The Earth Summit at Rio

THE EARTH SUMMIT AT RIO

Politics, Economics, and the Environment

RANEE K. L. PANJABI

Northeastern University Press
BOSTON

Northeastern University Press

Library of Congress Cataloging-in-Publication Data

Panjabi, Ranee K. L.

    The Earth Summit at Rio : politics, economics, and the environment / Ranee K.L. Panjabi.

        p. cm.

    Includes bibliographical references and index.

    ISBN 1-55553-283-7 (cloth : alk. paper)

     1. Economic development—Environmental aspects.

    2. Environmental policy—International cooperation.

    3. Environmental law, International.   4. United Nations Conference on Environment and Development (1992 : Rio de Janeiro, Brazil)   I. Title.

    HD75.6.P36   1997

    363.7'0526—dc21   97-11423

Designed by Ann Twombly

Composed in Bodoni Book by Coghill Composition Co., Richmond, Virginia. Printed and bound by Thomson-Shore, Inc., Dexter, Michigan. The paper is Glatfelter Supple Opaque Recycled, an acid-free stock.

MANUFACTURED IN THE UNITED STATES OF AMERICA

01 00 99 98 97   5 4 3 2 1

This book is dedicated to my wonderful father,
Khooshie, author, journalist, and diplomat, who taught me
that life is so great a gift, even the Gods cannot give it back. I
also dedicate this to my loving mother, Lata, poetess and
diplomat, who one day casually suggested that I write
"something academic but interesting about the Rio Summit"
and got me started on this project.

This book could not have been written without the help and
inspiration of all my B.V. friends, whom I love very much.

Contents

Foreword

I am pleased to have been asked to recommend the book entitled *The Earth Summit at Rio: Politics, Economics, and the Environment* by Dr. Ranee K. L. Panjabi. This book is essentially a compendium of five articles previously published (at least three as lead article) in various distinguished journals of international law. The articles, which carefully examine particular aspects of the Earth Summit (United Nations Conference on Environment and Development) as well as two International Environmental Conventions which were negotiated in parallel process and opened for signature at the Summit, are genuine masterpieces which are most deserving of broader consideration by the academic and legal audience for which they were intended.

Rarely is it possible for an author who has not been directly involved in the preparatory process for a major international conference, and/or the conference itself, to capture the essence and true dynamics which prevailed. This book combines the product of thorough research blended with interesting and relevant expressions of diverse international opinion drawn from participants both from governments and from the media.

I have no hesitation in recommending this book. Readers will gain useful insight into the Earth Summit as well as the Conventions relating to Climate Change and Biological Diversity. Indeed, the book would undoubtedly serve as useful reference material if not as a primary source for students of the intricacies and dynamics of international politics.

I commend this important contribution to promote a greater appreciation of the Earth Summit, its successes and failures. I am unaware of any other writings which are as insightful and, in the result, objective.

ARTHUR H. CAMPEAU
*Canadian Ambassador for Environment
and Sustainable Development (retired)*

Preface

To manage our common future on this planet, we will need a new global legal system based essentially on the extension into international life of the rule of law, together with workable mechanisms for accountability and enforcement that provide the basis for the effective functioning of national societies. We are a long way from this today. The United Nations Conference on Environment and Development, the "Earth Summit" of 1992, defined many of the needs for continued development of international law, including the strengthening of existing instruments and agreements on new ones. But even this would move us only marginally towards establishing an effective international legal regime.

The global nature of many environmental issues—climate change, acid rain, ozone depletion, shared water, and nuclear accidents, to name a few—has virtually made an anachronism of national boundaries. In a world crying out for international solutions, a vitally important mission for all concerned with global environmental protection is to ensure that our leaders are alerted to the challenges confronting our shared planet, sensitized to their complexity, and encouraged to think creatively in finding answers to these problems. Law forms an important part of our daily life, whether or not we are conscious of the fact, and international environmental law has a crucial role to play in helping us all to meet the global environmental challenges.

The Earth Summit at Rio de Janeiro in 1992 demonstrated how the superstructure of global environmental protection and sustainable economic development is significantly undergirded by international legal instruments such as the Conventions on Climate Change and Biological Diversity. I have witnessed the growing importance and vitality of international environmental law; yet I have also noticed the dearth of comprehensive treatment of the subject. It is with great pleasure, therefore, that I welcome this work by my fellow Canadian, Dr. R. K. L. Panjabi.

Dr. Panjabi's material not only reflects those vital issues illuminated at Rio, but also adds fresh insights into ways in which the human community might go about meeting the challenges we currently confront. It is refreshing to see a book dealing with a vitally important, yet difficult, subject living up to the highest expectations of a scholarly offering, yet rendered with an ease and fluency that will make it understandable to lawyers and non-laywers alike.

MAURICE F. STRONG
Senior Advisor to the Secretary-General of
the United Nations, and Liaison to the
International Monetary Fund and the World Bank

Acknowledgments

A book like this is inevitably the product of many minds, and I have many
people to thank for providing me so generously with their time and their
ideas. Although I take full responsibility for any failings of the book, I must
share the credit with all the wonderful people who made this project a reality.
First, I would like to thank Ambassador Arthur Campeau, solicitor, environmental activist, and former Canadian Ambassador for Sustainable Development. Ambassador Campeau was closely involved in the negotiations for the
Biodiversity Convention and was Personal Representative of the Canadian
Prime Minister to the Earth Summit. His office provided me with the masses
of documents which made this research project feasible, and he gave me an
insider's insight into the dynamics of the Rio Summit. Without his help this
book would not have been possible.

Ambassador Campeau also introduced me to Mr. Maurice Strong, Secretary-General of the United Nations Conference on Environment and Development, which met at Rio in June 1992. Mr. Strong read my original articles
and gave them his approval for publication in book form. His continuing
encouragement has been a source of great inspiration. It was with great de-
light that I learned that he has been presented the U Thant Peace Award
for his dedication to the United Nations and to the cause of environmental
protection.

The material in this book was originally published as a series of articles
(three of them lead articles) in the following academic journals of international law: *Dickinson Journal of International Law, California Western International Law Journal, Denver Journal of International Law and Policy*, and
North Carolina Journal of International Law and Commercial Regulation. I
wish to thank the editors-in-chief of those journals, all of them, for their
encouragement and help with this project: Florian Kogelnik of the University
of Denver, Jamison K. Shedwill of California Western School of Law, Henry
M. Perlowski of the School of Law at the University of North Carolina, and
Michelle Martin and her colleague Jason Blavatt of the Dickinson School of
Law. The present book is a revision of those articles, and I appreciate their

kind permission to reprint the articles in this compilation. I thoroughly enjoyed working with these dedicated editors and wish them all the best in their legal careers.

I also express my deep gratitude for the information and introductions provided so generously by Dr. Robert W. Slater, Assistant Deputy Minister, Environment Canada; Ms. Wendy Parkes, Manager of Operations, Biodiversity Convention Office; Mr. Denis Jolette, Senior Advisor to the Ambassador for Environment and Sustainable Development, Government of Canada; Mr. Michel Archambault, Director of Educational Institutions Programme, Canadian International Development Agency; Mr. Howard Mann, Legal Counsel, Environment Canada; Ms. Julie Wagemakers, Publications Officer, International Institute for Sustainable Development; Mr. Douglas Russell, Director of International Policy, Environment Canada; Mr. Ross Glasgow, Deputy Director Environmental Relations, External Affairs, Canada; Mr. Richard J. Kinley, Deputy Director, International Issues Branch, Environment Canada; Ms. Sid Embree, Policy Advisor, Environment Canada; Ms. Marlene Roy of the International Institute for Sustainable Development in Canada; and Mr. John Robinson, Vice-President, Policy Branch of the Canadian International Development Agency.

I thank as well the staff of the Office of the Prime Minister of Canada, who provided me with so much guidance and also sent me considerable quantities of material for my research. I owe a similar debt of gratitude to the office of Mr. Jean Charest, former Environment Minister of Canada. The External Affairs Ministries of the Governments of the United Kingdom, the Netherlands, Sweden, India, France, Malaysia, and Japan were generous in responding promptly to my innumerable queries. A particular thank-you to His Excellency Mr. Christian Krepela, Minister-Counsellor, Embassy of Austria in Ottawa, who spared so much of his busy time to answer my questions.

The project would not have materialized into book form without the encouragement of Mr. Scott Brassart, Editor, Northeastern University Press. His unfailing good humor and his patience have been a tremendous help, spurring me to carry on with the project and complete it.

Finally, I thank my wonderful mother, Lata Panjabi. This entire Rio project was initially her idea, and there have been times when she must have regretted her inspiration! She has valiantly read and corrected the manuscript and has been both unpaid research associate and constructive critic. Although Rio involved a great deal of work for us both, she has never once complained about the onerous tasks I have assigned her.

To say thank you to so many people seems small recompense for the gift of their encouragement, their time, and their insights. I hope, however, that this book fulfils in some small measure their collective dedication to the cause of saving our great planet.

RANEE K. L. PANJABI

CHAPTER ONE

Idealism and Self-Interest in International Environmental Law: The Rio Dilemma

This world of ours has changed dramatically in the past few years. A world superpower, the USSR, has dissolved into numerous component countries. Ethnic rivalries have forced the break-up of Yugoslavia and Czechoslovakia. Separatist groups worldwide are clamoring for recognition and a nation state of their own. The end of the Cold War has created an international situation filled with promise but fraught with peril. The absolutes of Cold War politics have been replaced by an international atmosphere of insecurity, aggression, and tension. It is as if the Cold War acted as a curb on all the nations of the world which bowed to its dynamic. Now released from the threat of total annihilation, the nations of the world appear to be indulging in an orgy of pent-up emotions which have simmered under the surface since the end of the Second World War. There seems little doubt that future historians will characterize this period as a turning point, an era when the peoples of the world selected or were forced into new directions, novel attitudes, and inno- vative ideas. Whether these directions, attitudes, and ideas are ultimately for the good of mankind or contribute to its eventual extinction will probably be determined in the present decade. The sheer exuberance of the era, the energy with which former totalitarian states are embracing democracy, the enthusiasm for "newness"—all indicate that this is the appropriate moment to channel these tendencies into positive directions which will benefit the entire planet.

Idealism and self-interest have converged and conflicted at various times in history. These two concepts have played a dramatic role, the one propping up and supporting the other, and, on occasion, being at odds with each other. Idealism has been defined as "the pursuit of high or noble principles" and goals.[1] On the other hand, self-interest refers to a "regard for one's own

Reprinted with kind permission from an original version published by the *California Western International Law Journal*.

interest or advantage especially with disregard for others."[2] Frequently these two concepts appear to be in serious conflict with each other. Indeed it could be argued that the history of mankind mirrors the great struggle between the higher priorities—idealism—and the harsher realities dictated by pragmatically inclined self-interest. Perceptions of idealism and self-interest are often induced by circumstances, personalities of leaders, and by prevailing conditions, both economic and political. When idealism and self-interest blend, and direct and determine the nature of political and economic action, profound beneficial change can occur. This chapter suggests a new focus on these two timeless concepts in the search for solutions to the problems which plague and threaten the survival of our planet.

It is unfortunate that the termination of the Cold War has resulted in a resurgence of nationalism because this makes the future of the entire planet more uncertain. Nationalism, fuelled as it is by self-centered notions of "us" against "them," has already exercised considerable influence on the history of mankind. It was in Europe that this idea first emerged in its modern form as medieval universalist philosophy grounded in the Christian faith gave way to a more secular, more exclusivist outlook. Language, geography, culture, history—all of these were channelled in the service of a new philosophy which incidentally suited the aims and ambitions of monarchs in England, Spain, and France. The concept of the nation state forced a new focus of loyalty, loyalty circumscribed not by heavenly boundaries but by geographic lines demarcating the areas within which each sovereign nation exercised its supremacy. Within his nation the monarch sought peace, security, and conformity to his views, both political and religious. Churches were controlled (as in Tudor England), and trade was encouraged largely to swell the coffers of the national government so that the ruler could exercise that most characteristic of national actions—the acquisition of territory by war. And so through the centuries the monarchs of Europe fought and the map of the continent changed like a kaleidoscope as geographic boundaries moved hither and thither and smaller, weaker entities were gobbled up by first one and then another sovereign nation. It was the merging of the monarch's personal ambition and self-interest with the concept of national glory which became the means of promoting war among his people—who, after all, had to pay the ultimate price in human sacrifice for those ever changing border lines.

It was nationalism which sparked the greatest adventure of them all—imperialism. The insatiable ambition for national glory could hardly be satisfied by acquisition of the limited territories provided by Europe. It was so much easier to take over non-European areas, where the people were neither afflicted with the concept of national prestige nor even conscious of the "us" and "them" syndrome which prompted so much of this action. Historians may not agree on the time frame of the age of nationalism. Some may concede, however, that from approximately the period of the European religious Reformations the dynamic of politics and economics has propelled most of the peoples of the world in this direction. Nationalism resulted in both unification and disruption. It brought together disparate peoples and gave them a new linkage, a binding force to a sovereign entity called the nation state. Yet it also disrupted the cultural and economic systems which had guided the lives of millions of people for centuries in the old civilizations of Africa, Latin America, and Asia.

Nationalism was perceived as being the ultimate form of self-interest and was projected as the greatest type of idealism. The pillage and devastation of entire civilizations were justified in the name of religion—the spreading of Christianity to the "heathen" of the world—and excused by the need of each European imperial power to outdo its neighboring rivals in territorial acquisitions and economic enrichment. In the Hobbesian state of nature which prevails among nations to this day, whenever self-interest is exercised on the international stage idealism is often utilized (rather cynically) by governments to explain, rally the people of the nation, and popularize and gain acceptance for the prevailing policies. The peoples of Europe were encouraged to go forth and colonize the world, to spread the benefits of Christianity and civilization to the "savages" of America, Africa, and Asia, to bring European ways of business and industry to the rest of the world. If, in the process, individuals were able to enrich themselves beyond their wildest expectations, this was small payment for the great sacrifice they were making!

Ironically, nationalism, which had propelled the nations of Europe into imperial ventures, would ultimately destroy all their empires. It was not long before the populations of the colonies began to resent the Europeans even as they adopted their ways, their politics, and their notions of national sovereignty. If national sovereignty could lead Europe to conquer the world, what might it not do for the peoples of Asia and Africa? In India, China, and the colonial territories of Africa, the imported concept of nationalism was eagerly seized by dedicated idealists and projected as the ultimate form of both idealism and self-interest. This idea focussed the energies of entire generations of Asians and Africans who fought for and eventually won independence from colonial rule. India was the first in this group to use the ideas of the West to win its independence, though its nonviolent freedom struggle drew on centu-

ries-old indigenous spiritual and religious traditions. The Indian struggle of
freedom was also the longest, deemed by some nationalist historians of India
to have begun in 1857—the year of the mutiny, or revolt—and to have ended
in 1947 when Britain finally left India. China, carved up like a melon by
several European powers, found its independence in the process of a long
civil war between the Nationalist forces of Chiang Kai-shek and the Commu-
nists led by Mao Zedong. Africa, also chopped up by the European powers
at the Conference at Berlin in 1884–85, went through a process of upheaval
as imperial rule drew to a close in the twentieth century; often, as in the
Congo, the former rulers left too abruptly and chaos ensued. In Latin
America, the decimation of the indigenous populations did not prevent widespread resentment against foreign and local elites, whose grip on the economies of these resource-rich areas was hard to break. The combination of
multinational companies and local tinpot dictators proved to be a real challenge for nationalist and rebel groups, and instability has characterized the
politics of several Latin American nations for decades.

Nationalism in the newly emerging nations displays many of the characteristics which marked the early phases of national sovereignty in Europe some
centuries ago. A sense of cocky self-assurance combined with a fierce competitive spirit motivates the international actions of the Afro-Asian world.
Having seen their interests subsumed for so long under the larger priorities of
imperial politics, these nations are very sensitive to any resumption, however
inadvertent, of imperial attitudes or colonial tendencies on the part of either
Europeans or Americans. To Asians, Africans, and Latin Americans, self-
interest dictates that the horrors of the past never recur; that political sovereignty be protected at all costs; and that economic independence be maintained even at a price which burdens their people. This self-interest has
become the ultimate form of idealism, for it seeks to cherish national identity,
protect an ancient culture, and ensure the survival of a unique way of life.

An understanding of this historical background is essential for anyone
seeking to gain some insight into the reason why there is so much controversy
about the threat to our planet and so much resistance to the attempts to
rectify the damage to the air, water, and land which have been degraded at
an alarming pace since the Industrial Revolution began in Britain and Europe. The environmental issue has become bogged down in a morass of historical antipathies and political misunderstandings. Nations of the North,
long accustomed to dictating their will to the South, are now finding their
wishes thwarted by an assertive, almost aggressive Southern reaction that
often surprises and mystifies them. The developing nations, now called the

South, always remember the imperial past and frequently react in an almost prickly manner to any suggestions which might impinge on their sovereignty.

Though Europeans and Americans as individuals now project a greater awareness of and sensitivity to global problems and share a keen appreciation of the fragility of the planet, their governments still appear to formulate policies motivated more by immediate self-interest than by long-term concerns. In fairness, it must be stated that this approach is now changing. Populations in Europe and North America are now light years ahead of their governments in globally oriented patterns of thought. Idealism in environmental concerns has almost acquired the comfortable prominence of a sacred cow, a deity to be ignored or slighted at one's peril. Governments are catching up or, more appropriately, are being dragged into environmentalism by popular concern. Idealism in environmental matters is now perceived as the ultimate self-interest. The problem for governments in Europe and North America is how to translate this idealism into a pattern of action which does not cause undue economic or social disruption. European and American nations, with a centuries-long enjoyment of popular sovereignty, economic and political independence, and an enviable lifestyle can afford now to view the larger picture, the planetary outlook, the global rather than the narrowly national map as pre-eminent in their perceptions. As they do so, they necessarily clash with the perceptions of the new nations of the developing world. This conflict of perceptions of what is idealism and what is in one's self-interest posed the great dilemma in the extravaganza played out at the Rio Summit in June 1992.

It has to be remembered that the developed nations enjoy a pre-eminence in the global economy both because they reaped the fruits of their own endeavor within the nation state and because they could command at will and even at whim the enormous resources of the colonies. There would be no First World today, with a minority of the Earth's population enjoying a comfortable lifestyle if the majority of the world's people in what used to be called the Third World had not contributed the incredibly cheap rubber, cotton, minerals, oil, and labor (including slave labor) which enabled the Europeans and Americans to industrialize, produce cheap factory goods, and sell these back to the areas of their empires which had provided the raw materials. In the process of creating what is today called the developed world, the European and American governments, driven as they were by an ethic which is not regarded as favorably now as it was then, destroyed the economic self-sufficiency of Asian and African countries under their control; converted varied agricultural systems into a precarious reliance on cash-crop production (rubber in Malaya, cotton and tea in India); forced colonial exports to bow to the dictates of fluctuating world trade pricing; and all but wiped out local crafts and ancillary production which provided supplementary income for hard-pressed farmers. At the same time, the Western nations introduced modern communications systems and brought railroads, telegraphs, and telephones, which ironically would eventually serve to knit the various ethnic groups of colonies like India into a greater national awareness and consciousness. While it serves no purpose to harp on past exploitation, it is important to understand that present perceptions in both the developed and developing nations have been forged by the past and that the future cannot be shaped without an awareness of the history shared by the thirty thousand diplomats, environmentalists, journalists, and businessmen[3] who converged on Rio in an ostentatious attempt to undo and halt the decline of the planet.

Although in retrospect the United Nations Conference on Environment and Development (UNCED), at Rio, may be deemed only a limited success, it did generate awareness and enhanced consciousness around the world about the fact that nationalism may have to chart a new course and make some concessions to international concerns. Popular idealism now dictates that air be purified, that water be cleansed of pollutants, that food be free of toxic contaminants. This is now perceived to be in the ultimate self-interest of the survival of the species, indeed of all life on this planet. For weeks before Rio, the media poured forth a torrent of facts emphasizing the unhealthy state of our environment—the poisons in the air, the depletion of the ozone layer, the destruction of lakes and wilderness areas, the ravaging of centuries-old forests, the human encroachment on the Amazon. No one can now dispute that the planet is in serious trouble. The problem is how to repair the damage already caused.

The concerns of people in all nations have increased considerably in the past decade, and with ample justification. "Worldwide, carbon emissions could rise from 6 billion tons to 20 billion tons by 2100."[4] Approximately two thirds of the carbon dioxide arises from the use of fossil fuels. "They released about 5.3 billion tons of carbon in 1986."[5] William Cline, an economist with the Institute of International Economics in Washington, D.C., suggests that the warming of the Earth's surface temperature caused by these emissions is irreversible.[6] Although the impact on the economy of the United States of America could be minimal (on a conservative estimate, a lowering of economic output by 1 per cent by the year 2050),[7] in certain areas of the developing world the damage and flooding could be extensive.[8] Scientists have speculated that the average surface temperature could rise by 1.5 to 4.5 degrees.[9] Although "[t]he science on this has the clarity of a mudpie,"[10] there is "plenty of scientific evidence that greenhouse gases are increasing in the atmosphere. Meteorological records show that, since comparative records began in the 1850s, the past decade has been the warmest yet."[11] On the assumption that by the beginning of the next century greenhouse gases will have doubled from their levels before the Industrial Revolution, "most scientists assume there will be a 2.5 degree increase in average temperatures by the year 2040."[12] Estimates of losses to the economy of the United States hover around the staggering figure of $60 billion, including $7 billion in land loss and preventive measures against oceans rising and $18 billion in agricultural losses.[13]

It has been estimated that by the year 2000 carbon dioxide emissions from the burning of fossil fuels will contain about 7 billion tons of carbon,[14] a figure that could rise to between 13 and 23 billion tons by 2050.[15] Still more gases are released by deforestation, other forms of burning, and general exploitation of land.[16] "The developed market economies of North America, Western Europe and Japan produce 49 per cent of global carbon dioxide emissions, the economies of Eastern Europe 25 per cent and the developing countries 26 per cent."[17] However, it is likely that the nations of the so-called South will replace the developed world (the North) in the next century if present pollution trends continue unabated. The developing nations are already and rapidly assuming the dubious distinction of being significant polluters in the field of greenhouse gas emissions.[18]

Global warming is not the only threat to our planet. The much lauded discovery in 1928 of chlorofluorocarbons (CFCs, the chemicals which are found in an assortment of products from refrigerators to rubber pillows)[19] has resulted in a chemical attack of unprecedented proportions on the layer of ozone which protects the Earth from the sun's ultraviolet radiation. The ozone consists of a thick belt of "triple-oxygen atoms (ozone) that encircles the Earth between 15 and 50 kilometres from the ground."[20] It has been estimated that one CFC molecule can destroy approximately 100,000 ozone molecules.[21] "A decrease of only 1 per cent in ozone concentration can lead to an increase of approximately 2 per cent in the effective irradiance of ultraviolet B (UV-B) light at the earth's surface."[22] The effect on human beings will be a higher incidence of skin cancer and cataracts;[23] a reduction in agricultural output and in food supplies generally is also possible.

Given its present population of 5.8 billion,[24] Spaceship Earth may just barely survive the rigors of global warming and ozone depletion. Population expansion, however, exacerbates all the other aspects of the environmental crisis. The population of the world increases by about 1 million every four days.[25] The figures vary according to the source one consults. The year of the Earth Summit the United Nations was reported to have estimated that world population would increase to 6 billion by 1998;[26] projections for 2100 are 11.2 billion.[27] At present more than 4.5 billion of the world's people inhabit developing countries, the area collectively labelled the South,[28] and most of the estimated growth will occur in the poorest regions of Africa and southern Asia.[29] "The growth rate will be the equivalent of one new country the size of Mexico every year."[30] According to one estimate, of the 92 million people added to the population of this planet annually, approximately 88 million are born in the developing countries.[31] This growth in numbers continues to be the most serious of the many challenges facing mankind. The consequences for the next century of such unprecedented growth are staggering.

If current population levels and increases continue, "[t]wo-thirds of the people of Africa—about 1.1 billion people—will be without adequate water supplies by 2025."[32] The World Health Organization has estimated that each year food, clothing, and shelter will have to be provided for an additional 80 to 100 million people.[33] In a twenty-year period, this could mean a need for a 36 per cent increase in food, drinking water, and agricultural products.[34] Even today malnutrition and preventable disease kill 10 million children every year.[35] The effects of human depredation of the environment are daily killing 40,000 of our Earth's most vulnerable people, children whose deaths (according to the United Nations) are related to hunger and its health consequences.[36] Given a huge increase in population, the tragedy of such deaths will become even more catastrophic.

It would be naive to assume that the developed world can somehow immunize itself against these crises. "The main leverage that the poor now have over the rich is the threat that they may drop in—indefinitely. As disparities among nations grow, pressures to migrate are rising. According to the United Nations, 75 million people are migrating illegally every year, skipping from country to country in search of work or food."[37] Poverty and political unrest have generated huge refugee flows as desperate men, women, and children flee from their homes in search of a safe haven and sustenance to keep them alive.[38] The search for a better life drives 2 million people to emigrate from the poor nations to the richer countries annually.[39]

Were the problem confined only to the climate and population, it would be serious enough. However, the land, the oceans, all that sustains life is now imperilled by the actions of the human species. The outlook is worse than grim. The land is rapidly losing its fertility because of human activity which is literally working the soil to death. "In the past 45 years, an area approximately the size of China and India combined has experienced moderate to extreme soil degradation as a result of human activities, according to UN studies."[40] This area of almost 3 billion acres is approximately 11 per cent of the vegetated surface of the planet.[41] Deforestation, animal grazing, and heavy agricultural usage have combined to degrade 1.1 billion acres in Asia and 793 million acres in Africa.[42] The agricultural abuse of land has resulted in extensive erosion and the loss of 25 billion tons of topsoil each year,[43] this at a time when the world desperately needs more food: in the past two decades the ranks of the chronically hungry have increased by 90 million to a frightening total of 550 million.[44] Some estimates of annual soil loss are even higher, approaching 28 billion tons.[45] The International Soil Reference and Information Centre in the Netherlands has concluded that human activity has resulted in the degradation of approximately 15 per cent of the land area of this planet.[46] Hence, when more space and more food-producing land will be required, these vital necessities are likely to be in short supply.

The oceans of this planet serve in vital ways to sustain and nourish human and animal life. Human beings have responded to this beneficence by annually dumping six and a half million tons of garbage into the world's oceans.[47] The United Nations estimates that normal oceanic shipping activities result in oil spillage or seepage of about 600,000 tons each year.[48] In a 1985 review the National Research Council estimated that, considering various points of origin, between 1.7 and 8.8 million metric tons of petroleum enter the oceans each year.[49] Arthur Kent, an NBC correspondent, on assignment covering Greenpeace, commented on pollution in the Mediterranean Sea: "The Med. is a vast open sewer for the nations of Europe. Greece pumps 98 per cent of its sewage raw into the Mediterranean, Italy 89 per cent."[50]

Besides oil, sewage, and garbage, the oceans have also been used as dumping grounds for nuclear waste.[51] It is certainly true that "[p]olluters will dump on the site that offers the least resistance."[52] These acts of sheer "ecostupidity"[53] or worse threaten the world's fisheries, which provide 16 per cent of the animal protein consumed by the people of this planet.[54] Pollution threatens the breeding grounds of this vital food supply, and overfishing to meet present needs promises to destroy it altogether. "Intensive fishing has mined many coastal and open-ocean fisheries. Catches of Atlantic cod and herring, Southern African pilchard, Pacific Ocean perch, King Crab, and Peruvian anchovies have all declined over the past two decades."[55] At the very moment when the human species is expanding beyond control, we are destroying the other planetary species which alone could sustain this human population explosion in the future. Brian Mulroney, Prime Minister of Canada at the time of the Earth Summit, emphasized this problem: "Overfishing, especially off of Canada's east coast, requires the urgent attention of the world's fishing nations, particularly the nations of the European Community."[56] A moratorium on cod fishing off the coast of Newfoundland in eastern Canada has been the result of overfishing by several nations.

The resilience and durability of this planet for so many thousands of years ought not to lull us into believing that it will all sort itself out somehow. The variety of plant and animal life on this Earth of ours has also to be protected. The destruction of entire species is one of the most tragic consequences of human activity run amok. Though no one is precisely certain, it appears that there are at least 10 million species of life on this planet.[57] The United Nations estimates the number at 30 million.[58] At least 140 plant and animal species become extinct every day.[59] So serious a loss of biological diversity has aroused global concern among environmentalists, who predict serious consequences for the future sustenance of human life if this trend continues. It is almost impossible to estimate the problems which will arise with the expected loss of about one quarter of 30 million species (U.N. estimate) projected to become extinct in the next two or three decades.[60] The problem affects every form of life, from plants and insects to birds, mammals, reptiles, and primates. Declining numbers and near extinction threaten three fourths of the world's species of birds and two thirds of the 150 known primate species.[61] Cautioning against this unprecedented destruction of life on Earth, the United Nations has pointed out:

> Biological diversity must be viewed as a global resource, like the atmosphere or the oceans. New uses for it are being discovered that can relieve human suffering and environmental destruction. Only a tiny fraction of species with potential economic importance have been utilized; 20 species supply 90 per cent of the world's food, and just three (wheat, maize, and rice) provide more than half. In most parts of the world, these few crops are grown in monocultures that are particularly sensitive to insect attacks and disease. Yet tens of thousands of edible species—many possibly superior to those already in use— remain unexploited. . . . The maintenance of biological diversity is a precondition for sustainable development.[62]

This incredible and wonderful variety of life represents "Earth's genetic insurance policy,"[63] and there is a possibility that at least 7 per cent of this natural wealth could be lost in the next quarter century.[64] There can be little doubt that the problems of this planet are reaching catastrophic proportions and, further, that there is very little time to solve these crises and yet sustain life at an acceptable level in the future. Dr. Mostafa Tolba, Director of the United Nations Environment Programme, highlighted the serious nature of the problem: "Humans continue to alter in a few decades precise ecological balances that have evolved over billions of years." Dr. Tolba continued, "The facts show again and again—in dwindling fish stocks, projected shortfalls in fuel wood, quickening soil erosion and millions of tons of greenhouse gases spewed into the atmosphere—that time is running out."[65]

The very fact that representatives from 167 countries,[66] led by more than 150 world leaders,[67] attended the Conference on Environment and Development held in Rio in June 1992 is indicative of the nearly universal interest and concern about the fate of this planet. The presence of so many heads of state at Rio demonstrates an awareness at the highest levels that the environment is The Issue of our time. While cynics would argue that association with environmental concerns is good public relations for any politician, there is definitely a dawning consciousness that "[u]ltimately all environmental protection is in everybody's self-interest."[68] Whether such consciousness will result in a spate of national legislation to clean up the damage remains to be seen. As with most such international gatherings, the level of rhetoric at Rio surpassed the extent of action taken. This dilemma was explained by Brazil's Acting Environment Minister, Jose Goldemberg: "There's a big difference between rhetoric and action. Rhetoric is very easy, action is very difficult."[69]

The action required at Rio was the expansion of the parameters of international environmental law to meet the challenge posed by the catastrophic destruction and degradation of the elements of sustainable life on Earth. The challenge was only barely met, and time alone will tell whether the initial steps taken at Rio will bear fruit in more stringent regulations and tighter controls on pollution. Whether nations take extensive action to clean up their own air, land, and water will also depend on the extent to which public pressure can be brought to bear on governmental leaders. Although they beamed during the Rio photo opportunities, they may now balk at the expensive implementation of environmental programs. In an apparent effort to goad the dignitaries at Rio to live up to their commitments, in his farewell address United Nations Secretary-General Boutros Boutros-Ghali reminded them that "[t]he function of the United Nations is not to mask general inaction with verbiage, speeches, reports and programs."[70]

The actual achievements of the Rio Conference were not insignificant, given the numerous obstacles placed in the path of progressive, effective action to cope with the environmental crisis. Global warming, raised as an internationally vital issue at the Toronto Conference on the Changing Atmosphere (1988),[71] was now tackled at Rio in a treaty limiting global emissions of carbon dioxide and other greenhouse gases.[72] The so-called Climate Treaty (reprinted as Appendix 4 to this book) aims to stabilize emissions of these gases at a safe level soon enough to "allow ecosystems to adapt naturally to climate change, to insure that food production is not threatened and to enable economic development to proceed in a sustainable manner."[73]

The Treaty, criticized for its weakness,[74] does take some important steps to deal with the problem. First, national programs have to be created to mitigate changes in the climate. The developed nations are required to "limit their emissions of greenhouse gases, to protect forests and other systems that absorb greenhouse gases and to demonstrate that they are taking steps toward meeting these objectives."[75] All parties are required to report bi-annually on measures taken to reach their 1990 levels of gas emissions and to conduct periodic reviews of progress taken to implement the provisions of the Treaty.[76] There is also a commitment by signatories from the developed world partially to fund efforts by developing nations to control emissions of these gases.[77] Developed nations are required to assist in technology transfers to the developing world in its efforts to implement measures aimed at stabilizing emissions.[78]

Although some of the negotiators in favor of a treaty on climate change were quite moderate in their requirements—parties "agree to try to hold greenhouse-gas emissions in the year 2000 to 1990 levels"[79]—achieving any specific binding goals proved to be a real challenge, particularly in view of the strong negotiating position taken by the Government of the United States of America. Though the White House staff denied the charge, the American Government was accused both before and during the Rio Summit of having "watered down the summit's central global warming treaty."[80] Lester Brown, President of the World Watch Institute, criticized the United States' position: "[t]he U.S. used to be in a leadership position in terms of the environment; now it has become the problem."[81] Brown's concerns center on the fact that stabilizing emissions to 1990 levels by 2000 will result in only an 11 per cent reduction in carbon dioxide emissions by the year 2050.[82]

Motivated by the possible economic consequences of drastic environmental clean-up programs, particularly when unemployment rates in his country were at an eight-year high,[83] President George Bush refused to allow "the extremes of the environmental movement to shut down the US on science that may not be as perfected as we in the US could have it."[84] As his nation was accused of emitting the largest amount of carbon dioxide (approximately 23 per cent of the world's total),[85] the cost of implementing binding time deadline targets was unacceptable to the U.S. President. Eventually the "binding commitments were dropped"[86] to satisfy the apprehensions of the United States, and President Bush agreed to attend the Rio Summit. The *Times* of London called the compromise a "toothless treaty"[87] and stated that "[t]he compromise explains the differing accounts of the treaty as historic break-through and sell-out."[88]

Unfortunately, the spirit of compromise which enabled the world's only superpower to sign the Climate Change Treaty did not extend to its participation in the second major achievement of the Rio Summit—the Convention on Biological Diversity, popularly referred to as the Biodiversity Treaty (reprinted as Appendix 5 to this book). The Treaty seeks "to reconcile the goal of preserving species and ecosystems with that of economic development and reduction of poverty."[89] Signatories promise to attempt a blend of conservation and economic development with less resource exploitation, establishment of protected areas, preservation of ecosystems and species, and restoration of damaged ecosystems. Again, technology transfers and funding from the North may well be the only means for developing nations to implement their treaty obligations.[90] "Under its [the Treaty's] terms, developing countries agree to preserve areas of biological importance in return for financial aid from developed countries, in the form both of foreign aid and of royalties from companies that use material gathered in Third World countries."[91] There was less inclination on the part of all parties to compromise on this Treaty. The stakes for both sides of the dispute were perceived as being much higher than was the case with the Climate Change Treaty. The Biodiversity Treaty binds not merely governments but businesses as well. It has been called a "two-way street."[92] Not only is this Treaty a significant step forward in the protection of the global environment but it also ensures that countries in the South still rich in biodiversity get a fair share in the utilization of this resource. "Developed nations have agreed to pay developing countries for conserving and controlling the use of their plants and animals. And developing nations are agreeing to let other countries purchase samples of those species to make products such as improved food crops, medicines and cosmetics."[93] There is now a clear incentive for developing nations rich in biodiversity to conserve this resource and to protect it from short-term schemes which call for rapid deforestation and consequent species destruction. Although environmentalists at Rio argued in favor of a stronger legal framework to protect the Earth's genetic diversity, the present Treaty, while far from perfect, "does provide a solid framework on which to build."[94] There appears to be wide recognition that the Treaty has expanded the boundaries of environmental law and alerted nations in both the North and the South to the fact that a new order now governs the use of this resource. Environmental groups in England have concluded that "[t]he convention is likely to become the most effective instrument for promoting global conservation in the coming decades."[95]

The Government of the United States of America expressed its opposition and refused to sign the Biodiversity Treaty because of its apparent desire to protect the patent rights of American companies working in the expanding biotechnological field.[96] Given the dominance of the United States in this field of research and development, the position of the Bush Administration is understandable in terms of American self-interest. Unfortunately, the consensus at Rio was highly critical both of the American President and of his commitment to economic priorities over the cause of environmental protection. The Americans were also opposed to some of the financing provisions of the Biodiversity Treaty.[97] The *Times* (London) expressed some of the apprehensions shared by the rich nations: "developed countries could be obliged to contribute whatever sum the majority of signatories—which will be Third World countries—decide is needed to meet the aims of the convention."[98] Even as he voiced his very strong objections to the Treaty, President Bush committed his nation to protecting biodiversity. "We come to Rio," he said, "prepared to continue America's unparalleled efforts to preserve species and habitat. . . . [O]ur efforts to protect bio-diversity itself will exceed . . . the requirements of the treaty. But that proposed agreement threatens to retard biotechnology and undermine the protection of ideas. . . . [I]ts financing scheme will not work."[99] Clearly, the environmental ideal had to be trimmed to a perceived self-interest concerning American economic circumstances. Environmentalist groups at Rio retaliated by unanimously declaring the United States the worst of the environmentally offending nations.[100] The United States signed the Biodiversity Treaty after Bill Clinton became President.

Even though the United States Government under George Bush was opposed to the third major document to be propounded at Rio, it did eventually sign it, albeit with some misgivings. Agenda 21, a lengthy blueprint for the clean-up of the environment, concerns "almost every environmental issue from overpopulation to overfishing."[101] Labelled an "absurd project,"[102] Agenda 21 has been criticized for not presaging "any fundamental change in the economic relationship between the rich nations of the developed world and the poor nations of the developing world, a change that many feel is necessary if poor countries are not to slip ever deeper into poverty and environmental degradation."[103] The underlying problem with Agenda 21 is not so much its ideals, which have wide support, but the potential difficulties which will arise when the world attempts to implement these ideas as concrete programs of action. Proponents of the document suggest that the cost of implementation could run as high as $125 billion per year.[104] The economies of the North, reeling from the recession of the 1990s, balk at the massive cost of Agenda 21, an expense which would involve at the very least a doubling of their foreign aid budgets.[105]

On 11th June 1992, the delegates at the Rio Summit unanimously accepted the Rio Declaration, the fourth significant document of the Conference. (The Declaration is reprinted as Appendix 2 to this book.) It is basically a statement of environmental principles which are not legally binding.[106] Though some member nations at the Summit had serious reservations about some of the principles, the acceptance of the Declaration provides a direction for all nations, provided they have the will to implement a program of sustainable development. That there are serious flaws in the Declaration cannot detract from its environmental principles, however limited the scope of the latter may appear to be. Criticism has been severe. The Canadian Participatory Committee of non-government agencies which advised the Canadian Government commented that "some of the principles are so gaseous as to dissolve upon examination."[107] Briefly, the Rio Declaration endorses the primacy of human beings as the center of concerns for sustainable development. It blends the priorities of development and environment, and it tries, somewhat half-heartedly, to urge implementation of the right to development through exchanges of science and technology. States are urged to resolve environmental problems nationally and internationally in a peaceful manner in accordance with the principles of the United Nations Charter.[108]

Having explored both the serious nature of the global environmental crisis and the steps taken at Rio to meet some of these problems, it would be worthwhile now to analyze the achievements of the Earth Summit to determine whether this great event was indeed a significant landmark in the creation of international environmental law, or whether, as some observers suggest, it was nothing more than a "gabfest."[109] The Rio Summit has to be assessed from a variety of perspectives, and any fair analysis must include consideration of the very real and compelling conflict between idealism and self-interest evident throughout the Conference.

There was a definite theatricality to the entire Rio event, an atmosphere akin to that of a gala performance or a Hollywood opening night. The vast numbers of the immediate audience, approximately thirty thousand, makes it one of the largest extravaganzas ever.[110] Innumerable world leaders— potentates and prime ministers, governmental delegations and representatives of non-governmental organizations—all vied for media attention and for a voice in the formulations being presented to the Conference. The entire world was watching on television and reading about the event in its newspapers as some six thousand journalists from every corner of the Earth covered the Summit.[111] Though the entire world had gathered to avert a tragedy, the atmosphere at Rio was almost that of an elaborate carnival, and one wonders whether this was really the ideal way to formulate principles of environmental law. Though it is true that similar spectacles have often resulted in significant treaties, as at the Congress of Vienna in 1814–15 following the Napoleonic Wars, the sheer size and impressive aura of the gathering at Rio may have led nations to posture more, to harden their positions, and to exaggerate their differences, factors which could have had a significant negative impact on the achievements of the Conference. One would like to think that legal principles, particularly when these involve several nations, will not be hammered out in the glare of global publicity and worldwide interest. It was perhaps fortunate that the basic elements of the Treaties and of the Declaration were formulated in a series of meetings preceding the Summit. It was also prudent of most supporters of the compromise Treaties which emerged from those earlier sessions to resist the urge to renegotiate to any great extent at Rio what had already been agreed to in earlier meetings. If Rio was largely an event to bring global attention to the tragedy of environmental degradation, then it can be deemed an unqualified success.

It was also successful in exposing the positions of various governments and in revealing which world leaders were sincere in their commitment to the environment. Rio was ultimately an exercise in global idealism, the vision of its creator, Maurice Strong, who firmly believes that the planet must be revitalized if the human species is to survive. This great and noble vision collided headlong with the national self-interest of a variety of political players at the Conference. Ironically, the positions of the United States of America and the more assertive of the developing nations like Malaysia were remarkably similar in terms of the primacy of self-interest over environmental concerns. The American Government's refusal to risk jobs in an economy battered by recession was interpreted at Rio as selfishness and callousness. That President Bush was standing for re-election that year was frequently alluded to in an explanation of the American reluctance to play the role of environmental world leader.

Some developing nations also presented a position based largely on self-interest. Malaysia, rich in forests, a resource it is logging at a rate which alarms most environmentalists, resisted efforts to deem this natural treasure of wood and biodiversity part of the global heritage of mankind. Malaysia was both strident and defensive in its stance at Rio, and the country's delegates argued very strongly that if the North wants the world's forests saved then the North must be willing to subsidize the preservation of this resource. A Malaysian diplomat, Ting Wen-Lian, commented assertively that if developed countries want developing countries to conserve their forests, they should attend to "the poverty, famine and crushing burden of external debt" which compel the poorer nations to fell their trees in order to survive.[112] It is significant that though the United States of America and Malaysia seemed to represent opposite poles in the environmental dialogue at Rio, both nations, in emphasizing the primacy of national self-interest over global idealism, were on a remarkably similar plane of thought. Both countries are committed to environmental improvement, but both resisted the economic sacrifice inherent in its implementation. Though the rhetoric highlighted the North-South divergence, the primacy of immediate national priorities in the policies of both nations demonstrates how akin they really are in their underlying attitudes. Both Malaysia and the United States of America figured in the list of the five worst environmental nations.[113] The list was prepared and widely publicized by the various environmental activists gathered to share ideas in the unofficial people's summit, labelled the Forum, which also convened in Rio at the same time as the more formal Earth Summit.

Although the nations on the extremes of the North-South drama as played out at Rio were the major hindrance to the passage of strong measures to clean up and preserve the environment, most nations in both North and South overcame their nationalistic apprehensions to sign the compromise agreements which were passed at the Conference. The endorsement by so many nations of each of the Treaties and the unanimous acceptance of the Rio Declaration presage a new global consciousness on many levels, indicating that the pre-eminence of national self-interest must now give way to the ideals of sustainable development. It is as if the majority of the nations of the world have finally realized that there is a clear difference between short-term self-interest which would resist environmental legislation in order to safeguard economic concerns and long-term self-interest in which the entire planet survives and perhaps even thrives because timely measures have been taken to implement the ideals expounded by environmentalists. Among the nations of the North, the American position was regarded as extreme. It was in marked contrast to that of the European nations, some of which argued emphatically for firm timetables for curbing carbon dioxide emissions in the Climate Change Treaty—a provision not acceptable to the United States and therefore dropped in the final document. Austria, the Netherlands, and Switzerland initiated a proposal to commit signatory nations to immediate stabilization of emission levels, a move which, according to one Austrian delegate, resulted in a threatening letter from the U.S. Government.[114] Clearly, the American position had led to its isolation in the sphere of environmental diplomacy; regrettably, an opportunity to assume the mantle of world leadership in this important facet of international relations was forsaken. It is impossible to speculate on the probable consequences of a vigorous American assumption of leadership in this attempt to expand the parameters of environmental law. Certainly, the inclusion of firm timetables in the Climate Change Treaty and U.S. endorsement of the Biodiversity Treaty would have made a profound difference.

The fact that countries like Canada and the United Kingdom signed the Treaties,[115] propounded at Rio underscored the isolation of the United States of America. On 12th June 1992, Canadian Prime Minister Brian Mulroney promised at the Earth Summit that his country would "undertake to ratify both agreements this calendar year."[116] In Canada and the United Kingdom, public pressure in favor of environmentalism has been very strong, and Governments have acceded, despite reservations, to the principles and formulations of Rio largely in response to the will of their constituencies. As regards the United States of America, Albert Gore was the Chairman of the American Senatorial delegation to the Earth Summit. A few months later, he was elected Vice-President of the United States. At Rio, Gore was highly critical of the stance adopted by the White House,[117] deploring the failure of his country to provide leadership in this matter.[118] So intense was the hostility to the position of the American Government that few environmentalists or official delegations were prepared to give due credit to President Bush for agreeing to provide $150 million in new aid for the protection of forests.[119]

It is clear from the plethora of speeches, policy papers, and official statements which poured out of the Earth Summit that most nations of the world have awakened to the impending crisis in the environment and are anxious to do something about it. While there is no dearth of motivation or inclination, resolute action is another matter altogether. First, both North and South plead that no matter how ideally desirable it is to clean up the planet, realistically and in economic terms this is likely to be a prohibitively expensive venture, one which no country can envisage undertaking at the present time. So while subscribing with enthusiasm to the ideals of environmentalism, few nations are willing to turn those ideals into concrete plans of action which would fund the clean-up of rivers and lakes or stop the felling of forests. At the point where rhetoric has to be translated into action, the factor of self-interest, clothed in the language of the priority of saving jobs or encouraging development, takes over and rivers continue to be contaminated and forests disappear before our very eyes. It remains to be seen whether all the signatories to the two Treaties will in fact take adequate measures to implement the provisions within their own nations.

Part of the problem at Rio was the indulgence by all sides in the rather useless exercise of flinging responsibility and blame on the other party. The North blamed the South for exacerbating an already severe problem by over-populating the world. The South retaliated by flinging in the North's face its wasteful, extravagant lifestyle which causes most of the global pollution. The North, or rather its American leadership, refused to consider vital changes to the way of life of its people. The South countered by insisting on its right to develop in much the same resource-wasteful manner that the North had earlier relied on for its industrialization. The bickering and quibbling before and during Rio demonstrate clearly that for a number of countries self-interest is still the guiding principle of foreign and domestic policy and that there is not as yet a clear realization that environmental idealism has to be adopted to ensure the survival of the planet.

If the nations of the world, East, West, North, and South, could view the long-term perspective of present policies, they would see a world in which the deserts keep expanding because agricultural abuse and deforestation have destroyed the soil which sustains and nourishes plant life. If the pollution of Earth's rivers, lakes, and oceans continues at present levels, we can expect to lose fish and shellfish as a normal food source for land animals and man. The elimination of this source of protein will not merely result in greater hunger on Earth; it will also wipe out thousands of jobs for those who have depended on the water to earn their livelihood. If the air becomes dirtier because of industrial and automobile emissions, we can expect higher rates of illness worldwide; this is likely to affect children and the elderly, who are most susceptible to the ailments associated with pollution. A reliance on present policies sacrifices the future to an extent which ought to be unacceptable to any government in any nation. This form of immediate self-interest, gratified at enormous risk to future, long-term considerations, hardly seems worth the price. If nations fail, at this crucial juncture, to subsume their own perceived self-interests under the greater interest of the entire planet, they risk not merely diplomatic isolation but economic loss and environmental degradation.

There is evidence that some governments and business leaders have already understood that environmental action is likely to be not only idealistic but very profitable in the near future. "U.S. corporations are taking the initiative in getting rid of their ozone-reducing chemicals. The Hughes Corp. now uses a chemical derived from lemon juice (yes, lemon juice) instead of CFCs in its weapons-manufacturing program."[120] 3M has reduced pollution and increased profits in its U.S. operation.[121] Northern Telecom has taken measures to end its use of CFCs.[122] Executives from Chevron, Volkswagen, and Mitsubishi, among others, have banded together to form the Business Council for Sustainable Development in an attempt to encourage the translation of environmental ideals into pragmatic business practices worldwide.[123] The Canadian Chemical Producers Association has initiated the Responsible Care program, which blends business interests with environmental responsibility. As of early June 1992, the program had affiliates in twenty countries, including the United States, Japan, Australia, and a number of European nations.[124] It is clear that in a number of cases, "[b]usiness is moving faster than the laws require."[125]

Environmentalism is itself becoming a major industry, worth in the United States alone about $150 billion.[126] The present myopic self-interest which persuades some governmental leaders that jobs rely on polluting industries has to be re-examined in light of the possibilities opening up for employment in the clean-up of the planet. Nations like the United States, gifted with inventive genius and a tradition of individualism and innovative entrepreneurship, could stand to gain considerably by becoming active participants in this real "new world order" which has already arrived. By clinging to an outdated perception of self-interest, the world's one remaining superpower risks isolating itself and being unable to compete in a rapidly changing economic and environmental attitude now sweeping the planet. If the United States commits itself to the implementation of sustainable development globally and evinces an interest in assisting the poorer nations to develop in an environmentally safe manner, it will not only enhance its international image but safeguard its political pre-eminence in the world. It would be naive to assume that with the conclusion of the Cold War there are no serious threats to the Free World. While the prospect of world war is now remote, the likelihood of global pollution on a massive scale is an ever growing threat. The consequences of death by war or death by foul air are the same. If the American Government can devote the same energy to the environment that it was able to bring to the expulsion of Saddam Hussein from Kuwait, much can be done to protect future generations of every nation and to ensure that the legacy we leave will not lead our children to curse us for befouling their world. Although the Treaties signed at Rio are only a small step in the expansion of international environmental law, a sincere commitment to implement these provisions would galvanize the developing world to follow suit and could well lead to an era of greater understanding and harmony. We all need a global recognition that the ideals of environmentalism are ultimately the most realistic form of self-interest. All nations can only benefit from an awareness that idealism and self-interest are not at odds with each other, that in fact they complement and sustain each other. What is at stake in acting on this new awareness? Only the future of our species and of our home, Earth.

Notes

1. *Webster's Encyclopedic Unabridged Dictionary* (New York: Portland House, 1989), 707.
2. *Ibid.*, 1294.
3. *Globe and Mail* (Toronto), 19th May 1992, A1:4.
4. *Christian Science Monitor*, 4th June 1992, 6:5.
5. *Global Outlook 2000* (New York: United Nations, 1990), 79.
6. *Globe and Mail*, 3rd June 1992, D10.3.
7. *Ibid.*
8. *Ibid.*
9. *Ibid.*, A6:1.
10. *Ibid.*
11. *Ibid.*, B1:4.
12. *Ibid.*, 7th May 1992, A2:3.
13. *Ibid.*
14. *Global Outlook 2000*, 79.
15. *Ibid.*
16. *Ibid.*
17. Stewart Boyle, "Global Warming—A Paradigm for Energy Policy?" *Energy Policy* (February 1989), cited in *Global Outlook 2000*, 79.
18. Ivan L. Head, *On a Hinge of History* (Toronto: University of Toronto Press, 1991), 94.
19. *Evergreen Foundation*, 15th April 1992, 7:3.
20. *Globe and Mail*, 4th June 1992, A6:4.
21. *Ibid.*, A6:5.

22. Head, *On a Hinge of History*, 95.

23. *Ibid.*

24. World Resources Institute, *World Resources, 1994–1995: A Guide to the Global Environment* (Washington, D.C.: WRI, 1994), diskette.

25. *World Monitor*, June 1992, 31.

26. *Globe and Mail*, 29th April 1992, A1:6.

27. *Ibid.*, 6th June 1992, A4:2.

28. World Resources Institute, *World Resources, 1994–1995*.

29. *Globe and Mail*, 29th April 1992, A2:6.

30. *Ibid.*

31. Sandra Postel, "Denial in the Decisive Decade," in Lester R. Brown et al., *State of the World, 1992* (New York: W. W. Norton, 1992), 3.

32. *Globe and Mail*, 29th April 1992, A2:6.

33. *Christian Science Monitor*, 2nd June 1992, 4:3.

34. *Ibid.*, 3.

35. *Ibid.*, 10th June 1992, 3:4.

36. *Ibid.*, 2nd June 1992, 1:4.

37. *Globe and Mail*, 26th May 1992, B18.

38. See R. K. L. Panjabi, "The Global Refugee Crisis: A Search for Solutions," *California Western International Law Journal*, vol. 21, no. 2 (1990–91): 247–63.

39. *Globe and Mail*, 29th April 1992, A2:6.

40. *Christian Science Monitor*, 2nd June 1992, 10:2.

41. *Ibid.*

42. *Ibid.*

43. *Ibid.*, 4:3.

44. *Ibid.*

45. *World Monitor*, July 1992, 8.

46. *Globe and Mail*, 20th May 1992, A2:3.

47. *Christian Science Monitor*, 2nd June 1992, 4:3.

48. *Ibid.*, 10:1.

49. National Research Council, *Oil in the Sea: Inputs, Fates, and Effects* (Washington, D.C., 1985), cited in Head, *On a Hinge of History*, 82.

50. Arthur Kent, *Dateline*, NBC Television, 11th August 1992.

51. *Christian Science Monitor*, 2nd June 1992, 10:1.

52. *Newsweek*, 1st June 1992, 30.

53. *Ibid.*, 37.

54. *Christian Science Monitor*, 2nd June 1992, 10:2.

55. John C. Ryan, "Conserving Biological Diversity," in *State of the World, 1992* (New York: W. W. Norton, 1992), 13.

56. Prime Minister Brian Mulroney, Address at the Canadian Museum of Civilization, Hull, Quebec, 1st June 1992 (Office of the Prime Minister).

57. *Christian Science Monitor*, 2nd June 1992, 10:1.

58. *Globe and Mail*, 20th May 1992, A2:5.

59. Postel, "Denial in the Decisive Decade," 3.

60. *Globe and Mail*, 20th May 1992, A2:5.

61. Ryan, "Conserving Biological Diversity," 13.

62. *Global Outlook 2000*, 95.

63. *Times* (London), 3rd June 1992, 15:1.

64. *Ibid.*

65. *Christian Science Monitor*, 2nd June 1992, 4:3.

66. *Ibid.*, 8th June 1992, 3:1. The *New York Times* (14th June 1992, 10:4) estimated that 178 nations attended the Rio Conference.

67. *Times* (London), 4th June 1992, 1:5.

68. *Newsweek*, 1st June 1992, 33.

69. *Christian Science Monitor*, 11th June 1992, 5:4.

70. *New York Times*, 15th June 1992, A8:1.

71. *Times* (London), 1st June 1992, 12:1.

72. *Christian Science Monitor*, 2nd June 1992, 10:4.

73. *New York Times*, 13th June 1992, 4:1–2.

74. *Globe and Mail*, 3rd June 1992, A6:3.

75. *New York Times*, 13th June 1992, 4:1–2.

76. *Ibid.*

77. *Christian Science Monitor*, 2nd June 1992, 10:1–5.

78. *New York Times*, 13th June 1992, 4:1–2.

79. *Globe and Mail*, 5th June 1992, A4:6.

80. *Times* (London), 2nd June 1992, 16:6.

81. *Globe and Mail*, 3rd June 1992, A6:3.

82. *Ibid.*

83. *Ibid.*, 6th June 1992, B2:1.

84. *Times* (London), 2nd June 1992, 16:6.

85. *Ibid.*, 1st June 1992, 12:3.

86. *Ibid.*

87. *Ibid.*

88. *Ibid.*

89. *New York Times*, 6th June 1992, 6:3.

90. *Ibid.*, 6:3–4.

91. *Globe and Mail*, 4th June 1992, A6:1.

92. *Christian Science Monitor*, 26th May 1992, 3:2.

93. *Ibid.*

94. *Times* (London), 4th June 1992, 15:4.

95. *Ibid.*

96. *Globe and Mail*, 4th June 1992, A6:1.

97. *Ibid.*

98. *Times* (London), 3rd June 1992, 15:2.

99. *New York Times*, 13th June 1992, 5:6.

100. *Globe and Mail*, 9th June 1992, A8:1; *Times* (London), 10th June 1992, 12:2.

101. *Globe and Mail*, 2nd May 1992, A7:1.

102. *Times* (London), 1st June 1992, 15:2.

103. *Globe and Mail*, 15th June 1992, A8:5.

104. *Times* (London), 9th June 1992, 27:1.

105. *Globe and Mail*, 2nd May 1992, A7:1.

106. *Times* (London), 12th June 1992, 1:7.

107. *Globe and Mail*, 2nd June 1992, A3:4.

108. *Ibid.*, A3:5–6.

109. *Ibid.*, 19th May 1992, A1:4.

110. *Ibid.*

111. *Ibid.*, 3rd June 1992, A23:1.

112. *Ibid.*, 6th June 1992, A4:6.

113. *Ibid.*, 9th June 1992, A8:1–2.

114. *New York Times*, 8th June 1992, A5:4.

115. *Times* (London), 13th June 1992, 1:2–4; *Canada and the Earth Summit*, Press Release by the Government of Canada at the Rio Summit, 3–14 June 1992.

116. Notes for an Address by Brian Mulroney, Prime Minister of Canada, to the Earth Summit, Rio de Janeiro, Brazil, 12th June 1992 (Office of the Prime Minister).

117. *Globe and Mail*, 6th June 1992, A4:2–4.

118. *Ibid.*, 13th June 1992, A4:3–5.

119. *New York Times*, 22nd June 1992, A10:2.

120. *Time*, 17th February 1992, 45:2.

121. *Ibid.*, 1st June 1992, 41:2.

122. *Ibid.*, 17th February 1992, 45:2.

123. *Ibid.*, 1st June 1992, 41:2.

124. *Responsible Care*, 5th June 1992.

125. *Time*, 17th February 1992, 45:2.

126. *Globe and Mail*, 6th July 1992, B2:4.

CHAPTER TWO

From Stockholm to Rio:
A Comparison of the Declaratory Principles
of International Environmental Law

Introduction

Environmentalism is the major political, economic, and social issue of the 1990s. This cause commands the type of grassroots support around the planet which few other ideas have evoked in the long history of mankind. Environmentalism has achieved the status of a philosophy few would dare to question or oppose. It may well become the primary ideology of the post–Cold War world, and if governments can respond adequately to the demands of their people, the human species may yet be able to salvage the only viable home it has. The problem lies not in the acceptance of environmental concerns but in the implementation of environmentalists' programs for clean-up. A planet facing debilitating recession in its richest nations and grinding poverty in its poorest cannot spare the funds and the resources to clean up the environment. An emphasis on short-term necessity may push aside the long-term interest which dictates that the Earth must be cleansed of its foul air and water if the human species is to survive.

The global concern about the problems of the environment has prompted two major world conferences in the past two decades. The first was held at Stockholm in June 1972 and the second at Rio de Janeiro in June 1992. These two conferences have been successful in generating global concern about environmental deterioration but have hitherto not resulted in action to match the level of international awareness and concern. The problem is that rhetoric about the environment is far easier to produce than action, and international forums tend on occasion to degenerate into "rhetoric-fests" where

Reprinted with kind permission from an original version published by the *Denver Journal of International Law and Policy*.

world leaders spout all the proper phrases but then go home and often fail to implement their internationally formulated promises.

Because environmentalism has now become a political issue, it has fallen prey to the vicissitudes of political expediency. It is often a victim of the fact that democratic governments—those most attuned to environmental awareness—are frequently short-lived and subject to electoral changes. Thus, the very situation which encourages free choice in democracies appears to militate against the implementation of long-term solutions which alone can save the planet. This is why it is so important to raise environmentalism above the level of party politics and make it an issue of national and international significance, not only in terms of promise and commitment but in terms of action and implementation. The numerous nations which participated in the United Nations Conference on the Human Environment in Stockholm and the United Nations Conference on Environment and Development (UNCED) in Rio were aware of the chasm between rhetoric and real action which is an inevitable aspect of environmentalism. It was assumed at both conferences that the creation of blueprints of principle would emphasize the significance of environmental concern. The acceptance by all nations of a body of principle could be an initial step to encourage not merely national measures to curb pollution but international treaties to improve the global environment. Both conferences paid particular attention to the formulation of these principles, which were to guide and chart a path for mankind. The Stockholm Conference produced the Declaration of the United Nations Conference on the Human Environment (reprinted as Appendix 1 to this book), consisting of a preamble and twenty-six principles. The conference at Rio resulted in acceptance of the Rio Declaration on Environment and Development (reprinted as Appendix 2 to this book), which has a brief preamble and twenty-seven principles.

This chapter will attempt an in-depth comparison of these two international documents, with a view to discerning the significance of each and pointing out the differences in emphasis between the two decades, the 1970s and the 1990s. As the Rio Declaration specifies the significant directions for the next few years, the emphasis will be on this instrument of international endeavor. World opinion on the priorities, successes, and weaknesses of the Rio Declaration is relevant to a full understanding of its significance. It has to be pointed out that length constraints limit the degree to which every aspect of the comparison can be explored. Nor can every principle in each Declaration be analyzed. The focus will be on the Rio Declaration as it stands and not on the process by which it was created. For the sake of convenient comparison, the Declaration of the United Nations Conference on the Human Environment will be referred to as the Stockholm Declaration.

Although both international conferences produced other instruments for acceptance and signature,[1] the formulation of bodies of principle was significant in establishing the fact that environmentalism is not merely an issue of political and economic concern. The Stockholm and Rio Declarations have brought a new dimension to the tone of environmentalism by generating an awareness that these issues override the more mundane considerations which prompt ordinary political action. The Declarations signified a universal acceptance of and acquiescence in the primary, fundamental fact that environmentalism is a matter of values. It may not be too early to call it the new ethic of the 1990s. Inspired by the pain of millions who suffer daily from the effects of pollution, the Declarations serve to remind the leaders of the planet both about the seriousness of the problem and about the need for dramatic measures to alleviate the lot of those who have been victimized. Even though both Declarations fall short of reflecting the moral authority which initially inspired them, they form a useful charter and benchmark for future direction.

Although cynics may argue that such pronouncements of principle are hopelessly naive, an exercise in futility, more honored in the breach than in the observance, the very fact that so many nations could after intense deliberation formulate these principles is indicative of the universality of awareness concerning the many facets of the environmental problem. The skepticism of those who witnessed the failure to implement most of the Stockholm principles is akin to the pessimism of those who daily decry the fact that mankind appears not to heed or follow the teachings of the great religious leaders. However, few of those who lament the religious deficiencies of modern citizens would be in favor of discarding the great body of religion and philosophy that has been produced for the benefit of mankind. Similarly, the inadequate implementation of environmental principle ought not to daunt or deter us from realizing its inherent significance and relevance.

An emphasis on legal principle can be very useful in the international realm, where there is as yet no supranational sovereign power to control states. While acknowledging that nation states are still in a virtual state of nature with respect to the exercise of power, few would dispute the fact that international law daily erodes the realm of that natural sovereignty, circumscribing it with treaties, trade agreements, United Nations resolutions, and a plethora of good-faith accords. William Thorsell has written about the "erosion of sovereignty that has been accelerating since the 1970s."[2] Iraq learned the hard way that unilateral aggression simply does not pay when its invasion of Kuwait was challenged by international military intervention. States which choose to behave like renegades risk being treated as pariahs. The people of Iraq continue (at time of writing) to suffer the consequences of the actions of their dictator, Saddam Hussein. By a curious twist of fate, it was left to Saddam Hussein—the champion of irresponsible national sovereignty—to demonstrate the importance of and dire necessity for an international approach to environmental concerns. By flooding the Persian Gulf with oil and by blowing up over seven hundred Kuwaiti oil wells[3]—an act which befouled the air all the way to India—he demonstrated the international impact of ecoterrorism and underscored the need for universal measures to save the planet. If governments are not yet willing to surrender their sovereign powers for the environmental benefit of all inhabitants of this planet, it is imperative that at the very least they be reminded constantly of both their obligations and their responsibilities in this regard. A short body of easily comprehensible principle serves a useful purpose as an environmental beacon guiding nations and their citizens toward greater environmental awareness. Hopefully, realization of the dimensions of the problem will generate further efforts to practice environmental protection.

Nationalism is today the single biggest obstacle to environmentalism. The "us" against "them" mentality that it generates breeds a reliance on narrow perspectives based on self-interest and the immediate expedient requirements of a sovereign state. The international outlook, on the other hand, favors a long-term, less self-centered approach. The problem is that while environmental agreements are internationally formulated, they have to be nationally implemented. Hence, what should be achieved with universal benefit in view is often subsumed under the self-interest of the sovereign state. The fact is that today, despite the plethora of international agreements and treaties, and the existence of the United Nations and its numerous bodies, sovereign states are still essentially in a state of nature in their relations with each other. This unfortunate reality makes decisive action in the environmental field extremely difficult. Given the existence of so many nations on this planet, the best that one can hope for is that many of them will be inspired by their own environmental rhetoric and by their international commitments to take the steps necessary to clean up their own area of the planet. The principles of environmentalism may be formulated on the large international stage. Regrettably, the materialization of these principles takes place in the narrower national sphere. Therein lies the biggest obstacle and the greatest challenge for environmentalists.

Another facet of the problem was eloquently explained by Frances Cairncross in the *Economist*:

> Nature is no respecter of national boundaries. Across those dotted lines on the globe, winds blow, rivers flow and migrating species walk or fly. The dotted lines may carve up the earth, but the sea and the atmosphere remain open to all, to cherish or plunder.

Cairncross laments the inevitable consequences: "When people in one country harm that bit of the environment they assume to be theirs, many others may suffer, too. But how, and how much, can countries make their neighbours change their ways?"[4]

The limitations of the present political structure of the planet, with its multitude of small, medium, and large nation states, certainly slow the pace of environmental activity. The vast disparities of resources, wealth, and population among these many states adds to the problem. But as Maurice Strong, Secretary-General of UNCED, commented:

> We do not have a central world government, and if we are going to insist that you set up a central world government and a central world gendarmerie before you deal with environment problems, the planet will be dead. We have to work with the system we've got, which is nation states working together through the United Nations, which is the only global organization that can perform that function.[5]

The World Commission on Environment and Development proposes in its report *Our Common Future* that,

> [b]uilding on the 1972 Stockholm Declaration . . . there is now a need to consolidate and extend relevant legal principles in a new charter to guide state behaviour in the transition to sustainable development. . . . The charter should prescribe new norms for state and interstate behaviour needed to maintain livelihoods and life on our shared planet, including basic norms for prior notification, consultation, and assessment of activities likely to have an impact on neighbouring states or global commons.[6]

The call to create new normative structures was timely and valid. However, the implementation of this proposal would not be easy. As Dennis Lloyd commented: "Law . . . cannot but be a reflection—however partial and imperfect—of the society in which it operates, and if that society contains inherent contradictions these will be manifested in the fabric of the law itself."[7] Prior to the publication of *Our Common Future*, the United Nations General Assembly had accepted the World Charter for Nature in 1982.[8] The World

Charter for Nature and the Stockholm Declaration have been appropriately called "decennial touchstones in the area" of international environmental law.[9]

Ten years after the World Charter for Nature and five years after the publication of *Our Common Future*, the task of creating a body of principles of international environmental law was still formidable, given the diversity of interests, both national and regional, which sought representation for particular points of view. The environmentally conscious developed world envisaged an Earth Charter which, in Howard Mann's opinion, "would be readable, understandable and accessible to everyone, i.e., a document that was not a typical UN resolution but that would provide an important tool for the shaping of public opinion on, and support for, the concept of sustainable development."[10] According to A. O. Adede, writing in *Environmental Policy and Law*, during the process that led up to the Earth Summit, the United Nations Working Group's priorities were to make the principles short and concise, with an inspiring and appealing text which would be "easily understood by the general public."[11] If the Stockholm and Rio Declarations appear less than perfect in their inspirational value, it could equally be suggested that they at least address the primary concerns of the majority of governments represented at the two environmental conferences.

Although the Stockholm and Rio Declarations have not been categorized as legally binding, their acceptance was by no means easy in either case. The discussions and debates over principle drew forth the full panoply of argument on the North-South divergence, the developed-versus-developing conflict, and the contest of the rich nations and the poor nations. That so much thundering rhetoric accompanied the formulation of declarations never meant to be legally binding indicates that delegates at both Conferences were keenly aware of the ultimate significance of creating a body of environmental principles. Principle can be the precursor of law, provided it evokes sufficient public support and generates a type of moral authority of its own. A body of international legal principle can also, over the course of time, be assumed by some of its proponents to be acceptable as customary law.[12] Even though the Stockholm and Rio Declarations may not bind nations in the legal sense, they do oblige them morally to respect their ideas as indicators of a universal consensus about the priorities of environmentalism. Representing Thailand at UNCED, Dr. Chulabhorn Mahidol reminded delegates that the Declaration would "have a very strong political and moral force."[13]

The differing priorities of the developed world, for environmentalism, and the developing world, for development, were reflected in the long and rather bitter debates which resulted in the somewhat choppy and less than inspirational text of the Rio Declaration. As Frank McDonald reported in the *Irish Times* (Dublin), "The original idea among the developed countries was to produce a ringing declaration in Rio which 'kids all over the world could hang on their bedroom walls.' But then the developing countries rather unhelpfully pointed out that many of the children in their part of the world don't have bedrooms."[14] Lucia Mouat noted in the *Christian Science Monitor* that the proposed Earth Charter became "a graphic symbol of the North-South divide"[15] and was converted into a more pedestrian, rather wordy Declaration which the world accepted with a somewhat resigned sense of inevitability. Bowing to the inescapable force of international political reality, the Head of the Canadian Delegation and Personal Representative of the Canadian Prime Minister at Rio, Arthur Campeau, described the final Declaration as "a document suitable for bureaucrats."[16]

The delegates even argued at Rio about the status of the Rio Declaration. The United States of America was in favor of placing the body of principle as a preamble to the detailed Agenda 21 document.[17] As Agenda 21 is also not legally binding, this idea would appear to have some logic. However, Agenda 21 is also unlikely to be read widely or perused by the general public. It has been criticized in the *Times* (London) as a "750-page document of unsurpassed UN verbosity, intended to be the world's work programme for sustainable development."[18] Whereas attaching the Rio Declaration to Agenda 21 might well have ensured its consignment to oblivion in future years, assigning independent status to the Declaration guarantees it a wider audience. It has also been suggested that this status gives it "the effect of what many describe as 'soft law.' "[19]

The Stockholm Conference had raised public awareness about our ailing planet. The Rio Summit extended this interest worldwide as television and radio carried its message to the far ends of the earth. The Stockholm Conference attracted only two Heads of Government, Indian Prime Minister Indira Gandhi and Swedish Prime Minister Olaf Palme.[20] The Earth Summit at Rio drew over one hundred Heads of State and Government.[21] Dr. Michael Oppenheimer of the Environmental Defense Fund explained that "You can't be treated as a world leader on any issue without being a player on the environment."[22] And as Brad Knickerbocker commented in the *Christian Science Monitor*, "Since the last gathering convened 20 years ago, both the seriousness of global environmental problems and general awareness about them have increased dramatically, as has the level of human suffering due to related poverty."[23]

In 1972, the USSR and its Eastern European allies boycotted the Stockholm Conference because the German Democratic Republic could not participate equally with other member nations.[24] At Rio there was more global participation both at the governmental and non-governmental level, reflecting the seriousness of environmental degradation and possibly also illustrating the effectiveness of the Stockholm Conference in heightening international awareness of the problem.

The following comparison of the two documents of principle will include quotations in italics from both, with appropriate references.[25] The clauses of the two documents have been grouped under subheadings which explain the essential subject matter of the various principles. Those portions of the principles which are relevant to the theme of the subheading will be quoted. Occasionally, an important principle may be repeated, as it may require analysis under different subheadings. It was felt that this method of proceeding was preferable to following one or the other document seriatim, in view of the fact that the two documents merge and diverge continually. For the reader's convenience, the full texts of the two Declarations are reprinted as Appendix 1 and Appendix 2 to this book.

Population

The Stockholm Declaration underlined the significance of the population issue. Its preamble states:

> *The natural growth of population continuously presents problems on the preservation of the environment, and adequate policies and measures should be adopted, as appropriate, to face these problems.* (*Stockholm Declaration*, Preamble 5)

The body of the Stockholm text was, however, rather vague with respect to specific solutions for a problem which was already recognized as fundamental by most member states of the United Nations.

> *Demographic policies, which are without prejudice to basic human rights and which are deemed appropriate by Governments concerned, should be applied in those regions where the rate of population growth or excessive population concentrations are likely to have adverse effects on the environment or development, or where low population density may prevent improvement of the human environment and impede development.* (*Stockholm Declaration*, Principle 16)

In a very real sense, it could be argued that the implementational failures of Stockholm in a variety of areas propelled and galvanized the frenzied activity before Rio. The intervening two decades had not been kind or benign for either the planet or its dominant species. The world's population, 3.5 billion at the time of the Stockholm Conference,[26] had risen to approximately 5.5 billion by the time delegates gathered in Rio for the Earth Summit.[27] It is interesting to note that the first United Nations Conference on Population, held in Rome in 1954, had projected a world population figure of 3.5 billion by 1980.[28] In light of those predictions, the actual figure is even more alarming. "Between 1950 and 1985, world population grew at an annual rate of 1.9 per cent compared with 0.8 per cent in the half-century preceding 1950."[29] In the 1990s, it is estimated that the Earth has to feed, house, and clothe an additional 92 million people each year, 88 million of them inhabiting the developing world.[30] The United Nations estimates that world population will reach 7 billion by the year 2010.[31] Although the crisis had escalated in seriousness over the intervening twenty years, the delegates at Rio were even less capable of decisive action than their predecessors at Stockholm.

The failure of the Rio Declaration to mention the population issue in precise and clear terms was perceived as a fundamental flaw. Hence, what was excluded from the body of principles was as important as what was ultimately included. James Brooke of the *New York Times* suggested that "[c]lauses in documents concerning population growth were watered down after closed-door lobbying by delegates from the Vatican and Saudi Arabia, both for religious reasons."[32] This criticism is echoed by Christopher Young, who asserted: "The Vatican, with support from Muslim fundamentalist countries, fought successfully to draw the teeth from any declaration about the need for population control, on which other attempts at worldwide environmental progress may well depend. Heavy lobbying by the Catholic Church has managed to remove any direct mention of family planning or population control from the relevant Earth Summit documents."[33]

The position of the Catholic Church was ably presented at the Earth Summit by His Eminence Angelo Cardinal Sodano, Secretary of State for the Vatican, and by Archbishop Renato R. Martino, Apostolic Nuncio and Head of the Holy See Delegation to the United Nations Conference on Environment and Development. Asserting that the position of the Holy See "regarding procreation is frequently misinterpreted,"[34] Archbishop Martino insisted that the "Catholic Church does not propose procreation at any cost."[35] He specified that "[w]hat the Church opposes is the imposition of demographic policies and the promotion of methods for limiting births which are contrary to the objective moral order and to the liberty, dignity and conscience of the human being."[36]

While conceding that "[e]veryone is aware of the problems that can come from a disproportionate growth of the world population,"[37] Cardinal Sodano emphasized the linkage between the poverty of the many and the wastage of resources by the few. Echoing Pope John Paul II, the Cardinal reminded delegates at UNCED that "the pollution of the environment and risks to the ecosystem do not come primarily from the most densely populated parts of the planet."[38] Statistical analyses support the Cardinal's position. At the present time, 85 per cent of the world's income is enjoyed by a mere 23 per cent of its population.[39] Maurice Strong, who organized UNCED and was Secretary-General at both international conferences, highlighted the serious inequity between the rich and poor nations by pointing out that a child born in the developed world would consume twenty to thirty times more of the planet's resources than a child born in a developing nation.[40] The per person energy consumption of Europeans is ten times that of Africans. North Americans consume twenty times the energy utilized by Africans.[41] It was not unnoticed that the position of the religious leaders was sympathetic to and in tune with the attitude of many delegates from the developing nations. Glenn Godfrey, Attorney General of Belize, denied that overpopulation is one of the main causes of environmental degradation and insisted that "poverty is caused not by a growing population, it is caused rather by the failure of the developmental process to distribute the wealth . . . in a socially just manner."[42] In an editorial, the *Times* (London) was critical of the Holy See:

> What the Vatican has done, with a hint of mischief making for its own purposes, is to orchestrate the voice of Third World resentment towards such Western demographic arrogance. But it would have been far more honest to have let the argument come to the surface at Rio than to try to forestall it by diplomatic pressure.[43]

In fairness it has to be noted that developing nations are keenly aware of the serious nature of their population problem. Nawaz Sharif, Pakistan's Prime Minister at the time of the Rio Conference, agreed that "[d]eveloping countries must assume their full share of responsibility in limiting population growth to manageable levels."[44]

There was considerable finger-pointing and mutual recrimination before and during the Earth Summit. The consideration of world population control fell victim both to the influence of religious tradition and to North-South wrangling. Eugene Linden, writing in *Time*, put the matter in appropriate if gloomy perspective: "If human numbers and consumption continue to rise unabated, there is little hope for the other creatures with whom we share the earth and a high probability of catastrophe for humanity itself."[45] British Prime Minister John Major emphasized the consequences of ignoring the population crisis by stating that the Rio process "has no chance of success if we do not do much better in our efforts to slow the growth of population." He added that if humanity failed to take such measures the Earth would destroy itself.[46] Despite the lengthy discussions which preceded Rio, the best the delegates could agree on was:

> *To achieve sustainable development and a higher quality of life for all people, States should . . . promote appropriate demographic policies. (Rio Declaration*, Principle 8)

Clearly, in tackling the most basic and fundamental social, economic, and environmental problem facing the planet, the delegates at Rio had regressed from the ambiguities agreed to at Stockholm. *Time* referred to this as "perhaps the worst example of bureaucratic obfuscation."[47]

Political Causes

The historic chasm between North and South—a chasm of their shared imperial past—haunted delegates at both Stockholm and Rio. As the victims of decades of economic deprivation, the developing nations inevitably focussed on the legacy of colonial rule and its relation to environmental degradation. Their former imperial masters, now enjoying the status of developed countries, were less anxious to bring these political issues of the past to the surface in what had been termed an environmental discussion. In the end the South won, to an extent. Both Declarations referred to the political issue uppermost in the minds of the delegates from the developing world.

> *Man has the fundamental right to freedom, equality and adequate conditions of life, in an environment of a quality that permits a life of dignity and well-being, and he bears a solemn responsibility to protect and improve the environment for present and future generations. In this respect, policies promoting or perpetuating* apartheid, *racial segregation, discrimination, colonial and other forms of oppression and foreign domination stand condemned and must be eliminated. (Stockholm Declaration*, Principle 1)

The concern is expressed again in Principle 15 of the Stockholm Declaration:

> *Planning must be applied to human settlements and urbanization with a view to avoiding adverse effects on the environment and obtaining maximum*

social, economic and environmental benefits for all. In this respect projects
which are designed for colonialist and racist domination must be abandoned.
(*Stockholm Declaration*, Principle 15)

In the 1970s, colonialism with its consequences was still clearly a priority on the agenda of international problems insofar as developing nations were concerned. As Dr. W. K. Chagul, Tanzania's Minister of Economic Affairs and Development Planning, commented, "The evils of *apartheid*, racial and colonial oppression, far from being irrelevant, are at the very core of environmental problems in Africa due to the degradation they cause to the human resources by taking away the rights of the many and thereby bringing benefits to only a minority."[48] Latin American nations echoed the complaints of the Africans by arguing that "the 'economic imperialism' of multi-national corporations, based in the U.S. and elsewhere, was depriving them of effective control of their economies, and resulting in the rapaciously wasteful spoliation of their resource bases, carried out under absentee managers who had no real concern for the local environment."[49]

By the time the Rio Conference convened in 1992, the emphasis of the political concerns of the developing nations had adjusted to a more marked concentration on the fate of the Palestinians, particularly those in the Occupied Territories under Israeli rule. In the preliminary planning phases for UNCED, it became evident that "the conference was to be as much about the North-South dialogue between the rich and the poor as it was about the environment."[50] The North-South dialogue zeroed in on the Israeli occupation of Arab land to highlight an issue which is a sore point with proponents and opponents of Israeli actions with respect to its Middle East neighbors. The Arabs, and the Palestinian representatives in particular, were anxious to incorporate some form of condemnation of oppression and occupation into the Rio Declaration. Farouk Kaddoumi, Head of the Political Department of the Palestine Liberation Organization, stated at UNCED that "[p]rograms aimed at environmental upgrading are closely related to the necessity of removing all forms of oppression."[51] Mr. Kaddoumi emphasized the consequences of Israeli control over Palestinians: overcrowding, deforestation, reduction of the area of cultivable land, overgrazing, reduction of available water supplies, use of herbicides by settlers in the Occupied Territories, soil erosion, and desertification.[52] The Palestinian delegate concluded that the condition of occupation precluded the fulfilment of sustainable development.[53] The Syrian Vice-President, Abdul-Halim Khaddam, included Israeli activities in Lebanon in his denunciation of Israel's environmental record.[54]

During the preparatory process for UNCED, Israel objected strenuously to the attempt to castigate its policies in the Occupied Territories in the drafts of the statement of principle.[55] At Rio, however, the Israelis were more positive in tone. Conceding that "conflicts and disputes all over the world prevent genuine cooperation,"[56] Dr. Uri Marinov, Director-General of Israel's Ministry of the Environment, reminded delegates of the environmental discussions which formed part of the ongoing Middle East peace process and suggested that "[t]he environmental negotiations of the current peace talks should be used first for environmental purposes; but the opportunity they present as a basis for overcoming political controversy should not be ignored."[57]

According to the *Times* (London), the Americans brokered a deal on this Middle Eastern controversy over the Rio Declaration: "A deal was struck by the Americans on Israel's behalf . . . by which the reference will be removed from the summit's giant work programme, Agenda 21, while remaining in the declaration itself and the document was agreed unchanged."[58]

The controversial clause in the Rio Declaration reads:

> *The environment and natural resources of people under oppression, domination and occupation shall be protected. (Rio Declaration,* Principle 23)

The satisfaction of the developing nations was expressed by the Minister of Foreign Affairs of the United Arab Emirates. Rashid Abdullah al-Noaimi informed colleagues at UNCED that his nation welcomed "the principle contained in the agreed declaration of the Conference pertaining to the need to provide protection for the environment and natural resources of the peoples suffering from oppression, domination and occupation."[59]

National Sovereignty

> *States have, in accordance with the Charter of the United Nations and the principles of international law, the sovereign right to exploit their own resources pursuant to their own environmental policies . . . (Stockholm Declaration,* Principle 21)

It is evident that most developing nations, having suffered years of foreign rule and exploitation, are still wary of international phraseology which they believe may restrict the exercise of their recently acquired independent status. This hesitancy was evident at Stockholm. During the lengthy process which resulted in the formulation of the Stockholm Declaration, the delegation from the People's Republic of China (then a new member of the United Nations)[60] proposed its own series of principles, which included the following rather significant clause: "Any international agreement should respect the sovereignty of all countries. No country should encroach on another under the pretext of environmental protection."[61] The Chinese suggestion is reflected in the Stockholm Declaration.

The emphasis on national sovereign rights was by no means exclusively a concern of the developing world. The Canadian delegation presented a draft of principles for incorporation in the Stockholm Declaration. The first of those principles reads: "Every state has a sovereign and inalienable right to its environment, including its land, air and water and to dispose of its natural resources."[62] It is apparent that Canada "played a central role in the drafting of Principle 21."[63]

Although the United Nations Charter bases the organization "on the principle of the sovereign equality of all its Members,"[64] one wonders whether the frequent restatement of national sovereignty is really necessary in light of the growing realization that environmental catastrophe is a global concern and will require international efforts to alleviate the plight of its victims. It would appear that all nations, rich and poor, are anxious to preserve their national rights, even while they concede the global nature of the problems they are tackling.

The rather wordy tribute to national sovereign rights that was approved at Stockholm was repeated almost verbatim in the Rio Declaration. The only difference was the addition of two significant words in the Rio version:

> . . . *the sovereign right to exploit their own resources pursuant to their own environmental* and developmental *policies* . . . (*Rio Declaration*, Principle 2; emphasis mine)

With the inclusion of developmental policies, the Member States of the United Nations underscored the twin objectives of UNCED—namely, environment and development. They also raised this principle to the second in ranking order, possibly signifying its importance.

In view of the fact that the Rio Declaration in its Preamble reaffirmed the Stockholm Declaration and specifically sought to "*build upon it*" (*Rio Declaration*, Preamble, para 3), the restatement verbatim of the principle of sovereign rights in the 1992 document would appear to be redundant. Its very presence in both formulations suggests the significance of the issue of national sovereignty to Member States of the United Nations.

As the world has collectively intervened militarily to free Kuwait from Iraqi aggression and has intervened humanely to feed the starving in Somalia, a declaration justifying and emphasizing the primacy of national sovereignty would appear to be almost regressive in terms of the universalist ideals of the Earth Summit. Moreover, the European and North American member states of NATO came under increasing public pressure to intervene in Bosnia to force the Serbs to stop killing the residents of Sarajevo. As William Thorsell commented in the *Globe and Mail* (Toronto), "At the UN, the principle of self-authorized intervention in the 'domestic affairs' of nations for 'higher' purposes is emerging *ad hoc*."[65]

Simultaneously, there is a definite backlash from both developing and developed nations which find safety in clinging to the trappings of nationalism and are loath to concede to the new internationalism which accompanies environmental concerns. Wolfgang Burhenne and Marlene Jahnke, editors of *Environmental Policy and Law*, explained the situation at UNCED very lucidly:

> The problems of *national sovereignty* came again to the fore. Clearly, when world responsibility is accepted for environmental danger, some States have difficulty in accepting that this will also entail giving up some of their rights and accepting new responsibilities. On the one hand, this is understandable. On the other, all states have to accept that we have only one earth and that such sensitive aspects can only be solved with compromise on both sides.[66]

The addition of "developmental" considerations in the Rio Declaration and its placement in tandem with environmental policies also signalled a new effort at integration of the two concepts—a perception developed in the light of events arising from the Stockholm Conference, when it became apparent from the plethora of subsequent discussions, analyses, and conferences involving numerous facets of these issues that the interconnection had to be stressed. It is now obvious that overdevelopment in some countries has resulted in environmental degradation while underdevelopment in others has also caused environmental decline. The World Commission on Environment and Development expressed this realization:

> Environment and development are not separate challenges; they are inexorably linked. Development cannot subsist upon a deteriorating environmental resource base; the environment cannot be protected when growth leaves out of account the costs of environmental destruction. These problems cannot be treated separately by fragmented institutions and policies. They are linked in a complex system of cause and effect.[67]

The difficulty lies not so much in realizing this interconnection in theory but in keeping it at the forefront in practice. The practice of such integrated thinking will require considerable effort and adjustment of policies in the realm of resource management and business in every nation.

International Cooperation

Having deferred substantively in both documents of principle to the primacy of national sovereign rights, the delegates at Stockholm and Rio were equally eager to demonstrate their commitment to the concept of international coop- eration. Realizing that environmental restoration cannot proceed exclusively in the national sphere, the Member States of the United Nations pledged themselves to seek universalist solutions to the serious crisis of planetary decline. A study of both Declarations illustrates the tacit adherence to the idea of international cooperation in environmental matters. The movement from Stockholm to Rio is indicative of mounting world apprehension because of environmental degradation, and it therefore reflects not so much a new trend as a more specific sense of direction.

> *A growing class of environmental problems, because they are regional or global in extent or because they affect the common international realm, will require extensive co-operation among nations and action by international organizations in the common interest. The Conference calls upon Govern- ments and peoples to exert common efforts for the preservation and improve- ment of the human environment, for the benefit of all the people and for their posterity. (Stockholm Declaration, Preamble 7)*

The Stockholm Declaration called for

> *[i]nternational co-operation . . . in order to raise resources to support the developing countries in carrying out their* [environmental] *responsibilities. (Stockholm Declaration, Preamble 7)*

These appeals for international effort and cooperation were echoed in Prin- ciple 24:

> *International matters concerning the protection and improvement of the envi- ronment should be handled in a co-operative spirit by all countries, big or small, on an equal footing. Co-operation through multilateral or bilateral arrangements or other appropriate means is essential to effectively control, prevent, reduce and eliminate adverse environmental effects resulting from activities conducted in all spheres, in such a way that due account is taken of the sovereignty and interests of all States. (Stockholm Declaration, Princi- ple 24)*

It is interesting to note that in the very principle stressing the need for cooperation, mention had to be made once again of the importance of State sovereignty.

At Rio, delegates included a brief mention of the concept of cooperation in their Preamble:

> With the goal of *establishing a new and equitable global partnership through the creation of new levels of cooperation among States, key sectors of societies and people,* (*Rio Declaration*, Preamble, para 4)

and also in the Preamble:

> Working towards *international agreements which respect the interests of all and protect the integrity of the global environmental and developmental system, . . .* (*Rio Declaration*, Preamble, para 5)

The concept of international cooperation was spelled out in far greater detail in the Rio document, possibly because the experience of two decades between Stockholm and Rio had made this matter of critical importance. Hence, while governments around the world reiterate their commitment to national rights, they are now keenly aware of their mutual vulnerability when facing ecological crises. "Major disasters—Valdez, the Brazilian forests, fisheries, Chernobyl, 3-mile Island, the Gulf War and droughts in Africa—have added sharpness and urgency to the world's concern."[68] The pace of these disasters appears to have intensified. At Seveso, Italy, 193 people were injured in a dioxin leak in 1976.[69] An accident at a chemical plant in 1979 in Novoslbirsk, USSR, killed 300 people.[70] Leakage in 1984 at a pesticide plant in Bhopal, India, resulted in the death of approximately 2,500 people.[71] Although disasters of this nature occur within national boundaries, as the Chernobyl example shows the consequences can be international. The world is waking up to the fact that although we may not be under nuclear threat at the moment, there are other potentially serious and life-threatening dangers lurking in every nation of this planet. According to Frances Cairncross, writing in the *Economist*, there is more consensus now than there was in 1972 that "[i]nternational agreement is the best way to solve environmental problems that transcend national borders."[72]

Despite the many environmental agreements and treaties which have marked the two decades between Stockholm and Rio, few would dispute the fact that "[t]he world's environment is more degraded and is less stable than it was 20 years ago."[73] The World Watch Institute has estimated that Earth lost approximately 500 million acres of trees in the intervening two decades,

"an area roughly one-third the size of the continental U.S."[74] Equally catastrophic for the growth of food crops, the world lost about 500 million tons of
topsoil, "an amount equal to the tillable soil coverage of India and France
combined."[75] Food production per capita declined in ninety-four countries
between 1985 and 1989.[76] Vice-President Al Gore commented that "[m]od-
ern industrial civilization, as presently organized, is colliding violently with
our planet's ecological system."[77]

Environmental degradation is proceeding at an alarming pace in both developed and developing nations. No country is immune to the impact of damage to ecosystems. The shared problem makes cooperative effort the only
viable approach. The World Commission on Environment and Development
recognized the new reality:

> Until recently, the planet was a large world in which human activities and
> their effects were neatly compartmentalized within nations, within sectors
> (energy, agriculture, trade), and within broad areas of concern (environmen-
> tal, economic, social). These compartments have begun to dissolve. This
> applies in particular to the various global "crises" that have seized public
> concern, particularly over the past decade. These are not separate crises:
> an environmental crisis, a development crisis, an energy crisis. They are all
> one.[78]

Cooperation at Rio was framed along fairly precise lines. Although the
language of the Declaration is more bureaucratic than visionary, it is probably more useful than the more nebulous pronouncements which were formulated at Stockholm. Inevitably, the theme of cooperation underlies both
documents. The more specific content of the Rio Declaration is indicative
of the more serious attention paid to environmental cooperation since the
1970s:

> *States shall cooperate in a spirit of global partnership to conserve, protect
> and restore the health and integrity of the Earth's ecosystem. (Rio Declara-
> tion, Principle 7)*

Having stressed the concept of global partnership—implying joint responsibility—the Declaration went on to outline the areas wherein this cooperation
should occur:

> *States should cooperate to promote a supportive and open international eco-
> nomic system that would lead to economic growth and sustainable develop-
> ment in all countries, to better address the problems of environmental
> degradation. (Rio Declaration, Principle 12)*

The recognition of the need for more justice in the distribution of economic benefits was not merely an act of deference to developing nations. It reflects a growing realization in the developed, richer nations that the so-called Third World is set on a developmental course regardless of the environmental consequences, simply because these poor nations have no other choice if they are to give their present populations some kind of decent life. As Al Gore has suggested: "Rapid economic improvements represent a life-or-death imperative throughout the Third World. Its people will not be denied that hope, no matter the environmental costs. As a result, that choice must not be forced upon them."[79] One of the great challenges of the post-Rio process already under way is the implementation of this principle in the cause of global sustainable development and, hopefully, a cleaner environment.

International Cooperation with Respect to Environmental Damage

The Rio Declaration urges States to cooperate to prevent transboundary pollution.

> *States should effectively cooperate to discourage or prevent the relocation and transfer to other States of any activities and substances that cause severe environmental degradation or are found to be harmful to human health. (Rio Declaration, Principle 14)*

This issue is of more than academic concern, especially since the signing of the North American Free Trade Agreement between Canada, the United States of America, and Mexico.[80] *Newsweek*, referring to a survey by the Government of the United States, stated that "four of five American companies operating plants across the border in Mexico admitted they were there to take advantage of weak environmental laws."[81] It will take an enormous amount of government-to-government persuasion to ensure that the economic boom now proceeding in Mexico does not result in an environmental catastrophe.

In 1986, an accident at a nuclear reactor in Chernobyl, USSR, killed at least twenty-five people and spread radioactive fallout across Europe.[82] "Current estimates predict anything from 14,000 to 475,000 cancer deaths worldwide from Chernobyl. No one will ever know for certain."[83] The reaction to this disaster at UNCED was strong and the language of the principles more firm than is normal for such documents:

> *States shall immediately notify other States of any natural disasters or other emergencies that are likely to produce sudden harmful effects on the environ-*

ment of those States. Every effort shall be made by the international community to help States so afflicted. (*Rio Declaration*, Principle 18)

and this further provision:

*States shall provide prior and timely notification and relevant information to
potentially affected States on activities that may have a significant adverse
transboundary environmental effect and shall consult with those States at an
early stage and in good faith.* (*Rio Declaration*, Principle 19)

The mandatory nature of the language is very significant both in terms of the
concerns expressed and with respect to the precise obligations placed on
Member States of the United Nations regarding their responsibilities should
such an event occur. The Rio language carries more punch than the milder
provision on toxic substances included in the Stockholm Declaration:

*The discharge of toxic substances or of other substances and the release of
heat, in such quantities or concentrations as to exceed the capacity of the
environment to render them harmless, must be halted in order to ensure that
serious or irreversible damage is not inflicted upon ecosystems. The just struggle of the peoples of all countries against pollution should be supported.*
(*Stockholm Declaration*, Principle 6)

The delegates at the 1972 Conference followed the approach of opting for the
further creation of international law to deal with transboundary pollution:

*States shall co-operate to develop further the international law regarding
liability and compensation for the victims of pollution and other environmental damage caused by activities within the jurisdiction or control of such
States to areas beyond their jurisdiction.* (*Stockholm Declaration*, Principle
22)

This provision was repeated in the Rio Declaration:

*States shall also cooperate in an expeditious and more determined manner
to develop further international law regarding liability and compensation for
adverse effects of environmental damage caused by activities within their
jurisdiction or control to areas beyond their jurisdiction.* (*Rio Declaration*,
Principle 13)

The problem of transboundary environmental damage was, interestingly
enough, linked both at Stockholm (Principle 21) and at Rio (Principle 2)
with the emphasis on national sovereign rights, already discussed above. The
following provision appeared in the Stockholm Declaration:

States have . . . the responsibility to ensure that activities within their jurisdiction or control do not cause damage to the environment of other States or of areas beyond the limits of national jurisdiction. (Stockholm Declaration, Principle 21)

This part of Principle 21 was repeated verbatim in the Rio Declaration (Principle 2). The linkage of sovereign rights and national responsibilities was largely credited to Canada's active participation at Stockholm. As the Canadian Government explained,

> Principle 21 of the 1972 Stockholm Declaration represented a watershed in ✗
> international environmental law in acknowledging the sovereign right of
> states to exploit their own resources and the responsibility of states "to
> ensure that activities within their jurisdiction or control do not cause dam-
> age to the environment of other states or of areas beyond the limits of na-
> tional jurisdiction." Canada played a central role in the drafting of Principle
> 21. The conceptual framework provided by Principle 21 has been applied
> successfully by Canada in other negotiations, and is the legal foundation of
> virtually every international environmental agreement and legal instrument
> concluded since Stockholm.[84]

The Right to Development

The provisions on development were of primary importance to delegates at both Conferences because, although the principles were not deemed to be legally binding, there was an implicit feeling that the more the South could wrest in developmental assistance promises from the North the more likely it was that the Declarations could be utilized to hold the rich nations to account for their lapses with respect to sharing the world's wealth equitably. Both Declarations highlight the importance of development for the poorer nations and acknowledge the primacy of that concept. Indeed, it could even be suggested with a degree of justification (albeit a trace of exaggeration) that the Rio Declaration is almost a Sustainable Development Charter for the South. There are at least six principles in the Rio Declaration which deal specifically with the concerns of developing nations. The Stockholm Declaration includes the subject in its Preamble and in approximately six of its principles. It is, furthermore, arguable that both documents are filled with indirect references which apply to the poor and rich nations.

It is important to stress that developmental priorities are reflected in the Declaration on the Right to Development adopted by the United Nations General Assembly in 1986, which specifies that "States have the primary responsibility for the creation of national and international conditions favourable to the realization of the right to development."[85] The South could argue at Rio that it was simply acting in line with the principles of the Declaration on the Right to Development. The following principle incorporated into the Rio Declaration illustrates this point:

> *The right to development must be fulfilled so as to equitably meet developmental and environmental needs of present and future generations. (Rio Declaration*, Principle 3)

The nations of the South, in their rush to development, threaten to destroy the global environment even more vigorously than did the North in its somewhat slower Industrial Revolution. Self-interest in both North and South now dictates an approach which would assist Southern development in a manner that is not environmentally degrading so that the atmosphere, the soil, and the water of the entire planet can ensure the future survival of people in all parts of the globe. The South now holds a crucial card in its favor and has played it to the utmost in continually reminding the North that developed nations contribute most of the pollution of the planet, consume most of the resources, and generate most of the waste, while developing nations shoulder most of its economic backwardness, its degrading poverty, and its enormous debt load.[86]

The shared imperial past of Northern and Southern nations has left a legacy of inequitable distribution of the world's limited wealth. Political subjugation resulted in economic exploitation. Southern resources and markets were critical to the industrial development of imperial nations like Britain and France. In the past, the South had no choice but to contribute to the prosperity of the North. In fact, its contribution to Northern affluence still outweighs the foreign aid it receives from developed nations. The World Bank estimated that at the end of 1991 the total debt stock of developing nations was $1,281 billion.[87] This crippling debt burden can be traced back to the 1970s, when sudden increases in oil prices and low interest rates induced many developing nations to borrow from the rich nations. As interest rates rose in the 1980s, these countries found it increasingly difficult to meet their debt obligations. "Between 1973 and 1980, Third World debt increased by a factor of four, to $650 billion."[88] The consequences for the environment are dramatic, as Ben Jackson explains:

> Rising poverty and the desperate attempts of Third World countries to earn hard currency carry not only a heavy human and economic cost but also an environmental one. Many indebted nations rely on their natural resources—

whether timber or minerals like copper—to raise foreign exchange from trade. . . . Environmental resources are being exhausted just to pay debt.[89]

Although official development assistance to the South totalled a significant $49.7 billion by 1988, there was still a "massive, perverse redistribution of income" because the outward flow from Southern nations exceeded these payments by about $8 billion.[90] UNICEF estimated that about half a million children in the developing world died in 1988 because "social progress in Third World countries has been stalled or reversed by crushing debts and falling revenues."[91] Clearly, the South is still subsidizing the high standard of living in the North. It was inevitable that the delegates at Rio would have to confront this issue—the most fundamental in the North-South dialogue—in formulating all the agreements which were produced at that Conference.

At Stockholm in 1972 the delegates, realizing that global environmental degradation was a consequence of industrialization in the North and under-development in the South, believed that narrowing the gap between the rich and poor nations was one important solution.

> *In the developing countries most of the environmental problems are caused by under-development. Millions continue to live far below the minimum levels required for a decent human existence, deprived of adequate food and cloth-ing, shelter and education, health and sanitation. Therefore, the developing countries must direct their efforts to development, bearing in mind their prior-ities and the need to safeguard and improve the environment. For the same purpose, the industrialized countries should make efforts to reduce the gap between themselves and the developing countries. In the industrialized coun-tries, environmental problems are generally related to industrialization and technological development. (Stockholm Declaration, Preamble 4)*

Unfortunately, the ideals of Stockholm failed to materialize. Jean Charest, Canada's Minister of Environment, reminded delegates at the Rio Earth Sum-mit that in the past "thirty years, income disparities between the North and the South have grown from twenty times to *sixty* times." He commented that "this trend is simply not sustainable."[92] By 1992, the escalation of the world's economic problems made for sharper reactions at Rio. This time actual blame was assigned to the rich nations:

> *. . . States have common but differentiated responsibilities. The developed countries acknowledge the responsibility that they bear in the international pursuit of sustainable development in view of the pressures their societies place on the global environment and of the technologies and financial resources they command. (Rio Declaration, Principle 7)*

The nations of the South could justify their somewhat rancorous mood at Rio with plenty of statistical evidence. To compare the situation with respect to two prominent nations, the United States of America and India: The U.S.A., with 5 per cent of the world's population, consumes 25 per cent of its energy, emits 22 per cent of all CO_2 produced, and accounts for 25 per cent of the world's GNP.[93] On the other hand, India, with 16 per cent of the world's population, uses a mere 3 per cent of its energy, emits three per cent of all CO_2 produced, and accounts for only 1 per cent of global GNP.[94] And India is by no means among the world's poorest nations, for in that type of comparison, the disparities would be far greater. From the perspective of the North-South divide, the North, "which has 25 percent of the world population, consumes 70 percent of the planet's energy, 75 percent of its metals, 85 percent of its wood."[95]

To understand the shift in mood over the two decades between Stockholm and Rio, one could compare the gentle suggestion in the Stockholm Declaration:

The non-renewable resources of the earth must be employed in such a way as to guard against the danger of their future exhaustion and to ensure that benefits from such employment are shared by all mankind. (Stockholm Declaration, Principle 5)

with the far more strident tone in the Rio Declaration:

To achieve sustainable development and a higher quality of life for all people, States should reduce and eliminate unsustainable patterns of production and consumption. . . . (Rio Declaration, Principle 8)

Critics derided the weakness of the Rio provision calling for the elimination of unsustainable patterns of development. Frank McDonald of the *Irish Times* referred to the "watered-down reference to the contentious issue of over-consumption by the rich North."[96] He also pointed to the use of the milder word "should" rather than "shall" in Principle 8 and commented that this "tells its own tale about the nit-picking at the core of UNCED."[97] It has to be remembered, however, that the United States of America, whose consumer society was uppermost in the mind of those who supported Principle 8, was itself strenuous in rejecting any condemnation of its affluent way of living. American delegates insisted "over and over that 'the American lifestyle is not up for negotiation.' "[98] Given that approach by the world's only superpower, the inclusion of any provision on lifestyle limitation may be deemed a bold and progressive development. The South may not have won entirely at Rio, but its rhetoric was strident. An Indian environmentalist, Maneka Gandhi, pointed out that a single Western child consumes as much as 125 Eastern children do. She concluded that "nearly all the environmental degradation in the East is due to consumption in the West."[99]

If the developing nations could be said to have had a "wish list" at the Earth Summit, there are indications of their desires in the Rio Declaration. The eradication of poverty was primary:

> *All States and all people shall cooperate in the essential task of eradicating poverty as an indispensable requirement for sustainable development, in order to decrease the disparities in standards of living and better meet the needs of the majority of the people of the world. (Rio Declaration, Principle 5)*

Among developing nations there are also disparities with respect to income, resources, and degree of development. Inevitably, the poorest of these nations have elicited universal compassion, as illustrated in the United Nations operation to feed the starving people of Somalia in 1993. The leading role played by the United States of America in Operation Restore Hope is in sharp contrast to the somewhat self-centered positions espoused by American delegates at Rio, positions which drew criticism from delegates and the media alike. At Rio, delegates agreed that

> *[t]he special situation and needs of developing countries, particularly the least developed and those most environmentally vulnerable, shall be given special priority. (Rio Declaration, Principle 6)*

At Stockholm, delegates assumed that one solution to such grinding poverty was price stability:

> *For the developing countries, stability of prices and adequate earnings for primary commodities and raw material are essential to environmental management since economic factors as well as ecological processes must be taken into account. (Stockholm Declaration, Principle 10)*

Unfortunately, the intervening two decades had seen little or no improvement in this regard. In fact, the poorest nations of the world have few alternatives to primary-commodity trading. A year after the Stockholm Conference, in 1973, oil-producing countries in the Middle East combined to raise the price of oil, which skyrocketed from a little over one dollar a barrel at the beginning of the 1970s to approximately forty dollars in 1979.[100] Nations involved in commodity production of tin, coffee, and cocoa[101] also attempted to cash in on what appeared to be a new global economic structure of more equity and fairness in trade. Their enthusiasm was to be short-lived. The lack of economic diversity in many developing countries—part of the imperial heritage which geared these economies to the particular needs of the colonial power—meant that often these nations, short of foreign exchange, tended to overproduce their major agricultural commodity to earn ready cash. This led to gluts and more problems as Northern nations, anxious to protect their own farming sector, protected themselves against commodity imports.[102] As Ivan Head explains:

> In the countries of the South . . . there is in all-too-many instances an overwhelming dependence on a single economic activity. Simply stated, most economic eggs are in one basket. And to make dependence even more keen, more often than not those eggs assume the form of a single agricultural commodity—coffee, cocoa, sugar, sisal, ground-nuts, etc.—or a single mineral—tin, copper, gold, as examples.[103]

Slower growth and a world recession also led to a decline in demand in the North for some commodities from the South.[104] Ivan Head describes the consequences of such undiversified economic systems:

> Should world prices collapse in one of these commodities, or should access to a major market be blocked, the results can be catastrophic. If the bulk of foreign exchange earnings is derived from that product, government income falls, social programs suffer, all related economic activity (transportation, processing, etc.) stagnates, and the country finds itself in desperate circumstances. This is the reality of an undiversified economy. This is the disadvantage of a seller in a buyer's market.[105]

Should the North succeed in developing a more equitable trading system to encourage developing nations to diversify economically and to export goods which the North no longer manufactures, there is a possibility that less Northern funding will have to be allotted for aid projects in those nations.

Developing nations in their wish list also included the following provision:

> *States should cooperate to promote a supportive and open international economic system that would lead to economic growth and sustainable development in all countries, to better address the problems of environmental degradation. (Rio Declaration, Principle 12)*

The developing nations also feared that environmentalism could become another means of justifying trade restrictions on imports into the North from Southern nations. This was a major concern at Rio. Malaysia put up a stiff resistance against all attempts at environmental labelling of Southern products. The Malaysians defeated these green-labelling plans, which would have enabled consumers to avoid buying timber from rain forest sources. As the South does not generally use eco-labelling on products from the North, the Malaysians were able to convince delegates that such practices by the North would establish a double standard.[106] In response to Southern apprehensions, the following terms constituted a second paragraph to Principle 12:

> *Trade policy measures for environmental purposes should not constitute a means of arbitrary or unjustifiable discrimination or a disguised restriction on international trade. Unilateral actions to deal with environmental challenges outside the jurisdiction of the importing country should be avoided. Environmental measures addressing transboundary or global environmental problems should, as far as possible, be based on an international consensus.* (*Rio Declaration*, Principle 12)

The last sentence of Principle 12 in the Rio Declaration is more precise than the following vague provision in the Stockholm Declaration:

> *International matters concerning the protection and improvement of the environment should be handled in a co-operative spirit by all countries, big or small, on an equal footing.* (*Stockholm Declaration*, Principle 24)

Some of the Rio formulations, though weak from an environmental point of view, have more clarity than the cloudy promises undertaken by Member States at Stockholm. With respect to world trade and environmental policies, for example, Stockholm's submission reads:

> *The environmental policies of all States should enhance and not adversely affect the present or future development potential of developing countries, nor should they hamper the attainment of better living conditions for all, and appropriate steps should be taken by States and international organizations with a view to reaching agreement on meeting the possible national and international economic consequences resulting from the application of environmental measures.* (*Stockholm Declaration*, Principle 11)

Development: Technology Transfer

High on the South's wish list is the issue of science and availability of technology. The developing nations are well aware that their economic situation precludes the investment in research and development which continues to afford a safely comfortable future for the economies of the North. In a world in which technology becomes obsolete so quickly, sustained commitment to scientific research has become the key to economic progress. The South has always demanded that the North accord it the benefits of this technology so that it can also develop at a pace acceptable to its burgeoning population. This issue was as important at Stockholm as it was at Rio, and the repetition and emphasis on this matter underscores the feeling among the poorer nations that they have been deprived by the rich nations not merely of the past and the present but of their future as well. Hence the issue of technology transfer was on occasion specified and sometimes implied in the Stockholm Declaration.

Environmental deficiencies generated by the conditions of underdevelopment and natural disasters pose grave problems and can best be remedied by accelerated development through the transfer of substantial quantities of financial and technological assistance as a supplement to the domestic effort of the developing countries and such timely assistance as may be required. (Stockholm Declaration, Principle 9)

The Declaration further provides that

[s]cience and technology, as part of their contribution to economic and social development, must be applied to the identification, avoidance and control of environmental risks and the solution of environmental problems and for the common good of mankind. (Stockholm Declaration, Principle 18)

followed by this surprisingly specific principle:

Scientific research and development in the context of environmental problems, both national and multinational, must be promoted in all countries, especially the developing countries. In this connexion, the free flow of up-to-date scientific information and transfer of experience must be supported and assisted, to facilitate the solution of environmental problems; environmental technologies should be made available to developing countries on terms which would encourage their wide dissemination without constituting an economic burden on the developing countries. (Stockholm Declaration, Principle 20)

Additionally, Principle 12 called for "*additional international technical and financial assistance*" for developing countries.

Unfortunately, emphasis in a body of principle, albeit an international code of principle, was no guarantee of performance by member nations which acceded to the Stockholm Declaration. In the two decades between Stockholm and Rio, developing nations found that access to Northern technological advances was neither easy nor cheap. In 1980, eight years after the Stockholm promises, developing countries paid approximately two billion dollars in fees and royalties, mainly to industrial countries.[107] Although there is widespread support for the sharing of technology, there is an important obstacle which prevents widespread fulfilment of this Stockholm aspiration. Industrial countries have evolved elaborate legal systems for protecting the patent rights of inventors and discoverers. The fact that many of these are in private hands makes it "difficult for governments to transfer them on noncommercial terms."[108]

Because of the importance of proprietary rights in industrialized societies, it is argued that patent protection acts as an incentive for the development of new technology.[109] The statistics bear out the significance of this point: in 1980, "industrialized market economies accounted for 65 per cent of the world total of patents granted, and the socialist countries of Eastern Europe held 29 per cent."[110] The total for developing countries was a mere 6 per cent, with many patents being granted to nonresidents.[111] There is a crying need for a sharing of the world's scientific and technological resources. As the World Commission on Environment and Development has explained: "The promotion of sustainable development will require an organized effort to develop and diffuse new technologies, such as for agricultural production, renewable energy systems, and pollution control. Much of this effort will be based on the international exchange of technology: through trade in improved equipment, technology-transfer agreements, provision of experts, research collaboration and so on."[112] Vice-President Gore of the United States of America called for the "rapid creation and development of environmentally appropriate technologies—especially in the fields of energy, transportation, agriculture, building, construction, and manufacturing capable of accommodating sustainable economic progress without the concurrent degradation of the environment" and for the quick transfer of this information to all nations, especially the developing nations.[113] However, he also stressed the importance of more secure protection for patents,[114] in view of the fact that "the dissemination of new, appropriate technologies will likely be critical to our success in saving the environment."[115]

The Rio Declaration emphasizes that

> *States should cooperate to strengthen endogenous capacity-building for sus-* should
> *tainable development by improving scientific understanding through ex-*
> *changes of scientific and technological knowledge, and by enhancing the*
> *development, adaptation, diffusion and transfer of technologies, including*
> *new and innovative technologies. (Rio Declaration, Principle 9)*

Although the operative word was "should" and not the stronger "shall," given the political climate resulting from the North-South divide at Rio, and the somewhat controversial stance adopted by the Bush Administration,[116] this provision for technological transfer was probably the best that could be achieved. E-Hyock Kwon, Minister of Environment for the Republic of Korea, reflected the South's perceptions at Rio when he insisted that "[t]he policies, information and technologies for sustainable development should be available and accessible to all countries."[117]

Implementation of Environmentally Safe Development

At Stockholm, delegates concluded that the key to successful performance of development projects with an environmental focus was through the concept of planning, mainly to be exercised at the national level.

> *Rational planning constitutes an essential tool for reconciling any conflict between the needs of development and the need to protect the environment.* (*Stockholm Declaration*, Principle 14)

The concept of "planning" was applied to the issues of human settlements and urbanization:

> *Planning must be applied to human settlements and urbanization with a view to avoiding adverse effects on the environment and obtaining maximum social, economic and environmental benefits for all.* (*Stockholm Declaration*, Principle 15)

and to resource management:

> *In order to achieve a more rational management of resources and thus to improve the environment, States should adopt an integrated and co-ordinated approach to their development planning so as to ensure that development is compatible with the need to protect and improve the human environment for the benefit of their population.* (*Stockholm Declaration*, Principle 13)

Planning for environmental improvement was to be a fundamental task of nation states:

> *Appropriate national institutions must be entrusted with the task of planning, managing or controlling the environmental resources of States with the view to enhancing environmental quality.* (*Stockholm Declaration*, Principle 17)

Although the concept of planning has now in some minds acquired a socialist tinge—social democracies like Nehru's India engaged in elaborate five-year economic plans, as did Communist states like the Soviet Union—there was clearly a need felt in the 1970s for each State to produce a coherent system for tackling environmental concerns. In the 1970s the concept of planning was the method resorted to by many of the newly independent nations, which utilized the idea to develop their fledgling economies. However, planning was easier than performance. As events of the 1970s and 1980s wreaked economic havoc on many developing nations, the guiding concepts of the Stockholm era had to be trimmed to the realities of the new economic world order which illustrated that economic success came not from massive governmental action plans but from the operation of free-market forces.

The concept of planning was not repeated in the Rio Declaration. Instead, more specific duties were assigned to national governments:

[1.] *States shall enact effective environmental legislation. (Rio Declaration,* Principle 11)

[2.] *States shall develop national law regarding liability and compensation for the victims of pollution and other environmental damage. (Rio Declaration,* Principle 13)

It is important to note that this latter obligation had also been included in Principle 22 of the Stockholm Declaration.

Transboundary waste disposal, now a serious threat to many societies, was prohibited—albeit somewhat mildly—at Rio:

[3.] *States should effectively cooperate to discourage or prevent the relocation and transfer to other States of any activities and substances that cause severe environmental degradation or are found to be harmful to human health. (Rio Declaration,* Principle 14)

[4.] *National authorities should endeavour to promote the internalization of environmental costs. (Rio Declaration,* Principle 16)

[5.] *Environmental impact assessment, as a national instrument, shall be undertaken for proposed activities that are likely to have a significant adverse impact on the environment and are subject to a decision of a competent national authority. (Rio Declaration,* Principle 17)

[6.] *States shall immediately notify other States of any natural disasters or other emergencies that are likely to produce sudden harmful effects on the environment of those States. (Rio Declaration,* Principle 18)

[7.] *States shall provide prior and timely notification and relevant information to potentially affected States on activities that may have a significant adverse transboundary environmental effect and shall consult with those States at an early stage and in good faith. (Rio Declaration,* Principle 19)

The obligations undertaken at Rio, if adhered to by Member States of the United Nations, do form a cohesive body of environmental guiding principle. It is clear that Rio moved ahead of Stockholm in specifying the duties of governments.

Development: The Apprehensions of Developing Nations

In 1972 the Brazilian delegation argued that "environment is a conspiracy of the rich to keep us in a state of happy savagery."[118] No nation reflects the change in mood from the 1970s to the 1990s as clearly as does Brazil. Brazil's transformation from the bête noire of Stockholm[119] to the welcoming host at Rio is a dramatic illustration of the fact that developing nations have now realized that environmental concerns are their problem and not merely some Northern fad.

Very briefly: confrontation erupted at Stockholm over Brazil's plan to construct a dam on the Pirana River, which it shares with Argentina. Brazil argued that developing nations could not afford the luxury of environmental protection, and it lobbied strongly against a draft principle which stated: "Relevant information must be supplied by states on activities or developments within their jurisdiction or under their control whenever they believe, or have reason to believe, that such information is needed to avoid the risk of significant adverse effects on the environment in areas beyond their national jurisdiction."[120] This principle, originally numbered 20, "was to be the only section of the Declaration on the Human Environment tabled and left for debate in the next meeting of the General Assembly."[121] When in late 1972 the United Nations General Assembly approved the Stockholm decisions, Principle 20 of the draft Declaration was not re-inserted into the final text and "had been effectively erased."[122]

The consultative principle was adopted at Rio:

> States shall provide prior and timely notification and relevant information to potentially affected States on activities that may have a significant adverse transboundary environmental effect and shall consult with those States at an early stage and in good faith. (Rio Declaration, Principle 19)

Developing nations were initially quite skeptical about the proposed environmental gathering at Stockholm. Development and escape from crippling poverty were uppermost in the minds of Southern governments, not environmental clean-up and conservation. As Wade Rowland comments eloquently on the range of perceptions of the South when the idea of an environmental conference was proposed, opinion among the developing nations ranged from an assumption that problems relating to the environment were a concern for the highly-developed nations alone . . . to a belief that the developed nations were using environmental doomsday predictions as a racist device to keep the non-white third world at a relatively low level of development.[123]

The convening of a conference of scientists and development experts from the South at Founex, Switzerland, in June 1971 served to bring the developing nations on board in support of the Stockholm agenda.[124] The meeting at Founex "had a major impact in expanding the international environment agenda beyond concerns about conservation and pollution to wider issues including flows of development assistance, trade and development."[125] The integration of development and environment had not yet occurred, but this linkage was inevitable once scientific research confirmed the serious nature of global environmental decline. The participants at Founex reported that "developing countries must view the relationship between development and environment in a different perspective. In their context, development becomes essentially a cure for their major environmental problems."[126] This awareness that the developing world had a different set of priorities was to become central to the formulation of both the Stockholm and Rio Declarations. It was this consciousness of differing roles which made the creation of both bodies of principle a rather delicate balancing act. If the ultimate instruments were not completely satisfactory to anyone, that in itself is a reflection of the level of compromise which had to be accepted for universal agreement.

The most serious apprehension of the South was that environmental concerns would impede development in the poor nations. Hence it was imperative to make certain that this did not happen. At Stockholm one way to ensure this was to keep restating the priorities of developing nations. Hence, in the Stockholm Declaration *"environmental deficiencies"* were attributed to *"conditions of underdevelopment"* (Principle 9). The requirements of developing nations for *"stability of prices and adequate earnings for primary commodities and raw materials"* were deemed *"essential to environmental management"* (Principle 10). Most important was the following caution:

> *The environmental policies of all States should enhance and not adversely affect the present or future development potential of developing countries.* (*Stockholm Declaration*, Principle 11)

and this provision:

> *Resources should be made available to preserve and improve the environment, taking into account the circumstances and particular requirements of developing countries.* (*Stockholm Declaration*, Principle 12)

Clearly, the Stockholm Declaration conceded much to the agenda of the developing nations in attempting to meet the apprehensions of these nations about the possible economic threat posed by Northern concerns about the environment.

At Rio, the process begun at Stockholm went even further. *"Developed countries acknowledge the responsibility that they bear in the international pursuit of sustainable development"* (Principle 7). The rich were put on notice to eliminate *"unsustainable patterns of production and consumption"* (Principle 8). The South has consistently stressed the fact that it is Northern wasteful processes which are leading to environmental degradation. This discussion was no less vehement at Rio than it was at Stockholm. Indeed, by 1992 the Southern delegates were armed with twenty years of depressing statistics to demonstrate both that the Stockholm measures had by and large failed and that the North had created even more environmental problems in the two decades between the Conferences.

The success of the South in including its agenda in the Rio Declaration can be gauged by the extent of criticism levelled at the final product emerging from the long and bitter negotiating sessions which occurred over the creation of this body of principle. According to Geoffrey York and James Rusk, reporting in the *Globe and Mail*, there was a clear feeling that environment had been subordinated to development.[127]

The South's apprehensions concerning its ability to keep up with Northern environmental standards (and possible resulting discrimination if it failed to do so) were to some extent palliated by recognition that

> *environmental standards, management objectives and priorities should reflect the environmental and developmental context to which they apply. Standards applied by some countries may be inappropriate and of unwarranted economic and social cost to other countries, in particular developing countries.* (*Rio Declaration*, Principle 11)

This principle may be compared with its Stockholm predecessor, which stated:

> *Without prejudice to such criteria as may be agreed upon by the international community, or to standards which will have to be determined nationally, it will be essential in all cases to consider the systems of values prevailing in each country, and the extent of the applicability of standards which are valid for the most advanced countries but which may be inappropriate and of unwarranted social cost for the developing countries.* (*Stockholm Declaration*, Principle 23)

The South also insisted at UNCED that *"trade policy measures for environmental purposes should not constitute a means of arbitrary or unjustifiable discrimination or a disguised restriction on international trade"* (*Rio Declaration*, Principle 12).

Finally, on this issue of development, from Stockholm to Rio, the world progressed from endorsing the essential need for development to accepting it as a human right.

> *Economic and social development is essential for ensuring a favourable living and working environment for man and for creating conditions on earth that are necessary for the improvement of the quality of life.* (*Stockholm Declaration*, Principle 8)

At Rio, nations agreed that

> *[t]he right to development must be fulfilled so as to equitably meet developmental and environmental needs of present and future generations.* (*Rio Declaration*, Principle 3)

The version above was accepted after amendment of the draft sponsored by Pakistan and China which referred to the inalienable right to development of poor nations.[128]

Despite the criticism levelled (often justifiably) at the vagueness of the Rio Declaration, there can be no doubt that for the world's poorest and most dispossessed, it at least affords recognition that their present state of economic deprivation has not gone unnoticed by the world. Teatao Teannaki, President of the Republic of Kiribati, stated that he was "particularly pleased to note that recognition is given to special needs of the least developed and those most vulnerable to environmental problems."[129] In inching its way toward recognizing the validity of the human right to development (with environmental safeguards), the world has taken an initial step towards alleviation of the plight of millions alive today and millions more yet to be born.

Conservation

The Stockholm Declaration, in its rather lengthy Preamble, expresses the concern which brought delegates from many nations to Sweden to deal with the problems of the environment:

> *The protection and improvement of the human environment is a major issue which affects the well-being of peoples and economic development throughout*

the world; it is the urgent desire of the peoples of the whole world and the duty of all Governments. (Stockholm Declaration, Preamble)

Both Declarations have ambiguous, if well-intentioned, principles which because of their vagueness appear to decline to the level of platitudes. The delegates at Stockholm held that

[t]he natural resources of the earth including the air, water, land, flora and fauna and especially representative samples of natural ecosystems must be safeguarded for the benefit of present and future generations through careful planning or management, as appropriate. (Stockholm Declaration, Principle 2)

followed by this equally vague pronouncement:

The capacity of the earth to produce vital renewable resources must be maintained and, wherever practicable, restored or improved. (Stockholm Declaration, Principle 3)

and by this rather nebulous principle:

The non-renewable resources of the earth must be employed in such a way as to guard against the danger of their future exhaustion and to ensure that benefits from such employment are shared by all mankind. (Stockholm Declaration, Principle 5)

Collectively, these appear to be classic examples of United Nations "internationalese," a language of promises without commitments. It was left to the delegation of the United States to present the positive aspect of Principle 2 (cited above) in the Stockholm Declaration. The American delegate declared the provision to be of "notable importance."[130]

Realization of the weakness of these principles prompted the delegates at Rio to attempt to be more specific in terms of actions to be undertaken to achieve the Stockholm principles. This attempt was not successful:

States shall cooperate in a spirit of global partnership to conserve, protect and restore the health and integrity of the Earth's ecosystem. (Rio Declaration, Principle 7)

and this provision:

In order to achieve sustainable development, environmental protection shall constitute an integral part of the development process and cannot be considered in isolation from it. (Rio Declaration, Principle 4)

The critical issue of conservation underlies all attempts at environmental improvement in every nation of the world. The failure of the Rio Declaration to produce a forthright pronouncement urging the world in the direction of conservation was a serious weakness. Conservation is undoubtedly the key to environmental success for the future. If we manage to conserve and use less of the Earth's resources now, we can possibly progress towards the universally accepted goal of sustainable development.

Polluter-Pays Principle

One of the unforeseen results of the Industrial Revolution was the fact that the movement to produce and sell inexpensive manufactured goods to growing numbers of consumers carried certain invisible costs that were borne by society as a whole. Minerals were often strip-mined and the ground destroyed in the frenzy to gain raw materials. Agricultural land was rapidly gobbled up by the ever growing urban and suburban concrete jungles which became the norm in much of North America and western Europe. Factories polluted the air and tainted the water. Frequently, they were situated next to rivers and lakes into which sewage was regularly dumped without much thought of future consequences. Municipal sewage systems were also constructed to dump raw waste into oceans. Eventually this decades-long activity caught up with mankind. The realization set in that industrialists in particular had made free use of air and water supplies, polluting and befouling them at no cost to themselves. These were hidden costs which the entire national community, and eventually the entire planetary community, would pay. As Ben Jackson comments: "The stress on economic growth as the goal for development also fails to take account of the future impact of present actions. It places the emphasis on profitable living now without considering the price—environmentally (and therefore probably economically)—to be paid later." Jackson highlights the consequence of this development-oriented approach: "For many environmentalists the greatest weakness of the 'more growth' solution is its failure to anticipate the future threat to the environment."[131]

In terms of past actions which have resulted in pollution, the clean-up costs are staggering. "Estimates of the cost of remedial steps worldwide to overcome the effects of pollution range as high as U.S. $300 billion annually."[132] The implementation of an environment-oriented agenda worldwide will entail not merely clean-up of past problems but action to prevent future pollution. Prior to the Earth Summit its Secretary-General, Maurice Strong, suggested in an interview that the costs to implement the ambitious Rio program would amount to an annual sum of $625 billion, with approximately 80 per cent of the costs being borne by developing nations, supplemented by an annual sum of $125 billion from the developed nations. Strong went on to explain that as development assistance already totalled approximately $55 billion, another $70 billion would be required.[133]

The irresponsible consumption of air and water has resulted in much of the damage which the world now has to remedy. However much nations may balk at Strong's figures, the harsh reality is that the world will pay for pollution, either now or later. The World Commission on Environment and Development explained the problem:

> Air and water have traditionally been regarded as "free goods," but the enormous costs to society of past and present pollution show that they are not free. The environmental costs of economic activity are not encountered until the assimilative capacity of the environment has been exceeded. Beyond that point, they cannot be avoided. They will be paid. The policy question is how and by whom they will be paid, not whether. Basically, there are only two ways. The costs can be "externalized"—that is, transferred to various segments of the community in the form of damage costs to human health, property, and ecosystems. Or they can [be] "internalized"— paid by the enterprise. The enterprise may invest in measures to prevent the damages and, if the market for its product allows, pass the costs along to the consumer. Or it may invest in measures to restore unavoidable damage—replanting forests, restocking fish, rehabilitating land after mining. Or it may compensate victims of health and property damage. In these cases too, the costs may be passed on to the consumer.[134]

A growing realization that costs must rise to pay for past pollution has been balanced with a determination to ensure that in the future polluters and not entire populations bear the burden of their actions. As Ivan Head comments, "The age-old practices of discharging wastes into flowing streams, and fumes into the atmosphere to be blown away by prevailing winds, are now subject not just to criticism but, increasingly, to censure and judicial constraint."[135] The polluter-pays principle is the logical consequence of such thinking. The aim of the policy is to force manufacturers to internalize environmental costs and transfer these to their prices for products.[136] Member countries of the Organization for Economic Cooperation and Development (OECD, based in Paris) agreed in 1972, the year of the Stockholm Conference, to adhere to the polluter-pays principle in their environmental policies.[137]

The success of incorporating environmental costs into product prices has been largely in the developed nations, where industries have been able to afford conversion to safer methods of manufacture and where consumers have been able to afford the resulting higher prices. The developing nations have not been as successful, particularly with respect to their exports, because environmental "costs continue to be borne entirely domestically, largely in the form of damage costs to human health, property and ecosystems."[138] Nor is this situation seen as a serious danger by a number of governments of developing nations.

At Stockholm, some developing nations demonstrated their unwillingness to shoulder the financial burden arising from environmentally unsafe production, possibly because this would blunt their competitive edge in world trade. The idea that they could be compensated for trade losses was rejected by leading developed nations like the United States of America, Canada, and Britain.[139] "Developing nations . . . argued loud and long that to expect the indigents of the world to accept financial responsibility in environmental trade upsets, on exactly the same basis as the wealthy industrial nations, was not only unfair but palpably absurd. The developing nations had their back to the wall as it was."[140] Given this defensive stance, it was almost inevitable for States at Stockholm to emphasize that

> *[t]he environmental policies of all States should enhance and not adversely affect the present or future development potential of developing countries, nor should they hamper the attainment of better living conditions for all, and appropriate steps should be taken by States and international organizations with a view to reaching agreement on meeting the possible national and international economic consequences resulting from the application of environmental measures.* (*Stockholm Declaration*, Principle 11)

Aside from a provision that "*States shall co-operate to develop further the international law regarding liability and compensation for the victims of pollution*" (*Stockholm Declaration*, Principle 22), there was no adherence to the polluter-pays principle in this instrument prepared in 1972.

By 1992 the mood had changed, as the developing nations became increasingly aware of the hidden costs of their competitive trading edge in certain commodities and products. The social consequences were visible from Africa to Southeast Asia. Hence at Rio the following principle was adopted:

> *National authorities should endeavour to promote the internalization of environmental costs and the use of economic instruments, taking into account the approach that the polluter should, in principle, bear the cost of pollution,*

with due regard to the public interest and without distorting international trade and investment. (Rio Declaration, Principle 16)

Delegates at Rio went further to prevent states from risking the environment and using the justification that there was scientific uncertainty about the dangers posed by their activities:

In order to protect the environment, the precautionary approach shall be widely applied by States according to their capabilities. Where there are threats of serious or irreversible damage, lack of full scientific certainty shall not be used as a reason for postponing cost-effective measures to prevent environmental degradation. (Rio Declaration, Principle 15)

Whether these provisions will be implemented globally remains to be seen. Certainly, in the post-Rio process now under way, it is critical that world public opinion monitor potential polluters in every nation to ensure that the polluter-pays principle is implemented. Although nations are not legally bound by the principles of Rio, in this instance the moral commitment might well be enforced by citizens whose vigilance may be the only way to ensure that all of society does not continue to suffer from the actions of a few.

Public Participation in Environmental Improvement

Environmental improvement is unlike any other major issue because it demands the commitment and action of every man, woman, and child on this planet. Because environmental decline is caused by people, the clean-up and prevention of pollution demand the active and continuing participation of citizens in every nation. Hence it is even more a people's issue than it is a governmental one. Although the restoration of major polluted sites requires government funding and action, the prevention of pollution involves a thoughtful consideration of the fate of the planet by each and every human being who inhabits it. Environmental consciousness has grown because of the dedication of thousands of men and women in every nation who have created, from a grassroots beginning, a movement which is now global and which can now compel governments to act according to its aspirations. Delegates at Stockholm were aware of the significance of the popular factor in raising environmental concerns:

To achieve this environmental goal will demand the acceptance of responsibility by citizens and communities and by enterprises and institutions at every level, all sharing equitably in common efforts. Individuals in all walks of life as well as organizations in many fields, by their values and the sum of their

actions, will shape the world environment of the future. . . . The Conference
calls upon Governments and peoples to exert common efforts for the preserva-
tion and improvement of the human environment, for the benefit of all the
people and for their posterity. (*Stockholm Declaration*, Preamble 7)

Environmentalism became a serious obligation for the individual:

Man . . . bears a solemn responsibility to protect and improve the environment
for present and future generations. (*Stockholm Declaration*, Principle 1)

This rather vague exhortation followed:

The just struggle of the peoples of all countries against pollution should be
supported. (*Stockholm Declaration*, Principle 6)

The most useful method specified was to be education about environmental
concerns:

Education in environmental matters, for the younger generation as well as
adults, giving due consideration to the underprivileged, is essential in order
to broaden the basis for an enlightened opinion and responsible conduct by
individuals, enterprises and communities in protecting and improving the
environment in its full human dimension. It is also essential that mass media
. . . disseminate information of an educational nature, on the need to protect
and improve the environment in order to enable man to develop in every
respect. (*Stockholm Declaration*, Principle 19)

In the twenty years between the United Nations Conference on the Human
Environment and the United Nations Conference on Environment and Development, the level of public involvement escalated to the point at which
governments appeared to be led—some might say dragged—towards environmental consciousness by their populations. Even though the high-minded
intentions of the Stockholm Conference did not materialize as successfully
as delegates had hoped, the sheer scope of environmental enthusiasm was
given a tremendous boost by the Conference. The level of popular support
for the cause of averting planetary decline spawned its own growth industry
of volunteer non-governmental organizations devoted to environmental concerns, media attention to the topic, investigative reporting of governmental
failures in safeguarding against pollution, and committees of concerned businessmen who became quickly aware of the need to cater to this growing
phenomenon or be boycotted and financially destroyed by it. Prime Minister
John Major of the United Kingdom commented at UNCED on this dramatic
development: "The environment is no longer the specialist concern of a
few—it has become the vital interest of us all."[141]

The galvanized public support for the environmental cause has also been a direct consequence of the fact that governments have dragged their feet over commitments made at Stockholm and that the problems of global pollution have escalated to almost catastrophic levels. His Majesty the King of Sweden, addressing the opening of UNCED, emphasized this concern when he reminded delegates about the uneven progress following Stockholm's great promise: "There has been great environmental improvement on the local, national and regional levels, while the global threats are more serious than ever."[142] The recognition that environmental success—or what there was of it—was largely because of local initiative also underscored the significance of involving the public not only in the debate but in the performance of environmental clean-up. At Rio delegates decided to forge a new global partnership:

> With the goal *of establishing a new and equitable global partnership through the creation of new levels of cooperation among States, key sectors of societies and people,* . . . (*Rio Declaration*, Preamble, para 4)

There was clearly a recognition, brilliantly stated by Maurice Strong, that "the 'carrying capacity' of the Earth could only sustain present and future generations 'if it is matched by the caring capacity of its people and its leaders.' "[143] Accordingly, delegates included the following important language in the Rio Declaration:

> *Environmental issues are best handled with the participation of all concerned citizens, at the relevant level.* (*Rio Declaration*, Principle 10)

It was important not merely to encourage public involvement. It was felt to be equally important to enable the people to gain information about environmental concerns from public authorities. There is a growing realization that governments might attempt to block public access to information about risks to health. The Soviet Government's handling of the Chernobyl disaster is only one of the more glaring and obvious examples of the manner in which governments seek instinctively to conceal rather than reveal their mistakes. "A secret Soviet decree prohibited doctors from diagnosing illnesses as radiation-induced."[144] Such cover-ups can have both national and international consequences. Probably with the Chernobyl example in mind, the Rio delegates decided that

> *at the national level, each individual shall have appropriate access to information concerning the environment that is held by public authorities, including information on hazardous materials and activities in their communities,*

and the opportunity to participate in decision-making processes. States shall facilitate and encourage public awareness and participation by making information widely available. Effective access to judicial and administrative proceedings, including redress and remedy, shall be provided. (Rio Declaration, Principle 10)

The mandatory tone of this provision makes it a prime example of the kind of language which should have been included throughout the Rio Declaration. It is in the forthright formulation of such provisions that the Rio Declaration moved the world in a progressive direction. Although the nonbinding nature of this instrument poses some problems for any individual or group challenging secretiveness in a government, the very fact that this principle exists in such strong language serves notice on governments that the new world order includes openness and free public access in matters of public interest. Had the entire Declaration been formulated in such pungent tones, it could have "provided a vision of how the people of the world could survive together in the next century."[145]

The concept of a global partnership attracted considerable attention at Rio. The Indonesian delegate, Dr. Emil Salim, made the point that the Rio Declaration should lead to the forging of "a new global partnership between nations and peoples, a partnership in which rights and obligations are equitably shared by all, a global partnership based on a renewed and improved division of labor between nations and an equally improved sharing of benefits and efforts between people."[146] The goal of the partnership would be sustainable development and proper management of the planet's resources.[147]

However worthwhile an exercise it may be, the framing of provisions about global partnership will not automatically guarantee success for the goals of environmentalism. As Vincent Perera, Sri Lanka's Minister of Environment and Parliamentary Affairs, explained, "UNCED . . . marks the *establishment* of the global partnership. But the test of the partnership is in its implementation."[148]

War as a Source of Environmental Concern

The environmental group Greenpeace produced a critique of the Rio Declaration in which it alleged that in the creation of the Declaration substance became immaterial as the desire to write a text became the driving force.[149] Nowhere is this point more evident than in the provision about warfare. Twenty years earlier, delegates at Stockholm had agreed that

> *Man and his environment must be spared the effects of nuclear weapons and all other means of mass destruction. States must strive to reach prompt agreement, in the relevant international organs, on the elimination and complete destruction of such weapons.* (*Stockholm Declaration,* Principle 26)

Frank McDonald, writing in the *Irish Times*, commented: "Twenty years later, under pressure from the US which fought for the exclusion of all 'Stockholm-type language' on this issue,"[150] the Rio Declaration merely and mildly suggested that

> *Peace, development and environmental protection are interdependent and indivisible.* (*Rio Declaration,* Principle 25)

and that

> *Warfare is inherently destructive of sustainable development. States shall therefore respect international law providing protection for the environment in times of armed conflict and cooperate in its further development, as necessary.* (*Rio Declaration,* Principle 24)

States were also urged to

> . . . *resolve all their environmental disputes peacefully and by appropriate means in accordance with the Charter of the United Nations.* (*Rio Declaration,* Principle 26)

Greenpeace commented, "If this is progress, we are in deep, deep trouble."[151]

An old African proverb states that when two elephants fight, it's always the grass that gets hurt.[152] Although the existence of a nuclear deterrent kept the two superpowers, the United States of America and the Soviet Union, from direct confrontation during the Cold War, each superpower patronized a number of client states, smaller, often poor nations whose petty conflicts with neighboring countries escalated and became part of Cold War politics. Each superpower armed its client states and fought its Cold War rival to the last drop of blood shed by the men and women of developing nations. No region was exempt from this war by proxy, which became the norm during the Cold War in Latin America, the Middle East, Africa, and Southeast Asia. Having a nuclear deterrent did not mean that there was less war and less misery in the world. War only became a game which rich nations played at the expense of poor countries. Between the Cold War years of 1945 and 1989, 127 wars were fought on this planet. "All but two of them have been in or between developing countries."[153] Occasionally, a superpower was itself dragged into the fray, as was the United States in Vietnam and the USSR in Afghanistan. As Ivan Head explains:

> The decade of the 1980s may well be recorded as one of the most brutal in this or any other century. It was a decade of wars. . . . The decade just past has seen the lengthy slaughter of the Iraq-Iran War as well as bloody civil wars—many of them surrogate conflicts supported or sponsored by super-power champions—in Afghanistan, Angola, Cambodia, El Salvador, Ethiopia, Guatemala, Lebanon, Mozambique, Nicaragua, Peru, Philippines, Sri Lanka and Sudan. It is estimated that in all these wars the death count exceeded 4 million.[154]

Whether the wars were civil or regional or international, the price in terms of human life and environmental destruction was always severe. Growing realization of the direct and indirect costs of war prompted delegates to urge UNCED in the direction of denunciation, if not renunciation, of warfare. Dr. Zvonimir Separovic, representing the new Republic of Croatia, told delegates that "[w]ar is highly detrimental to human wellbeing, to the human environment and development." He reminded delegates of the devastation war had caused in the former Yugoslav territory and urged the Earth Summit to consider the "aggressive nature of human behaviour expressed in violence against other human beings and the environment in the form of war." Indicating that his country's contribution to UNCED "is centred on the environmental impact of war," Dr. Separovic insisted that the Rio Declaration "include a condemnation of war and express concern for the irreparable consequences of war operations" as well as a call for "international action against a new kind of crime which might be called ecocide."[155]

Warfare also diverts the resources and funds of developed and developing nations away from health, education, and environmental clean-up—progressive, worthwhile avenues—to unproductive devastation which exacerbates human misery and environmental degradation. The World Commission on Environment and Development commented that "competitive arms races breed insecurity among nations through spirals of reciprocal fears. Nations need to muster resources to combat environmental degradation and mass poverty. By misdirecting scarce resources, arms races contribute further to insecurity."[156] The amounts involved in military research, production, and consumption, if directed to environmental and human concerns, could have a dramatic and immediate impact. Approximate estimates would suggest that global military spending had already reached $1 trillion by 1986.[157] "Annual military spending is still greater than the combined income of half of humanity."[158]

Developing nations, where most of the wars and consequent death and environmental degradation occur, continue to divert increasing amounts of their scarce resources to the military. "Since 1960, developing countries have increased their military expenditures at a rate that is double the rise in per capita income. As a percentage of GNP, developing countries dedicate 1.6 per cent to health care, compared to 5.2 per cent to military expenditures."[159] In human terms, military expenditure and debt servicing in the developing world (according to UNICEF estimates) cost each family in those poor nations approximately $400 per year.[160] Weapons imports from the North cost developing countries approximately $39 billion per year.[161] "In 1988, military spending in poor countries totalled $145 billion."[162] By the time the World Commission on Environment and Development contributed to the debate on this and related issues, the global arms trade, much of it in the developing world, had consumed over $300 billion.[163] Though there was much wringing of hands over this situation and the consequent loss of resources for more productive uses, delegates at Rio were unable to agree on a firm, principled stand against warfare and its consequences. With respect to this provision, the situation at Rio was clearly a regression from the formulation agreed to at Stockholm.

It was left to a delegate from the Middle East—an area which had recently witnessed the horrors of environmental warfare—to view the warfare provision in an optimistic light. Rashid Abdullah al-Noaimi, Minister of Foreign Affairs of the United Arab Emirates, informed delegates at Rio that his nation welcomed the principle "enunciated in the Declaration . . . which calls upon states to respect the rules of international law which provide for the protection of the environment in times of armed conflict."[164] One can only hope that the type of environmental havoc wrought on the Persian Gulf by Saddam Hussein will not be repeated in other conflict-ridden parts of the world.

Provisions Unique to the Stockholm Declaration

Although we have thus far compared and on occasion contrasted the principles which emerged from both environmental conferences, it is interesting to note the points where the two documents diverge completely. These areas are indicative of trends and tendencies of each era and also illustrative of specific weaknesses in the particular Declaration which failed to enunciate those principles. The present analysis will concentrate on the principles themselves rather than the preambles.

OCEANIC POLLUTION

The issue of oceanic pollution was considered at Stockholm and the following principle adopted:

> *States shall take all possible steps to prevent pollution of the seas by substances that are liable to create hazards to human health, to harm living resources and marine life, to damage amenities or to interfere with other legitimate uses of the sea.* (*Stockholm Declaration*, Principle 7)

Oceanic pollution continued for twenty years despite this high-minded principle. The issue is now so serious that it can be considered a global catastrophe in the making. There is hardly an ocean now which does not regularly receive sewage and waste. As Philip Elmer-Dewitt commented in *Time*:

> Anyone who has been near the seashore lately—or listened to Jacques-Yves Cousteau on TV—knows that the oceans are a mess, littered with plastic and tar balls and rapidly losing fish. But the garbage dumps, the oil spills, the sewage discharges, the drift nets and factory ships are only the most visible problems. The real threats to the oceans, accounting for 70% to 80% of all maritime pollution, are the sediment and contaminants that flow into the seas from land-based sources—topsoil, fertilizers, pesticides and all manner of industrial wastes.[165]

The most apparent form of pollution in the world's oceans occurs when ships carrying oil run aground, an event which is happening with alarming frequency. The whole world watched the horror of oil pouring from a wrecked tanker, the Liberian-registered *Braer*, which ran aground on 5th January 1993 on the southern tip of Mainland, the largest of the Shetland Islands, approximately 160 kilometers northeast of Scotland.[166] The ship was carrying about ninety-three million liters of oil on its journey from Norway to Canada. According to a contemporary Associated Press report, if all this oil spilled into the ocean it would be "twice the amount that was dumped when the Exxon Valdez hit a reef in Alaska on March 24, 1989."[167] Although these dramatic oil spills catch global attention, it must be remembered that oil pollution of the oceans goes on continuously. According to the United Nations, "about 600,000 tons of oil enter the oceans each year as a result of normal shipping operations."[168] It has also been estimated that about six and a half million tons of litter are cast into the world's oceans annually.[169]

There is no dearth of international agreements to protect the oceans. The Antarctica Treaty signed in 1959 and a subsequent Agreement of 1991 protect that fragile region with regulation of waste disposal and marine pollution.[170] Best known of the plethora of treaties is the Law of the Sea Treaty (1982), "the product of more than a decade of often contentious negotiations."[171] Participating States are expected to control diverse sources of ocean pollution, "including discharges and runoff from cities and agriculture, ocean dumping of wastes, releases from boats, oil exploration and drilling, mining, and air pollution deposited in the ocean."[172] This controversial convention also requires "express prior approval by the coastal state for dumping in the territorial sea, in the EEZs [exclusive economic zones], and onto the continental shelf. . . . States also have an obligation under the Law of the Sea to ensure that their activities do not injure the health and environment of neighbouring states and the commons."[173] A decade after its completion, the Law of the Sea Treaty had still not entered into force because its implementation required ratification by sixty signatory nations. Environmentalists like Hilary French suggested with respect to some provisions that "the position of the U.S. government was the biggest obstacle."[174] Despite this opposition, a number of the provisions of the Treaty became accepted as customary international law and were implemented by various nations,[175] "with positive effects on fish stocks, ocean pollution, and freedom of the seas."[176]

By 1992, the year of the Earth Summit, sixty-seven nations had signed on as contracting parties to the Convention on the Prevention of Marine Pollution by Dumping of Wastes and Other Matter (more commonly referred to as the London Dumping Convention). This Convention was established in 1972, the year of the Stockholm Conference, and entered into force in 1975.[177] The Convention prohibited the dumping of radioactive and other forms of dangerous waste into the ocean. It now "outlaws dumping of all forms of industrial waste by 1995," and provided for a ban on ocean incineration of wastes to take effect by the end of 1994.[178]

Clearly, the post-Stockholm process was not very successful in protecting the oceans from either deliberate dumping or accidental pollution. This is an area where very tough international law is called for, law which stiffens penalties and liabilities to such a point that shipping companies are deterred from using any but the most seaworthy vessels. A very stringent regime of fines for deliberate polluters would appear to be one possible solution. Nations have to be made to abide by their pledge to uphold the London Dumping Convention. This commitment states that Contracting Parties will "take all practical steps to prevent pollution of the sea by the dumping of waste and other matter that is liable to create hazards to human health, to harm living resources and marine life, to damage amenities or to interfere with other legitimate uses of the sea."[179] If the *Braer* disaster in Scotland and other such incidents of environmental damage increase public awareness, this could result in more "aggressive action to prevent oil spills, increase inspection of foreign ships and improve emergency preparedness."[180]

At the Rio Conference in 1992, there was awareness of the serious nature of global pollution of the oceans. Kinza Clodumar, Minister of Finance for the Republic of Nauru, in the central Pacific, called for "an immediate and permanent ban on the deliberate dumping of all toxic materials into the oceans, including especially radioactive wastes."[181] The delegate from Nauru reminded his colleagues at the Earth Summit that "dumping at sea accounts for only a fraction of ocean pollution. Three fourths of ocean pollutants enter directly from land, either in runoff or through the air. . . . Ocean pollution from land based sources is an issue that is central to the health of the biosphere. The global community can ignore the issue only at its increasing peril."[182] A delegate from a maritime nation thousands of miles from Nauru was similarly concerned. Thorbjorn Berntsen, Minister of Environment of Norway, emphasized the particular attention his nation has paid to degradation of the marine environment. He proposed curbing land-based pollution, strengthening rules and inspection routines applicable to global shipping, and imposing a permanent moratorium on ocean dumping of radioactive waste. The Norwegian delegate also pointed out that "contaminated nuclear production sites and potential runoff from more or less casually selected land based deposits represent a threat to the marine environment."[183] There was an evident and widespread interest in prohibiting ocean dumping and controlling the level of ocean pollution.

Unfortunately, this obvious level of concern did not translate into any specific recommendations in the Rio Declaration. There were no provisions in that instrument which specifically addressed the issue of the oceans, although it could be argued that the various principles on transboundary pollution might be applicable. Given the seriousness of the crisis, the escalation of the problem despite the existence of so many other international treaties, and the plea of the World Commission on Environment and Development ("the most significant initial action that nations can take in the interests of the oceans' threatened life-support system is to ratify the Law of the Sea Convention"),[184] inclusion of a precise formulation on the oceans in the Rio Declaration would have been not only appropriate but timely. In this very important matter, Rio demonstrated a regression from the stand taken at Stockholm.

To be fair to the delegates at Rio, the issue of the oceans does form an important chapter of the enormous document Agenda 21.[185] In early April 1992, delegates attending the Fourth Session of the Preparatory Committee for UNCED "agreed to strengthen existing global agreements aimed at controlling land-based sources of marine pollution—fertilizers, pesticides, and the like, which account for more than two-thirds of ocean pollution. They have also agreed to improve collection of data on both marine resources and damage from pollution."[186] Simultaneously with the global meeting in Rio, a smaller group of delegates representing twenty-nine countries attended a Vessel Traffic Service symposium in Vancouver, Canada. The symposium addressed concerns about and solutions for ship-source pollution. The remedies for this problem ranged from improved surveillance to enhanced management competence to better communications between nations.[187]

The necessity of including a provision specifically related to oceans in the Rio Declaration can hardly be underestimated. The Declaration is likely to be the most widely read of all the formulations to emerge from that enormous conference. If, as its creators hope, the Declaration serves as an inspiration for future action, exclusion of the issue of the oceans cannot be justified.

> In the Earth's wheel of life, the oceans provide the balance. Covering over 70 per cent of the planet's surface, they play a critical role in maintaining its life-support systems, in moderating its climate, and in sustaining animals and plants, including minute, oxygen-producing phytoplankton. They provide protein, transportation, energy, employment, recreation, and other economic, social, and cultural activities.[188]

WILDLIFE

In *Earth in the Balance*, a thought-provoking analysis of the planet's environmental problems, U.S. Vice-President Al Gore comments that "we are creating a world that is hostile to wildness, that seems to prefer concrete to natural landscapes."[189] In creating and spreading our cement jungles, the human species has exhibited not merely its primacy in the biological scheme of things but its capacity ruthlessly to destroy the natural habitat of countless thousands of species which share the Earth with us. Although environmental action plans propel us to clean and clear the air, atmosphere, water, and land which we have degraded, environmental awareness counsels us to remember the damage we do every day to the millions of large and small inhabitants of the planet whose very survival is unfortunately in our all-too-callous hands. The ultimate ethic of environmental consciousness is to learn again to perceive ourselves not as the masters of the Earth but as one of its life forms. Our survival now depends on a merging of the actions, awareness, and consciousness implicit in a commitment to environmental goals.

At the Stockholm Conference, delegates pointed us in the right direction:

Man has a special responsibility to safeguard and wisely manage the heritage of wildlife and its habitat which are now gravely imperilled by a combination of adverse factors. Nature conservation including wildlife must therefore receive importance in planning for economic development. (Stockholm Declaration, Principle 4)

By the time the Earth Summit met at Rio, this significant principle had matured into an international treaty—the Convention on Biological Diversity[190]—which was signed and is currently being ratified by a number of nations. The concept of wildlife has been enlarged to include life forms in their original pristine natural habitat and considerable attention is now being paid to the preservation of the homes of insects, animals, and plants, not merely for idealistic purposes, but because some of the genetic material carried by these varied species could provide innumerable medical and scientific benefits for human use in the future. "We do not know, to within the nearest 20 million, how many species there really are on earth," comments Paul Harrison.[191] It is incumbent on the human species to ensure the survival of at least the majority of the estimated four to thirty million species.[192] Yet "there is a growing scientific consensus that species are disappearing at rates never before witnessed on the planet."[193] If we estimate the existence of about 30 million species today, we may be destroying about 17,500 of these life forms annually.[194]

The destruction of species is caused by a combination of human need and greed. The burgeoning human population is pushing into the natural habitats of other species, altering the environment so drastically that few other species can co-exist in the same space. In Kenya, the population is "pressing so hard on parks that protected land [a mere 6 per cent of this territory] is steadily being lost to invading farmers."[195] As Mrs. Rahab Mwatha of the Greenbelt Movement in Kenya commented, "We are awakening to the fact that if Africa is dying it is because her environment has been plundered, overexploited, and neglected."[196] The rise in human population has generated extraordinary demands for food; it has almost destroyed the once abundant Atlantic fisheries off Newfoundland, to the east of Canada. In Latin America and parts of Asia, deforestation, brought on by expanding population and the urge to develop, takes a daily toll in species extinction.[197]

It was seemingly realized at the Stockholm Conference that "protection of habitat is the single most effective means of conserving diversity,"[198] hence the mention of habitat in Principle 4 (above). Although nations can take credit for preparing and signing an international Convention on Biological

Diversity at the Earth Summit, it is significant to note that there was no provision pertaining to this all-important issue in the Rio Declaration. Given the ramifications and implications of the threat of species extinction which is occurring relentlessly every day, it might have been wise to lay emphasis on this matter in the Declaration in order to bring it to public attention. The fact that there is now a treaty in existence and that numerous governments have signed it hardly solves the problem of the survival of species on this planet. The international convention is only an initial step in a very long process. The inclusion of a provision on biodiversity in the Rio Declaration could have popularized the concept of species conservation and made it more of a people's issue than it is at the present time. The issue of biodiversity may be too specific for the general principles incorporated into the Rio Declaration. The primacy of the issue, however, and the fact that the Declaration is a veritable hodge-podge of articles, some of which are more precise than others, would argue in favor of its inclusion.

While it could be suggested that the Biodiversity Treaty made the principle redundant, it is also important to remember that the Stockholm and Rio Declarations are intended to serve as signposts for all the people of the planet. It would have been worthwhile in such an instrument to include an issue which is of crucial importance environmentally, economically, and in the long term for the survival of our human species as well. The existence of a binding treaty in international law ought not to preclude adherence to the general principle of species conservation in a document such as the Rio Declaration. It is likely that in future years far more people will read the Declaration than the treaty on biodiversity. If the point of the Declaration was to inform, encourage, and create enthusiasm at the popular level, so important a feature of environmental concern ought to have been included with no prejudice to the legally binding commitments entered into by nations which signed the Convention on Biodiversity.

Principles Unique to the Rio Declaration

WOMEN

If international instruments can be considered a reflection of the times in which they are formulated, then the Rio Declaration bowed to the inevitable in recognizing the significance of women to society and to the environmental movement. The twenty years since Stockholm had witnessed the implementation in most areas of the developed world of the women's revolution, with its resulting shift in attitudes among both men and women about their roles and importance in society. With this revolution has come a growing awareness of the injustice of past treatment of women and a consequent determination to rectify the situation. Unfortunately, the improvement in the condition of women has been mixed, when the subject is viewed internationally. In developing countries, the poorer, uneducated women still suffer a lifelong burden caused simply by their gender. This can range from female infanticide in China to occasional widow burnings in India. Garrick Utley of NBC television reported in 1993 that in Somalia some women accused of adultery had been stoned to death.[199] Child marriage, lack of education, heavy domestic toil, and often grinding physical farm labor are the fate of thousands of women around the world. It was therefore timely and fitting that delegates at the Earth Summit endorsed the following principle for inclusion in the Rio Declaration:

Women have a vital role in environmental management and development. Their full participation is therefore essential to achieve sustainable development. (Rio Declaration, Principle 20)

There was widespread support for inclusion of a provision concerning women. As Margaret Shields, Director of the United Nations Research and Training Institute for the Advancement of Women, reminded delegates at UNCED:

Women are not new to environmental concerns. They are the people who must walk further to collect fuel if an area is deforested. They are the people most affected by pollution of water supplies when they or their families become ill as a result. In rural areas, women are often placed at greatest risk from the use of dangerous products and not only women but also their infant and newborn children. Most important, women are the people who could make an essential contribution to any debate about the fundamental realities of environmental degradation. They are the ultimate consumers of environmental management decisions.[200]

Although the problems of women in developing nations were mentioned at Rio and in the process leading up to the Earth Summit, these concerns were not specifically addressed in the Rio Declaration. In 1986 Mrs. King, representing the Greenbelt Movement in Kenya, told the World Commission on Environment and Development that "women are responsible for between 60 to 90 per cent of the food production, processing, and marketing. No one can really address the food crisis in Africa or many of the other crises that seem to exist here without addressing the question of women, and really seeing that women are participants in decision-making processes at the very basic all the way through up to the highest level."[201] The role of women as a vital labor force in Latin America, Asia, and Africa has long been recognized,[202] but the rather vague formulation produced about women in the Rio Declaration failed to address this issue in any specific manner.

Given the amount of publicity and media attention generated by the Earth Summit, it would have been worthwhile to emphasize—if only for popular consumption—certain key areas of environmental and social concern with respect to women. This is probably why Princess Sonam Chhoden Wangchuck, representing the King of Bhutan at UNCED, emphasized the fact that "the recognition of the role of women, particularly rural women, in sustainable development is crucial as women are motivated by their primary concern to improve their families' quality of life." Pointing out that "too few women are involved in decision-making processes of environmental management and policy making," Princess Sonam Wangchuck expressed her delegation's view that "women's role in leadership and community-based participation must be vital components of sustainable development strategies."[203] The delegation from Thailand was headed by a Princess who is also a professor, Dr. Chulabhorn Mahidol. It was fitting that in her statement to UNCED she stressed that women "must play an equal partnership role in the integration of environment and development."[204]

In March 1992, British Overseas Development Minister Lynda Chalker called for a "change in attitudes towards women in developing countries" and indicated that this would be "a high priority in the British aid programme."[205] Thorbjorn Berntsen, Norway's Minister of Environment, proposed particular efforts to include women in decision making.[206]

Bella Abzug of the United States, Co-Chair of the Women's Environment and Development Organization, reported to the Plenary Session at UNCED about the activities of women's groups at Preparatory Meetings, action which resulted in the inclusion of a principle about the role of women in the Rio Declaration. Ms. Abzug also emphasized the fact that UNCED's mandate directed the Conference's Secretary-General, Maurice Strong, to "ensure that women's critical economic, social and environmental contributions to sustainable development be addressed . . . as a distinct cross-cutting issue in addition to being mainstreamed in all the substantive work and documentation."[207] It is quite apparent that the women's issue aroused international concern and interest.

Not everyone, however, was satisfied with the resulting principle which became part of the Rio Declaration. The Canadian Participatory Committee of non-government agencies which advised on the Declaration was highly critical of the final document. With reference to the statement on women (Principle 20, quoted above), the Committee commented that "as written, the principle is vacuous."[208] This criticism is probably justified, but in evaluating the progress from Stockholm to Rio the existence of a reference to women's participation is illustrative of forward movement, however minimal the actual distance travelled.

INDIGENOUS PEOPLE

In a world of incredibly fast change and rapid destruction of the old in favor of the new, we have in the twenty years since Stockholm begun to realize both the value and the vulnerability of those groups whose preference is an alternative lifestyle more in tune with nature than the frenzied pace an industrialized civilization normally allows. The indigenous people of the world have suffered a cruel and undeserved fate at the relentless hands of majority cultures which have victimized them either by deliberate genocide or by economic deprivation. Indigenous people are

> found in North America, in Australia, in the Amazon Basin, in Central America, in the forests and hills of Asia, in the deserts of North Africa, and elsewhere. . . . The isolation of many such people has meant the preservation of a traditional way of life in close harmony with the natural environment. Their very survival has depended on their ecological awareness and adaptation. But their isolation has also meant that few of them have shared in national economic and social development; this may be reflected in their poor health, nutrition, and education.[209]

There is now a greater awareness of and sensitivity towards the needs of such cultures. It is now mainstream thinking in North America not to view them as obstacles to development, but to perceive them as important contributors to civilizational harmony. Indigenous people form part of the rich tapestry of human culture, and in protecting their right to life and their cultural heritage the world is finally recognizing its debt to its own past. At Rio, delegates endorsed the following principle:

> *Indigenous people and their communities, and other local communities, have a vital role in environmental management and development because of their knowledge and traditional practices. States should recognize and duly support their identity, culture and interests and enable their effective participation in the achievement of sustainable development. (Rio Declaration,* Principle 22)

The United Nations proclaimed 1993 the International Year for the World's Indigenous People.[210] However, despite the international action involved in formulating the universal declaration on indigenous rights,[211] tribal people continued to be slaughtered in countries like Guatemala. Rigoberta Menchú, winner of the 1992 Nobel Peace Prize for her work on behalf of indigenous people in Guatemala, revealed to the world the tragedy of her people: one hundred thousand have died in the past three decades, and over forty thousand have vanished.[212] Indigenous people in a number of nations from Canada to India are agitating for recognition of their status, for an end to their marginalized existence at the outer perimeters of society. Realistically, indigenous leaders like Menchú admit: "I don't think we can have an indigenous nation, alone in the world, at the end of the 20th century."[213] However, it is now widely recognized that extinction is not their inevitable fate nor is it their only alternative. In recognizing the worth of groups which live harmoniously with nature, we are only belatedly admitting that "[t]he Aboriginal viewpoint corresponds closely with the ecological perspective."[214] As Richard Falk comments, "In a fundamental sense, indigenous peoples preserve and embody alternate life-styles that may provide models, inspiration, guidance in the essential work of world order redesign, an undertaking now primarily associated with overcoming self-destructive tendencies in the behaviour of modern societies."[215]

It is significant that the relevant principle in the Rio Declaration is rather general and vague with respect to the role of indigenous peoples. States are not mandated to support or to recognize them; the milder word "should" is used. There is no mention of the inherent rights of indigenous people, nor is there any recognition of their right to political identity within the nation state. It is uncertain how states should support their identity, culture, and interests, so ambiguous is the commitment—if it can even be called that. However, despite its obvious weakness, the provision carries the world a step forward from Stockholm in according international recognition to the world's indigenous people. "Perhaps ironically, the growth of modern communications and transportation has internationalized the struggle of indigenous peoples in the last decade or so."[216] There is now a growing consciousness that we must look to our indigenous roots to find alternate ways of living and create lifestyles which are sustainable and healthy for our planet. Time is running out for humanity. By their suffering, indigenous people remind us of the fate of fragile life forms; by their resilience, they give us hope that all may not yet be lost. This point was made dramatically at UNCED by Ji Chaozhu, Under-Secretary-General of the United Nations Department of Economic and Social Development, who quoted from Native American Chief Seattle's statement of 1855: "Whatever happens to the earth, happens to the people of the earth.

Man did not weave the web of life: he is merely a strand in it. Whatever he does to the web, he does to himself."[217]

Conclusion

After the intense, often feisty negotiating sessions which resulted in the Rio Declaration, most delegates attempted to put a positive face on the result of their wrangling. Hans Alders, Minister of Environment of the Netherlands, commented:

> We see the Declaration as a sound basis for the much needed development of international law, and therefore endorse it as it stands. The Declaration not only reaffirms the Declaration of Stockholm, but it takes matters further, as indeed it should. New important elements in this Declaration are the principle of responsibility for future generations, the precautionary principle, the principle of informed participation in decision-making, the recognition of the rights of indigenous peoples and the importance of youth and the role of women in managing the environment. It is our task to ensure that these principles will be embodied in all future national and international legal and policy instruments.[218]

Not all delegates were willing to perceive progress from the brutal negotiating sessions which had preceded the final production of the Declaration. Giorgio Ruffolo, Italy's Minister for Environment, admitted that he was not happy with some of the formulations in the Declaration.[219]

At the end, the conflict between development and the environment was evident in the way delegates perceived the final product of their deliberations. The South had fought hard to achieve a development-oriented Declaration. The North, more fragmented and less unified than the developing nations, found its concept of an Earth Charter converted into a statement emphasizing the primacy of development in a number of provisions. This was precisely what the South wanted. As Dr. Emil Salim, Indonesia's State Minister for Population and the Environment, remarked:

> [B]ecause the objectives of our Conference include both the environment and development, both aspects must be reflected in the Rio Declaration. Therefore, while affirming the responsibility to undertake global environmental action, the Declaration must also affirm the right of nations to pursue development. Only in this way will be able to counter the destructive potential of environmental degradation and the equally terrible potential of global social and political upheavals.[220]

The Rio Declaration did not satisfy all Southern nations. The Prime Minister of Malaysia, Dr. Mahathir bin Mohamad, believed that it had been "watered down upon insistence from the powerful and the rich."[221]

For the North, there was much dissatisfaction with the Rio Declaration. In framing their initial concept for an Earth Charter, the Canadians sought inspiration from the Universal Declaration of Human Rights. "But with the watered-down Rio declaration, 'the visionaries came up against the lawyers and bureaucrats, and the lawyers and bureaucrats won.' "[222] The problem, as the Canadians saw it, was that developing nations "campaigned for a declaration that would have endorsed their right to develop with little environmental constraint, while blaming the rich for the world's environmental problems."[223] The Canadian Participatory Committee of non-government agencies commented that "some of the principles are so gaseous as to dissolve upon examination."[224] The Austrian delegate was more optimistic. Ruth Feldgrill-Zankel, Austria's Minister of Environment, agreed that "it may appear deplorable that the Rio Declaration did not turn out to be an inspired and inspiring document comparable to the Universal Declaration of Human Rights." However, she felt "confident that we will all live to see the elaboration of a true Earth Charter. In the meantime, we consider the Rio Declaration as an important cornerstone and will give serious consideration to it in our own decision making process."[225] Speaking on behalf of the European Community, Portugal's Minister of Environment, Carlos Borrego, acknowledged that the Declaration contains "many important principles recognized for the first time at the global level" and stated that it reflects "in a balanced way the various interests and concerns both of developing and developed countries."[226]

The fact that the Declaration drew fire from both North and South is indicative that it probably passed the test of a document reflecting true consensus. No one was totally happy with it, but it was more or less acceptable to all nations. The delegate from Sri Lanka, Environment Minister Vincent Perera, suggested that developed nations ought to view the developing countries with "empathy and understanding. Perhaps if they had done so during the preparatory process the lead up to UNCED would have been smoother and less tortuous."[227] While conceding that the Declaration had not lived up to expectations, the delegate from Nauru, Kinza Clodumar, saw the significance of "nations . . . for the first time sitting down together to consider collectively problems of environment and development at the global level."[228]

Although it appears at first glance to be a document geared to the Southern agenda, in a very real sense most Northern nations achieved their major

objectives in the Rio Declaration. The British hoped to enshrine as fundamental principles the precautionary approach, the polluter-pays principle,
and the idea of public access to information.[229] This was largely achieved,
albeit not in the stirring, fluent language initially envisioned by the developed countries. The Government of Sweden stressed the importance of the
polluter-pays principle, the precautionary approach, and the responsibility
of nations to ensure that their internal activities do not endanger the environment of other countries.[230] Norway favored the polluter-pays principle and an
open and fair international trading system, and it insisted that "environmental concerns must not be used as a pretext to introduce discriminatory trade
practices."[231]

It was left to the Japanese to place the Declaration in proper perspective
by calling it a "significant first step in our efforts towards sustainable development."[232] Chancellor Helmut Kohl of Germany suggested that it was part
of a solid foundation for further measures.[233] With an eye on the future,
Prime Minister Anibal Cavaco Silva of Portugal proposed that "[t]he Rio
Declaration will have to serve as a basis for the establishment of new relations between all parties concerned, whether public or private, which will, in
a responsible manner, have to provide answers appropriate to the challenge
facing us."[234] Nations were obviously already gearing up for the process that
would follow Rio even as the endless parade of dignitaries marched past the
eyes of delegates and cameras from around the world. "The principles in-
scribed in the *Rio Declaration* will serve as benchmarks for future progress,"
stated Denmark's Minister for the Environment.[235] King Hussein of Jordan
urged nations to exert "greater efforts to perfect" the imperfect Summit
agreements.[236] Singapore's Minister for the Environment, Dr. Ahmad Mattar,
declared confidently that "the adoption of the Rio Declaration will bring
about a new world ethic towards our living environment."[237]

Canada took the initiative in resuscitating the initial vision of an Earth
Charter, a vision which had faded when the competing interests of North and
South resulted in the somewhat disjointed body of principle which we have
been examining. At Rio, Prime Minister Brian Mulroney suggested that "the
idea of an Earth Charter of environmental rights and responsibilities, which
has slipped beyond our grasp at Rio, should be revived."[238] The Canadian
initiative sought to utilize the existing Rio Declaration as a basis for a "visionary 'Earth Charter' that would integrate these principles."[239] Speaking in
Hull, Quebec, during Environment Week 1992, the Canadian Prime Minister
declared, "Just as the Helsinki Accords set a point of reference for human

rights and responsibilities, so an Earth Charter would set benchmarks for environmental rights and responsibilities."[240]

The Canadian proposal recognized that it is imperative that the process of formulating principles of international environmental law continue in years to come. There is an obvious need for environmental law to become part of the everyday consciousness of men, women, and children around the world. Environmental regeneration cannot be accomplished without extensive public enthusiasm and participation. It is simply not a matter which can be left to governments. If anything, Stockholm and Rio both demonstrated that governments are often far behind their populations in terms of heightened awareness of the significance of this issue. By coming to Stockholm and Rio with the old baggage of national sovereignty and developing versus developed country conflicts, the bureaucrats and politicians ultimately betrayed the idealistic vision which ought to have been incorporated in both instruments of principle. Time alone will tell whether the people will be able to recapture the vision from the politicians and create an Earth Charter which will impel and enthuse all five and half billion of us to work to save this planet.

Notes

1. The Stockholm Conference resulted in the Action Plan for the Human Environment and in resolutions on World Environment Day, Nuclear Weapons Tests, and the Convening of a Second Conference on the Human Environment. (See United Nations Document A/CONF. 48/14).

The delegates attending the 1992 Rio Earth Summit accepted the massive, detailed formulation of specific environmental plans called Agenda 21; agreed to the *Non–Legally Binding Authoritative Statement of Principles for a Global Consensus on the Management, Conservation and Sustainable Development of All Types of Forests* (see United Nations Document A/CONF.151/6/Rev.1, 13th June 1992; reprinted as Appendix 3 to this book); and signed the *United Nations Framework Convention on Climate Change* (see United Nations Document A/AC.237/18 (Part II)/Add. 1, 15th May 1992; reprinted as Appendix 4 to this book), as well as the more controversial Convention on Biological Diversity (5th June 1994; reprinted as Appendix 5 to this book) which the Government of the United States of America refused to sign.

2. *Globe and Mail* (Toronto), 26th December 1992, D6:1.

3. *Newsweek*, 1st June 1992, 31.

4. *Economist*, 30th May 1991, 5.

5. Maurice Strong, interview by Pierre Haski, *Christian Science Monitor*, 29th May 1992, 3:1.

6. World Commission on Environment and Development, *Our Common Future* (Oxford: Oxford University Press, 1989), 332.

7. Dennis Lloyd, *The Idea of Law* (London: Penguin, 1985), 225.

8. The vote was 111 to 1 in favor of passage. See United Nations Document G.A. Resolution 37/7, U.N. GAOR, 37th Session, Supp. No. 51, U.N. Doc. A/37/51 (1983).

9. Howard Mann, "The Rio Declaration," in American Society of International Law, *Proceedings of the 86th Annual Meeting* (Washington, D.C., 1st–4th April 1992).

10. *Ibid.*

11. A. O. Adede, "International Environmental Law from Stockholm to Rio—An Overview of Past Lessons and Future Challenges," *Environmental Policy and Law*, vol. 22, no. 2 (April 1992): 100.

12. For comment on the Stockholm Declaration's customary-law status, see Mann, "The Rio Declaration."

13. Statement by Professor Dr. Her Royal Highness Princess Chulabhorn Mahidol, Personal Representative of the King of Thailand, UNCED, Rio, 5th June 1992.

14. *Irish Times* (Dublin), 11th June 1992, 7:4.

15. *Christian Science Monitor*, 27th March 1992, 7:2.

16. *Globe and Mail*, 2nd June 1992, A3:2.

17. *Christian Science Monitor*, 12th March 1992, 3:4.

18. *Times* (London), 1st June 1992, 15:1–2.

19. *Christian Science Monitor*, 12th March 1992, 3:4.

20. *Summary Report of the Seminar Convened by the Canadian Department of External Affairs and International Trade and the Department of Environment*, Meech Lake, Quebec, 8th–9th December 1991, 5. (Hereinafter cited as *Summary Report*, Meech Lake.)

21. *Vancouver (B.C.) Sun*, 15th June 1992, A9:3.

22. *New York Times*, 14th June 1992, 10:2.

23. *Christian Science Monitor*, 2nd June 1992, 1:5.

24. *Summary Report*, Meech Lake, 5.

25. For the *Stockholm Declaration*, see United Nations Document A/CONF.48/114; for the *Rio Declaration*, see United Nations Document A/CONF.151/5/Rev.1. The texts are reprinted as Appendix 1 and Appendix 2 to this book.

26. Wade Rowland, *The Plot to Save the World* (Toronto: Clarke, Irwin, 1973), 13.

27. *Globe and Mail*, 29th April 1992, A1:6.

28. Rowland, *Plot to Save the World*, 31.

29. Department of International Economic and Social Affairs, *World Population Prospects: Estimates and Projections as Assessed in 1984* (New York: United Nations, 1986), cited in World Commission on Environment and Development, *Our Common Future*, 99.

30. Population Reference Bureau, *1991 World Population Data Sheet* (Washington, D.C., 1991), cited in Sandra Postel, "Denial in the Decisive Decade," in Lester R. Brown et al., *State of the World, 1992* (New York: W. W. Norton, 1992), 3.

31. *Globe and Mail*, 20th May 1992, A2:3.

32. *New York Times*, 14th June 1992, 10:1.

33. *Edmonton (Alta.) Journal*, 5th June 1992, A13:3.

34. Statement by Archbishop Renato R. Martino, Apostolic Nuncio, Head of the Holy See Delegation, UNCED, Rio, 4th June 1992.

35. *Ibid.*

36. *Ibid.*

37. Statement by His Eminence Angelo Cardinal Sodano, Secretary of State of the Vatican, UNCED, Rio, 13th June 1992.

38. *Ibid.*

39. Sandra Postel, "Denial in the Decisive Decade," in Lester R. Brown et al., *State of the World, 1992* (New York: W. W. Norton, 1992), 4.

40. *Times* (London), 4th June 1992, 11:2.

41. *Christian Science Monitor*, 10th June 1992, 3:4.

42. Statement by Glenn Godfrey, Attorney General, Belize, UNCED, Rio, 5th June 1992.

43. *Times* (London), 19th May 1992, 13:1.

44. *New York Times*, 14th June 1992, 10:1.

45. *Time*, 1st June 1992, 30:1.

46. *Times* (London), 13th June 1992, 1:3.

47. *Time*, 1st June 1992, 29:1.

48. Quoted in Rowland, *Plot to Save the World*, 52.

49. *Ibid.*, 53.

50. *Christian Science Monitor*, 8th April 1992, 18:1.

51. Statement by Farouk Kaddoumi, Head of the Political Department, Palestine Liberation Organization, UNCED, Rio, 11th June 1992.

52. *Ibid.*

53. *Ibid.*

54. Statement by Abdul-Halim Khaddam, Vice-President, Syrian Arab Republic, UNCED, Rio, 12th June 1992.

55. *Times* (London), 12th June 1992, 1:7.

56. Statement by Dr. Uri Marinov, Director-General, Israeli Ministry of the Environment, UNCED, Rio, 3rd June 1992.

57. *Ibid.*

58. *Times* (London), 12th June 1992, 1:7. See also "UNCED: Rio Conference on Environment and Development," *Environmental Policy and Law*, vol. 22, no. 4 (August 1992): 224:2.

59. Statement by Rashid Abdullah al-Noaimi, Minister of Foreign Affairs, United Arab Emirates, UNCED, Rio, 9th June 1992.

60. See United Nations Document General Assembly Resolution 2758 (XXVI), 25th October 1971, concerning the recognition of the People's Republic of China and the expulsion of Taiwan from the world organization.

61. Quoted in Rowland, *Plot to Save the World*, 93.

62. *Ibid.*, 88.

63. *Summary Report*, Meech Lake, 7.

64. Charter of the United Nations, Article 2.

65. *Globe and Mail*, 26th December 1992, D6:5.

66. *Environmental Policy and Law*, vol. 22, no. 4 (August 1992).

67. World Commission on Environment and Development, *Our Common Future*, 37.

68. Geoffrey Bruce, "A Review of the Stockholm Conference of 1972, of Some Significant Developments in the Environment since 1972, and of Some Major Challenges and Opportunities for Canada in Preparing for UNCED in 1992," in *Summary Report*, Meech Lake, 24.

69. Helen Caldicott, *If You Love This Planet* (New York: W. W. Norton, 1992), 76.

70. *Ibid.*

71. *Ibid.*

72. *Economist*, 30th May 1992, 5:2.

73. Bruce, "Review of the Stockholm Conference," 24.

74. *Time*, 1st June 1992, 22:2.

75. *Ibid.*

76. *Newsweek*, 1st June 1992, 37:3.

77. Al Gore, *Earth in the Balance* (Boston: Houghton Mifflin, 1992), 269.

78. World Commission on Environment and Development, *Our Common Future*, 4.

79. Gore, *Earth in the Balance*, 279.

80. Notes for an Address by Brian Mulroney, Prime Minister of Canada, to the Kennedy School of Government, Harvard University, Cambridge, Massachusetts, 10th December 1992, 4 (Office of the Prime Minister).

81. *Newsweek*, 1st June 1992, 30:1.

82. Caldicott, *If You Love This Planet*, 76.

83. Nicholas Lenssen, "Confronting Nuclear Waste," in *State of the World, 1992* (New York: W. W. Norton, 1992), 49.

84. *Summary Report*, Meech Lake, 7–8.

85. United Nations, General Assembly Resolution 41/128, 4th December 1986. Art. 3.1, U.N. GAOR Supp. No. 53, U.N. Doc. A/C.3/41/SR (Dec. 4, 1986). Also see R. K. L. Panjabi, Review of James Crawford, ed., *The Rights of Peoples* (Oxford: Oxford University Press, 1988), in *Virginia Journal of International Law*, vol. 30, no. 1 (Fall 1989).

86. For a strong example of the South's position, see Centre for Science and Environment, "CSE Statement on Global Environmental Democracy to be Submitted to the Forthcoming UN Conference on Environment and Development," New Delhi, India.

87. Government of Canada, UNCED Reports, "Debt Conversion Initiative to Promote Sustainable Development," June 1992.

88. Caldicott, *If You Love This Planet*, 130.

89. Ben Jackson, *Poverty and the Planet* (London: Penguin, 1990), 96 97.

90. Ivan L. Head, *On a Hinge of History* (Toronto: University of Toronto Press, 1991), 42–44.

91. *Globe and Mail*, 21st December 1988, A9:1.

92. Statement by Jean Charest, Minister of Environment, Canada, UNCED, Rio, 11th June 1992.

93. *Time*, 1st June 1992, 22.

94. *Ibid.*, 23.

95. *Newsweek*, 1st June 1992, 39:1.

96. *Irish Times*, 11th June 1992, 7:6.

97. *Ibid.*

98. *Time*, 1st June 1992, 32:1.

99. *Newsweek*, 1st June 1992, 39:1.

100. Jackson, *Poverty and the Planet*, 65.

101. *Ibid.*

102. See Head, *On a Hinge of History*.

103. *Ibid.*, 46.

104. See Jackson, *Poverty and the Planet*, chapter 4.

105. Head, *On a Hinge of History*, 46.

106. Dr. Lim Keng Yaik, Malaysian Minister of Primary Industries, interview by Stephen Bradshaw, BBC News and Current Affairs, telecast on the Canadian Broadcasting Corporation's News Network, June 1992.

107. UN Centre on Transnational Corporations, *Transnational Corporations in World Development: Third Survey* (New York: United Nations, 1983); cited in World Commission on Environment and Development, *Our Common Future*, 87.

108. Hilary F. French, "Strengthening Global Environmental Governance," in *State of the World, 1992* (New York: W. W. Norton, 1992), 165.

109. *Ibid.*

110. Commonwealth Working Group, *Technological Change* (London: Commonwealth Secretariat, 1985); cited in World Commission on Environment and Development, *Our Common Future*, 87.

111. World Commission on Environment and Development, *Our Common Future*, 87.

112. *Ibid.*

113. Gore, *Earth in the Balance*, 306.

114. *Ibid.*, 320–21.

115. *Ibid.*, 318.

116. During the Bush Administration, the United States Government refused to sign the Biodiversity Treaty because of a perceived threat to the country's biotechnology industry. This action earned the Administration much criticism at Rio and around the world. After Bill Clinton became President, the United States signed this Convention.

117. Statement by E-Hyock Kwon, Minister of Environment, Republic of Korea, UNCED, Rio, 11th June 1992.

118. *Summary Report*, Meech Lake, 4.

119. Rowland, *Plot to Save the World*, 53.

120. *Ibid.*, 53–54.

121. *Ibid.*, 55.

122. *Ibid.*, 135–136.

123. *Ibid.*, 47.

124. *Ibid.*, 48.

125. *Summary Report*, Meech Lake, 4.

126. *Founex Report*, cited in Rowland, *Plot to Save the World*, 49–50.

127. *Globe and Mail*, 2nd June 1992, A3: 3–4.

128. *Ibid.*

129. Statement by Teatao Teannaki, President, Republic of Kiribati, UNCED, Rio, 13th June 1992.

130. Rowland, *Plot to Save the World*, 100.

131. Jackson, *Poverty and the Planet*, 180.

132. Head, *On a Hinge of History*, 91.

133. Maurice Strong, interview by Pierre Haski, *Christian Science Monitor*, 29th May 1992, 3:4.

134. World Commission on Environment and Development, *Our Common Future*, 220–21.

135. Head, *On a Hinge of History*, 91.

136. World Commission on Environment and Development, *Our Common Future*, 221.

137. OECD, "Guiding Principles Concerning International Economic Aspects of Environmental Policies," Council Recommendation C(72)128, Paris, 26th May 1972, mentioned in World Commission on Environment and Development, *Our Common Future*, 221.

138. World Commission on Environment and Development, *Our Common Future*, 83.

139. Rowland, *Plot to Save the World*, 57.

140. *Ibid.*

141. Statement by John Major, Prime Minister, United Kingdom, UNCED, Rio, 12th June 1992.

142. Statement by His Majesty, King Carl Gustaf of Sweden, UNCED, Rio, 3rd June 1992.

143. *Irish Times*, 15th June 1992, 1:4.

144. Lenssen, "Confronting Nuclear Waste," 49.

145. *Globe and Mail*, 15th June 1992, A8:5.

146. Statement by Dr. Emil Salim, State Minister for Population and the Environment, Republic of Indonesia, UNCED, Rio, 5th June 1992.

147. Statement by K. K. S. Dadzie, Secretary-General of the United Nations Conference on Trade and Development, UNCED, Rio, 5th June 1992.

148. Statement by Vincent Perera, Minister of Environment and Parliamentary Affairs, Sri Lanka, UNCED, Rio, 3rd–14th June 1992.

149. *Irish Times*, 11th June 1992, 7:4.

150. *Ibid.*, 7:5.

151. *Ibid.*

152. Cited in Jackson, *Poverty and the Planet*, 150.

153. Head, *On a Hinge of History*, 6.

154. *Ibid.*, 116–17.

155. Statement by Dr. Zvonimir Separovic, Minister in the Government of the Republic of Croatia, UNCED, Rio, 10th June 1992.

156. World Commission on Environment and Development, *Our Common Future*, 297.

157. *Ibid.*, fn. 14, 306.

158. Jackson, *Poverty and the Planet*, 174.

159. Head, *On a Hinge of History*, 135.

160. UNICEF, *The State of the World's Children, 1990* (Oxford: Oxford University Press, 1990), cited in Jackson, *Poverty and the Planet*, 174.

161. Head, *On a Hinge of History*, 6.

162. Jackson, *Poverty and the Planet*, 174.

163. World Commission on Environment and Development, *Our Common Future*, 300.

164. Statement by Rashid Abdullah al-Noaimi, Minister of Foreign Affairs, United Arab Emirates, UNCED, Rio, 9th June 1992.

165. *Time*, 1st June 1992, 24:2.

166. *Globe and Mail*, 7th January 1993, 1:6 and A6:1–2.

167. *Ibid.*, A6:1–2.

168. *Christian Science Monitor*, 2nd June 1992, 10:1.

169. *Time*, 1st June 1992, 26:1–2.

170. French, "Strengthening Global Environmental Governance," 156.

171. *Ibid.*, 158.

172. Douglas M. Johnston, "Marine Pollution Agreements: Successes and Problems," in Carroll, *International Environmental Diplomacy*, cited in French, "Strengthening Global Environmental Governance," 158.

173. World Commission on Environment and Development, *Our Common Future*, 272.

174. French, "Strengthening Global Environmental Governance," 159.

175. World Commission on Environment and Development, *Our Common Future*, 274.

176. French, "Strengthening Global Environmental Governance," 159.

177. British Information Services, Ottawa, "Minister Praises Contribution of London Dumping Convention in Protecting Marine Environment," 12th November 1992.

178. French, "Strengthening Global Environmental Governance," 156.

179. Cited in British Information Services, Ottawa, "Minister Praises Contribution."

180. *Globe and Mail*, 7th January 1993, 1:6.

181. Statement by Kinza Clodumar, Minister of Finance, Republic of Nauru, UNCED, Rio, 9th June 1992.

182. *Ibid.*

183. Statement by Thorbjorn Berntsen, Minister of Environment, Norway, UNCED, Rio, 4th June 1992.

184. World Commission on Environment and Development, *Our Common Future*, 274.

185. See chapter 17, *Agenda 21*, United Nations Document, UNCED, Rio, 1992.

186. *Christian Science Monitor*, 2nd June 1992, 10.

187. *Vancouver Sun*, 9th June 1992, D2:3.

188. World Commission on Environment and Development, *Our Common Future*, 262.

189. Gore, *Earth in the Balance*, 26.

190. See United Nations "Convention on Biological Diversity," reprinted as Appendix 5 to this book.

191. Paul Harrison, *The Third Revolution* (London: I. B. Tauris, 1992), 59.

192. *Ibid.*

193. World Commission on Environment and Development, *Our Common Future*, 148.

194. Edward O. Wilson, "Threats to Biodiversity," *Scientific American*, September 1989, 108–16, cited in Caldicott, *If You Love This Planet*, 96.

195. World Commission on Environment and Development, *Our Common Future*, 153.

196. Statement by Rahab Mwatha, Greenbelt Movement, WCED Public Hearing, Nairobi, 23rd September 1986, cited in World Commission on Environment and Development, *Our Common Future*, 154.

197. John C. Ryan, "Conserving Biological Diversity," in *State of the World, 1992* (New York: W. W. Norton, 1992), 2.

198. *Ibid.*, 24.

199. Garrick Utley, *NBC Evening News*, 9th January 1993.

200. Statement by Margaret Shields, Director, United Nations Research and Training Institute for the Advancement of Women, UNCED, Rio, 9th June 1992.

201. Statement by Mrs. King, Greenbelt Movement, WCED Public Hearing, Nairobi, 23rd September 1986, cited in World Commission on Environment and Development, *Our Common Future*, 124.

202. See World Commission on Environment and Development, *Our Common Future*, 125.

203. Statement by Her Royal Highness Princess Sonam Chhoden Wangchuck, Representative of the King of Bhutan, UNCED, Rio, 11th June 1992.

204. Statement by Professor Dr. Her Royal Highness Princess Chulabhorn Mahidol, Personal Representative of the King of Thailand, UNCED, Rio, 5th June 1992.

205. Statement by Lynda Chalker, Overseas Development Minister, United Kingdom, British Information Services, Ottawa, 10th March 1992.

206. Statement by Thorbjorn Berntsen, Minister of Environment, Norway, UNCED, Rio, 4th June 1992.

207. Statement by Bella Abzug, Co-Chair of the Women's Environment and Development Organization, UNCED, Rio, 9th June 1992.

208. *Globe and Mail*, 2nd June 1992, A3:4.

209. World Commission on Environment and Development, *Our Common Future*, 114.

210. United Nations Document, General Assembly Resolution 45/164, 18th December 1990.

211. See United Nations, *Agenda 21*, chapter 26.

212. *Globe and Mail*, 13th November 1992, A7:2.

213. *Ibid.*, A7:4.

214. Richard Falk, "The Rights of Peoples (In Particular Indigenous Peoples)," in James Crawford, ed., *The Rights of Peoples* (Oxford: Oxford University Press, 1988), 23.

215. *Ibid.*

216. *Ibid.*, 19.

217. Statement by Ji Chaozhu, United Nations Under-Secretary-General, Department of Economic and Social Development, UNCED, Rio, 11th June 1992. For a longer, slightly different version of Chief Seattle's words, see Gore, *Earth in the Balance*, 259.

218. Statement by Hans Alders, Minister of the Environment, the Netherlands, UNCED, Rio, 5th June 1992.

219. Statement by Giorgio Ruffolo, Minister of Environment, Italy, UNCED, Rio, 4th June 1992.

220. Statement by Dr. Emil Salim, State Minister for Population and the Environment, Republic of Indonesia, UNCED, Rio, 5th June 1992.

221. Statement by Dr. Mahathir bin Mohamad, Prime Minister, Malaysia, UNCED, Rio, 13th June 1992.

222. *Globe and Mail*, 2nd June 1992, A3:4.

223. *Ibid.*, A3:3–4.

224. *Ibid.*

225. Statement by Ruth Feldgrill-Zankel, Minister of Environment, Austria, UNCED, Rio, 5th June 1992.

226. Statement by Carlos Borrego, Minister of Environment and Natural Resources, Portugal, UNCED, Rio, 3rd June 1992.

227. Statement by Vincent Perera, Minister of Environment and Parliamentary Affairs, Sri Lanka, UNCED, Rio, 3rd–14th June 1992.

228. Statement by Kinza Clodumar, Minister of Finance, Republic of Nauru, UNCED, Rio, 9th June 1992.

229. Foreign and Commonwealth Office, London, United Nations Conference on Environment and Development, Issue 25/92, 9th June 1992.

230. *Sweden: National Report to UNCED, 1992*, Ministry of the Environment, Sweden, 1991, 12.

231. Statement by Thorbjorn Berntsen, Minister of Environment, Norway, UNCED, Rio, 4th June 1992.

232. Speech by Kiichi Miyazawa, Prime Minister, Japan, for UNCED, Rio, 13th June 1992.

233. Statement by Helmut Kohl, Chancellor, Federal Republic of Germany, UNCED, Rio, 12th June 1992.

234. Statement by Anibal Cavaco Silva, Prime Minister, Portugal, UNCED, Rio, 12th June 1992.

235. Statement by Per Stig Moller, Minister of Environment, Denmark, UNCED, Rio, 9th June 1992.

236. Statement by His Majesty King Hussein I of Jordan, UNCED, Rio, 5th June 1992.

237. Statement by Dr. Ahmad Mattar, Minister for the Environment, Singapore, UNCED, Rio, 11th June 1992.

238. Statement by Brian Mulroney, Prime Minister, Canada, UNCED, Rio, 12th June 1992.

239. "Canada and the Earth Summit: Achievements," UNCED, Rio, 3rd–14th June 1992.

240. Prime Minister Brian Mulroney, Address at the Canadian Museum of Civilization, Hull, Quebec, 1st June 1992 (Office of the Prime Minister).

CHAPTER THREE

The South and the Earth Summit:
The Development/Environment Dichotomy

*Introduction: The Right to Development and the
Right to a Safe Environment*

Although critics have for years derided the United Nations for its lack of
effectiveness in solving the problems of this world, they have frequently overlooked the resounding success of that world institution in arousing public
awareness on a global level about human rights and human responsibilities.
The many political failures of the United Nations should not obscure the fact
that peasants in India are aware of and fighting for environmental protection;
that in 1989 residents of Beijing acquired the courage to demand political
freedom; that the people of Lithuania agitated for and eventually obtained
sovereign right to self-determination; that the people of Somalia, by virtue of
their quiet misery, stirred the conscience of the world, which rushed them
food and medicine to fulfill their right to life. When the Member States of the
United Nations formulated the principles embodied in the Universal Declaration of Human Rights, the International Covenant on Civil and Political
Rights, and the International Covenant on Economic, Social and Cultural
Rights,[1] no one could have predicted how much enthusiasm these ideals
would generate globally on the popular level. Nor could one have forecast
how these formulations would inspire international debate and pressure for
reform across the world. Hardly anyone would dispute the fact that people
are more aware of their rights now than at any time in history. This is largely
due to the development of communications systems which have turned our
planet into a vast global village. However, technology alone could not have
achieved this awareness of the need for improvement in the human condition
and indeed of the right that human beings have to improve their lives.

Reprinted with kind permission from an earlier version published by the *Dickinson Journal of
International Law.*

While one would not wish to overrate the impact of the Covenants, there is a clear linkage between the process by which people awaken to their rights—a process often initiated by the acquisition of knowledge about other populations who enjoy such rights—and the agitation for reform consequent to that awareness. That such agitation seeks global support and internal justification goes without saying. The fact that foreign support is usually presumed and frequently given suggests that certain human rights are now viewed as fundamental and normal. We have witnessed considerable popular sympathy around the world for the struggle of the people of Croatia and Bosnia. The Kurds in Iraq have been the object of universal concern as they struggle against the tyranny of Saddam Hussein. The existence of United Nations Covenants pledging support to the implementation of human rights has been significant in popularizing the principles enshrined in those documents and in educating people to the possibilities of a better life. The gathering of so many nations at United Nations headquarters in New York cannot but rivet the attention of the whole world to the activities of that organization. Inevitably the process of human awareness and awakening has resulted in a demand for change and a clamor for reform. The strength of this process can be gauged by the rapid disintegration of the totalitarian system which ruled the Soviet Union for so many decades and its replacement by a more democratic economic and political ruling philosophy.

Indeed, human rights have now become so significant in international political forums that academics have been prompted to analyze and categorize the various rights and to explain and define them. Karel Vasak compared the development of the law of human rights by reference to the famous slogan of the French Revolution of 1789, "Liberty, Equality, and Brotherhood," and concluded that there are three generations of human rights.[2] Civil and political rights, enshrined in the Universal Declaration of Human Rights (1948), the European Convention on Human Rights (1950), the International Covenant on Civil and Political Rights (1966), and the American Convention on Human Rights (1969),[3] may be considered first-generation rights. The so-called equality rights comprising the second generation may be found in the Universal Declaration of Human Rights, the International Covenant on Economic, Social and Cultural Rights (1966), and the European Social Charter (1961).[4]

Civil and political rights, constituting the first generation, are now to be found in the constitutions of many countries. "Preservation of these rights in the municipal law and national constitutions of most democratic countries confirms the widely held opinion that these basic rights have gained near universal acceptance."[5] The second generation, namely economic, social, and cultural rights,[6] "require[s] states to act positively toward the creation of better economic and social conditions for their citizens. Whether these rights are binding on a government is debatable. It has been suggested that the rights are binding to the extent that a government must strive toward their fulfilment."[7]

It is the third generation of rights, the so-called solidarity rights, which are believed to "transcend national boundaries and may be thought of as rights against the international community of states as a whole."[8] These include the right to economic development, the right to peace, and the right to a healthy environment, as well as rights to humanitarian assistance and to the common human heritage.[9] Unfortunately, "[t]he third generation of rights even less firmly invokes any sense of obligation on the part of governments."[10] Although, at the popular level, there is widespread interest in and recognition of the validity of these rights and universal concern about both development and the environment, the global fulfilment of these ideals by the Member States of the United Nations continues to be a problem. It is difficult to implement third-generation rights, largely because they

> imply an interdependence of individuals and nations, and, by extension, of individuals in all countries. But this interdependence would be essentially negative if it involved only mutual obligations. Its positive aspect is that it also involves mutual interests insofar as every individual shares with every other one the need for a suitable international order.[11]

The basic challenge is to get governments around the world to appreciate the nature of these shared interests and to act to give concrete form to the universal global consciousness which has already manifested itself in both the developing and developed worlds. Governments are by their very nature protective of national sovereignty; in the international arena, they can, on occasion, cling tenaciously to their self-interest even at the expense of vital universal needs. Hence the formulation of international environmental law is a process in which nations appear to inch forward, far too slowly for environmental activists and with far too much time consumed in mutual recrimination, interstate bickering, and an apparent lack of commitment to the urgent requirement to save the planet.

The United Nations Conference on Environment and Development (UNCED), held at Rio in June 1992, marked another step taken along the route to salvaging our planet; like most other such steps, it was accompanied by a veritable drama of rhetoric and mutual fault-finding. At this great Earth

96 *The Earth Summit at Rio*

Summit, two third-generation rights—the right to development and the right to a healthy environment—clashed head-on. It is rather ironic that these two concepts formed the theme of the summit and were its focus. Without development in Africa, Asia, and Latin America, millions will die of malnutrition and its consequent diseases. But if environmental concerns are not recognized as crucial to the developmental process, the entire planet could be at jeopardy from global warming, increased greenhouse gases, pollution of water and land resources, and deforestation. The Rio process fell victim to nationalistic priorities in some developing and developed nations which publicly espoused positions largely because of their impact on domestic opinion. And so the world inches forward at the pace of a tortoise in the formulation of international instruments to protect the environment, while the destruction and degradation of the planet proceeds at a pace faster than that of the proverbial hare. In this race, the odds are regrettably against the survival of the planet. It is a sad irony that, with the end of the Cold War and democratization proceeding in the former Communist states, there has never been a more opportune moment to effect dramatic and positive change in matters of international concern.

Unfortunately, the political act of ending the Cold War is seen in retrospect as a considerably easier task than meeting the economic challenge of creating a fair distribution of the world's wealth. Although totalitarian ideologies have failed and "evil empires" have fallen, none of this has dramatically alleviated the plight of millions who continue to starve and die in horrifying numbers every day. Indeed, it could be argued that while the risk of an East-West confrontation has cooled, perhaps forever, the arguments on a North-South level have become more serious, more vocal, and more confrontational.

Nowhere is the agitation for a new world order so strong as in the debate between the developed and the developing nations. The confrontation between these two great blocs has been ongoing for some years now. Even though the dialogue is at times bitter and vehement, there is now a clear realization that the two halves of this planet have a desperate need for each other and that neither one can survive without the other. The Earth Summit provided symbolic recognition of that interdependence and of the vulnerability of both the developed North and the developing South. It also provided an important forum for the South to vent its frustration at what it believes to be serious economic injustice. As India's Prime Minister, P. V. Narasimha Rao, told delegates at the Earth Summit:

> We inhabit a single planet but several worlds. There is a world of abundance where plenty brings pollution. There is a world of want where deprivation

degrades life. Such a fragmented planet cannot survive in harmony with Nature and the environment or indeed, with itself. It can assure neither sustained peace nor sustainable development. We must, therefore, ensure that the affluence of some is not derived from the poverty of the many. As Mahatma Gandhi put it with characteristic simplicity, our world has enough for each person's need, but not for his greed.[12]

The Perceptions of the South

This chapter explores.some of the perceptions and viewpoints of the South in an attempt to establish a greater level of mutual understanding than exists at the present time. To explain the Southern position, this section concentrates on two subjects which are crucial to both North and South, forest development and preservation and the international debt crisis and its impact. The human rights to development and to a safe and healthy environment cannot be achieved while reckless development and crippling debt wreak their havoc both in the short and long term for North and South alike. These two topics, growth and the environment, provide us with a dramatic picture of the type of problem policy makers confront in both the North and the South as they struggle to better the lot of their people. As all readers are doubtless aware, there are several areas of dispute between the rich and poor nations, and no single study of this length could hope to explore them all with any degree of depth. This chapter also suggests that unless the North-South dialogue can concentrate more on activity and less on acrimony the entire planet will suffer irreversible damage. Just as in the past international measures aimed at armament limitation could only be implemented by agreement between East and West, now it has to be realized that any serious attempts at environmental protection cannot be formulated without the cooperation of North and South. The development of environmental law has now become largely dependent on the resolution of the North-South conflict, and solutions will only be possible if the new world order acquires an economic rather than a purely political emphasis.

There is an urgency about this situation, which may well be the most compelling political, economic, and social issue of the 1990s. With the end of the Cold War the entire world must now turn its full attention to the problems of development and the environment, the two aspects of life on this planet which have assumed crucial significance and fundamental importance simply because they affect the lives of every single man, woman, and child in every country. There is already a growing global consensus at the popular level that the amount of poverty, disease, and suffering of the majority of the Earth's inhabitants is almost obscene when compared with the wealth, well-being, and comfort of its more fortunate minority. The affluence of the developed nations is being challenged not merely because it is in such stark contrast to the poverty of the developing world. Such difference in lifestyle is being questioned on grounds of principle now, principle which seeks its intellectual and moral grounding in the rights embodied in the United Nations covenants. The wealth of the few is also being challenged because there is a growing awareness—thanks to the statistical information provided by technological systems of assessment—that the production and enjoyment of that Northern affluence is largely at the expense of the developing world. Further concerns have been raised because the extravagant resource utilization which has given a minority so comfortable a lifestyle has resulted in extensive degradation of the fragile ecosystems which are crucial to human survival on this planet.

The dialogue of dissension between North and South contains numerous areas of dispute, and all these varied threads can often become entangled with the pragmatic need for governments to ensure their domestic survival. This chapter will attempt, first generally and then more specifically, to examine some aspects of the North-South confrontation to highlight the perceptions and clarify the misconceptions which have generated so much unnecessary misunderstanding and mistrust. It is important first to understand the wider picture of this confrontation before focussing on any of its details. One might begin by examining the basic assumptions of both North and South about the responsibility each element bears for the future survival of the planet.

As a result of a shared imperial past, the developed nations and the developing world—their former colonies—begin the debate on opposite sides. The often violent struggle for freedom from colonial rule generated attitudes on both sides which have spilled over into present dialogues on matters pertinent to the entire planet. That history should form so significant a part of the current debate is unfortunate but inevitable. The attitudes of the nations of the South are conditioned by the memory of past exploitation and they come to the present debate with those preconceptions, always alert to and suspicious of Northern motives, which they fear are once again predatory. The North also suffers from a general disinclination to understand these Southern apprehensions, fears that are solidified by the activities of some multinational companies, which to this day are perceived in Africa, Asia, and Latin America as carrying out the imperial, exploitative activities of the past.

The Southern tendency, as their presentations at Rio demonstrate, is to oversimplify the issue and to visualize the North—or at least particular countries, specifically the United States of America—as hostile to Southern development. There is a refusal to see complexity and diverse opinion on both sides of this dialogue. It is very unfair to assume that the United States Government, U.S. multinational companies, and the American people are all cast in the same mold. Yet such is the nature of North-South misunderstanding that such assumptions are frequently made, at least at the public level. Accordingly, the entire debate begins on a foundation of mutual misunderstanding.

This base of misunderstanding colors the subsequent dialogue. When the North emphasizes the primacy of universal concerns, the South perceives this as another ploy to deprive it of developmental benefits by imposing Northern environmental priorities. When the South makes a desperate plea about the poverty of its population, the North counters by emphasizing the importance of population limitation. In a very real sense, the two parties are not talking to each other but almost posturing past each other. Although I have labelled the respective positions Northern and Southern for convenience, I must emphasize that there is considerable divergence within these two elements, as each group contains moderates and extremists. A study of that kind of divergence would require an entire book.

Another problem which frequently arises in this dialogue concerns the tendency of both North and South to spend time on mutual recrimination rather than on active reconciliation. No one in the North would deny that industrialization as it proceeded in Britain, Europe, and North America has been extravagantly wasteful in its use of resources and that it has resulted in widespread environmental destruction in the form of acid rain, the hole in the ozone layer, the death of thousands of lakes, the depletion of fisheries, and the devastation of thousands of species. As George Mitchell, former Majority Leader of the U.S. Senate, has commented, "We—mainly the industrialized nations—have become the Prometheus of the twentieth century, destroying our own habitat."[13] Although the evidence is clearly there of the price to be paid for such development, elements of the South are unwilling to absorb that lesson and seek instead the benefits of quick development regardless of the economic cost to their environment. The North perceives this approach as short-sighted and obtuse, refusing to recognize how severely the hunger, disease, and consequent death in the South weigh on the minds of policy planners in the developing nations.

Accordingly, the priorities of the North and South are different. Having destroyed so much of the environment in its rush to development, the North seeks to regain a balance in nature and allow the Earth to be revitalized and cleansed of its pollutants and contaminants. The nations of the South argue instead that development, rapid development, has to be a primary consideration in the minds of its governments and that the consequences of such development will have to be dealt with as a secondary factor. The Northern emphasis on the environment is perceived by some Southern governments as an insidious attempt to delay development and to keep their populations starving so that the developed world can continue to enjoy its exclusive lifestyle. The BBC correspondent Stephen Bradshaw commented on this aspect of the conflict: "But governments in the South aren't just asking for more handouts to improve squalid living conditions. They also say that if the North is asking them to develop more slowly to protect the planet from pollution, they'll lose billions of dollars and remain in poverty for decades to come."[14] Both sides of this dispute appear to miss the point that a shared interest in environmentally safe development promises enrichment for both North and South.

Ironically, even though both North and South take measures which daily degrade the planet, they adopt the high road of adherence to environmental principles in their public positions. There is no dearth of advocates of environmentalism today; what is now required is the will to implement the ideas which are accepted worldwide. In effecting such action, the North and the South confront each other on the most serious aspect of their argument— money. The cost of clean-up and of environmental protection is perceived as being beyond the resources of any one nation, or even of groups of nations. The South insists that the North pay for environmental action because it has created most of the pollution of the planet. The North is equally insistent on its right to control the distribution of any funds it authorizes for global environmental protection to ensure that proper use is made of the money. Such assertions are resisted as imperialistic by the South, which counters with references to its heavy debt burden to Northern banks, a financial squeeze which makes rapid development imperative and environmental protection a secondary consideration.

Having examined in broad terms some aspects of the confrontation, it would be worthwhile now to examine more specifically a few of the problems of the South, the very real dilemma facing developing nations as they confront the two priorities of environment and development in the critical circumstances of vast poverty and desperate living conditions for millions. Although it is not possible within the scope of this chapter to explore all the problems and consider all the nations of the South, a study of particular aspects of the global situation with a few pertinent examples should serve to illustrate the serious nature of the situation. Such issues as the debt crisis and the destruction of forests by development are indicative of the dimensions of the problem and also provide insight into the real clash of interests which underlies the North-South conflict. Interestingly, even as the Earth Summit at Rio served to focus international attention on the problems of the South, it also provided much-needed publicity in the North about the linkage between the need for a clean environment and the requirements of basic necessities to sustain life in the South.

Even though not all nations accept the idea of an inherent right to development, there is now widespread recognition that the implementation of this human right in the South is the best guarantee of the continued enjoyment of the right to development in the North. This recognition is not merely an increase of universal compassion or of humanitarian concern. It is a reflection of a vital self-interest which perceives the interdependence of all parts of the world and of the necessity of realizing that no area of Spaceship Earth can be written off or disregarded if the whole is to survive and thrive in the next century. Although the right to development "remains a putative right not fully accepted into the body of generally accepted international law,"[15] the nations of the South are no longer prepared to remain the permanent poor relations of the North. The issue of economic fairness underlies much of the debate and will continue to be the dominant theme in international affairs for some time to come. Because of the reluctance of some developed states to yield their economic advantages to meet the wider necessity of environmental benefit for all, and "[d]espite the overwhelming acceptance by the United Nations General Assembly of the Declaration on the Right to Development[16] . . . the acceptance of a right to development remains questionable."[17]

The recognition of this North-South divergence of opinion also affects governmental and non-governmental attempts to come to grips with and tackle the daunting and very depressing issue of environmental degradation, much of it a consequence of the Northern rush to development and industrialization. Even though there is now a widespread realization in the developed world that the entire planet has paid a terrible price for progress, convincing governments in the South not to emulate the North is no easy matter. First, Governments like those of the United States during the Bush Administration seemed reluctant to assume the role of world environmental leader and take the bold and brave decisions to clean up the planet and to protect it from further degradation. The energy brought by the United States of America to the creation of a new world order in international politics was unfortunately not matched by an equally enthusiastic commitment in the economic or environmental field. Consequently, there is no example to inspire the other nations of the world, particularly those in the South. Second, although the North preaches endlessly to the South about sustainable development, it does not practice to any great extent what it preaches. The reduction in the standard of living which has occurred in developed countries in the past few years has been more a consequence of global recession than of any serious attempt to curtail this way of life. Recessionary pressures have to some extent achieved a limited reduction in this most envied of lifestyles, but as the North eases out of its recession the old habits will probably return. According to one estimate, 23 per cent of the world's population controls 85 per cent of its income,[18] and most of this concentration lies in a few hands, largely in the North. In a report produced shortly before the Earth Summit, the World Bank explained the signficance of such economic disparity: approximately 1 billion people (out of the world's 1992 population of 5.48 billion)[19] had to survive on less than $1 per day.[20] The developing countries find it hypocritical of the North to control so much of the world's wealth, utilize so much of its resources, and create so much of its pollution and then attempt to lecture them on the observance of reasonable development. Some Southern leaders have been very vocal in demanding that the North curtail the massive consumption which eats away at the Earth's resources. As the Malaysian Prime Minister, Dr. Mahathir bin Mohamad, commented: "You're asking us to cut back on our standard of living by asking us not to develop. Why can't we tell you to do the same? If you want us to do it, you do it first."[21] The Prime Minister of India was somewhat gentler in his remark: "we cannot have conservation of the environment without the promise of development, even as we cannot have sustained development without the preservation of the environment."[22] The difference in tone between the two statements should not obscure the underlying similarity of the call for greater economic equity by these prominent Asian leaders. Michael Howard, Environment Secretary for the United Kingdom, expressed sympathy with this Southern concern: "Of course we recognize the need for them to improve their economic situation, to improve the lot of their people. What we must do, together, is to achieve economic development in a way that is sustainable and in harmony with the environment."[23]

Although we in the developed world can appreciate the planetary dangers of Southern development emulating the earlier Northern example, we cannot convince the South of our sincerity unless we are willing to demonstrate, not merely by talk but by action as well. Although Rio demonstrated that a number of Northern nations are dedicated to curbing the emission of greenhouse gases and to the protection of biodiversity, the fact that this dedication to environmentalism is not shared with the same degree of intensity by all Northern governments was noted by the Southern leaders, who did not fail to remark on it.

The conflict between developmental and environmental rights is worsened by a fundamental economic inequity which has first to be dealt with if any kind of serious action is to be taken to assure the future of our planet. For all their loud clamoring, the developing countries are well aware that they do not have equal status in the North-South dialogue. The North enjoys the wealth, the industries, the highly educated work force, the space, and the affluent lifestyle. The South is burdened with overpopulation, malnutrition, ever increasing poverty, lack of living space and arable land, and a lifestyle which is declining annually in comparison to that of the North. "In 1987, the average income per capita in the First World was $12,070, while it was $670, or 6 percent of $12,070, in the Third World. Ten years earlier, this number was 9 percent, so the income of the poor actually decreased."[24] It is an awareness of the growing disparity between North and South which fuels much of the assertiveness of the Southern governments in no longer asking or begging for economic equity but in demanding it as their right. In the process of seeking fundamental reform in the way the world's wealth is distributed, the South has to use all the levers it can bring to a very unequal dialogue.

One such lever was employed in the 1970s, when states in the Middle East realized that they had for years been selling their oil at bargain prices to the North and reaping little of the benefit. In 1973 these countries raised the price of oil dramatically and set off a severe economic crisis in Northern countries which had become dependent on cheap and abundant Arab oil.

Now, in the 1990s, the South has found another lever, its own development and the risks its industrialization process poses for the air and climate of the North. Further, in proceeding with the destruction of its forest resources, it deprives the North of the one great cleaner which can still absorb its own foul gases. Now that the South has found a new point of leverage, it has not been slow to exploit its significance in both the media and the negotiations before the Earth Summit. For the South, "[t]he essential question, in the context of the North-South Dialogue, is how the burden of adjustment is to be shared in an equitable manner."[25] There is now an awareness, stridently expressed, that "[t]he South has consistently subsidised the gargantuan consumption of the North,"[26] that in the future the "development of the South can in no way be compromised by the North's pre-emption of the global environment space,"[27] and that now "[a] reform of the world economic system is, therefore, vital so that all citizens of the world, including the world's poor, are empowered to take control of their environment while the rich are made to pay the ecological costs of their consumption."[28]

The Forest as an Issue in the Development/Environment Dichotomy

The dilemma of choosing between environmental and developmental priorities is most noticeable in the debate about the Earth's forests. It is this issue which most clearly highlights the Southern position and its suspicion of the motives of the North. Having denuded much of its forest cover in the process of industrialization and expansion of its urban population, the developed world has discovered that it has now become dependent on the developing nations, which contain approximately 90 per cent of the world's tropical forests.[29] Although by one estimate developed countries have a 34 per cent forest cover as against a 29 per cent forest cover for developing states,[30] between 50 and 90 per cent of the treasure trove of biodiversity is to be found in the tropical forests of the South.[31] "Half of Latin America is covered by forests, as is 33 per cent of Asia and 27 per cent of Africa."[32] Some Southern nations are very fortunate in the extent of this vital resource. As the Malaysian Prime Minister, Dr. Mahathir bin Mohamad, pointed out at the Earth Summit, the land of his nation "is almost 60 percent covered with self-regenerating tropical rain forest with an additional 15 percent covered by tree plantations."[33] By way of contrast at the time of the Rio Summit, Britain had only 10 per cent of its land under forest cover.[34] It is important to note that because of reforestation, forest acreage in the developed world continues to expand, but in developing nations the opposite is the case, and the rate of deforestation has accelerated.[35] Paul Harrison explains in *The Third Revolution* that "[i]n some countries the losses are dramatic. In Madagascar only 34 per cent of the original forest cover is left. In the Côte d'Ivoire and the Philippines four-fifths of the original forest has been cleared. Ethiopia's forest cover dwindled from 40 per cent of her land area in 1940 to only 4 per cent fifty years later."[36] Almost 40 per cent of the world's forest cover had disappeared by 1980.[37] The Earth's remaining acres of rain forest are found mainly in the Amazon region, in central Africa, and in or near Southeast Asia (specifically in Malaysia, Indonesia, and New Guinea).[38] Of these, the Amazonian forest is the largest, covering "nearly three million square miles, an area nearly as large as the entire United States. It spans nine South American countries from Brazil in the east to Peru in the west, and from Venezuela in the north to Bolivia in the south."[39]

The Earth's forests are important for more than just their timber. Besides providing shelter and a way of life for thousands of the world's indigenous people, forests are a treasure trove of biological diversity. "Those portions of the earth that are covered with forests play a critical role in maintaining its ability to absorb carbon dioxide (CO_2) from the atmosphere and are thus essential in stabilizing the global climate balance."[40] Ben Jackson, an environmentalist with the World Development Movement explains that "[t]he forest plays a key part in regulating climate by its effects on the reflectivity of the land surface and by the way in which it recycles water from the earth back into the atmosphere."[41] The Earth's forests are one important element in the present war against global pollution. With the loss of one and a half acres of forest every second to felling and land clearance,[42] the entire planet stands to lose a vital resource. "The destruction of a forest can affect the hydrological cycle (the natural water distribution system) in a given area just as surely as the disappearance of a large inland sea."[43] Forests contribute to the production of rain clouds[44] and are therefore vital in bringing moisture for agricultural crops. "More water is stored in the forests of the earth—especially the tropical rain forests—than in its lakes."[45] Other benefits of forest cover include cleansing of the air and water and balancing of the climate.[46] The Earth's forests "also stabilize and conserve the soil, recycle nutrients through the shedding of their leaves and seeds (and eventually their trunks when they die), and provide the most prolific habitats for living species of any part of the earth's land surface. As a result, when we scrape the forests away, we destroy these critical habitats along with the living species that depend on them."[47]

The nation with the largest concentration of tropical forest, Brazil,[48] is already suffering the consequences of earlier deforestation. "Logging and agricultural and urban expansion have destroyed more than 95 percent of the once-vast Atlantic coastal rain forests and the coniferous Araucaria forests of southern Brazil."[49] The poverty of the majority of Brazilians and the population's pressure on the rain forest can be attributed in part to serious economic inequity within that nation. "Just 2 per cent of Brazil's landowners control 60 per cent of the nation's arable land. At least half of this land lies idle."[50] Although there has been some reduction in the logging of the Amazon rain forest since 1990, before then the destruction proceeded at an alarming pace. The previous military regime in Brazil (1964–85) allowed and even encouraged (with tax incentives) wholesale destruction via slash-and-burn clearing of large sections of the Amazon forest. "In the past decade the forest has declined by 12 per cent,"[51] a victim of the situation created by the world's inequitable economic order. Since it provides a readily marketable commodity, wood, the logging continues, albeit not at the frantic pace of the past. Environmentalists around the world fear that economic "pressures have turned the Amazon into a battleground and may eventually turn it into a desert."[52]

There are varied statistical estimates concerning the extent of deforestation and even about the amount of forest cover left on Earth. One estimate suggests that since 1972 the world has lost approximately five hundred million acres of forests, "an area roughly one-third the size of the continental U.S."[53] The United Nations has calculated that by the year 2000 developing nations will lose approximately 40 per cent of their remaining forests if present trends continue.[54] Given an annual loss of fifteen million hectares, much of it in Africa, Asia, and Latin America, the outlook for the next century is bleak indeed.[55] The Food and Agriculture Organization has concluded that tropical deforestation accelerated during the 1980s to destruction of approximately 1 per cent of the forest cover every year.[56] "This amounts to two-thirds of the area of the United Kingdom. Every ten weeks an area the size of the Netherlands. A Barbados every day."[57] Vice-President Al Gore has commented that "[w]herever rain forests are found, they are under siege."[58] He explained the situation affecting the world's forests: "They are being burned to clear land for pasture; they are being clear-cut with chain saws for lumber, they are being flooded by hydroelectric dams to generate power. They are disappearing from the face of the earth at the rate of one and a half acres a second, night and day, every day, all year round."[59]

The disappearance of forests is also attributable to the great increase in the demand for land for cattle grazing. In India, the need to feed 196 million head of cattle has resulted in an unprecedented assault on forested acreage.[60] Forests are being felled daily and the land converted to pasture. The environmentalist Jeremy Rifkin has calculated that an average cow consumes about 410 kilos of vegetation per month and is "literally a hoofed locust."[61] As global meat production has nearly quadrupled since 1950,[62] the amount of land devoted to growing animal fodder has also increased, by as much as five-fold in Mexico between 1960 and 1980, to give only one example.[63] "Without question, ranching is a factor in tropical deforestation."[64] It has to be remembered that the reason for much of the emphasis on cattle rearing is directly related to the global demand for beef. Although beef consumption has fallen in some nutrition-conscious societies, this product is still a vital commodity in world agricultural trade.

It has been argued that the incentive to destroy tropical forests arises from consumer demand in the industrialized nations of the North.[65] "Markets in the North demand a wide range of tropical products, sometimes wood and wood derivatives, sometimes animal or plant products grown in the clearings. The hard currency earned by the export of these items is a powerful incentive to entrepreneurs and governments in the South alike. . . . The insatiable desire of consumers in the North for wood panelling, for hamburgers, and for cocaine is imperilling the future of forests in several tropical countries."[66] Whether the demand springs from the North or from the South itself, the destruction of forests is still the consequence.

The case of El Salvador is indicative of the possible future of a number of countries which are now richly forested. This Central American nation was once covered by forests over 90 per cent of its land.[67] After the depredations of population encroachment and agricultural expansion, these vast tracts have been reduced to "a single 20 square kilometre plot of cloud forest."[68] Even though political and economic instability have forced approximately one million Salvadorans to flee their homes,[69] the country can barely support its still growing population. Poverty, malnutrition, and consequent diseases are part and parcel of the life of the majority of those who still live in that country.

It is estimated that Malaysia, presently rich in forests, will deplete this resource by the year 2000.[70] India, Sri Lanka, Nicaragua, Honduras, and Guinea are only some of the developing nations which have almost lost or will soon lose their forest resources.[71] For Bangladesh, the loss of forest cover has resulted in extensive flooding.[72] Thailand lost almost half of its forests between 1961 and 1985.[73]

The area of land suffering an annual loss of forest cover is approximately equal to the size of the State of Washington.[74] As with all environmental problems, it would be naive to assume that the consequences are confinable within national borders. "The presence or absence of trees anywhere is a major factor in the world's environmental health everywhere."[75] It is now apparent that "deforestation of a watershed in one country can lead to flooding in a neighbouring country. Destruction of a forest habitat for migrating wildlife can lead to the loss of a species in another country. Loss of tropical forests, which act as major absorbants of greenhouse gases, can speed the warming of the world climate."[76] At the present time, forests cover only 7 per cent of the land surface of the Earth.[77] However, this small area is home to between 50 and 90 per cent of the ten million or more species which inhabit the planet.[78] Although it is internationally recognized that "[f]orest conservation is the key to preserving the Earth's heritage of biological diversity and harbours the secrets of new life-saving drugs and other products,"[79] acting on this awareness is another matter altogether.

Although it is now also clear that the threat to the world's forests constitutes a major environmental crisis, the will to solve this problem is lacking. Nations of the North continue to fell their own trees to reap immediate profits from the logging industry; then, with what developing nations regard as incredible hypocrisy, the same countries turn around and blame the nations of the South for following their example! Satellite photographs have demonstrated that the forests of the Pacific Northwest—the richest in the United States—show far more damage from clear-cutting than do the rain forests of Brazil.[80] The Bush Administration, which promised $150 million a year in aid to developing countries to preserve their forests,[81] simultaneously worked to open logging activities on four million acres of United States territory which contain old-growth forests.[82] Similarly, although the Canadian Government's forestry negotiators emphasized that "[a]s part of the global community, we have a responsibility to manage our forests so as not to impair the transboundary or global environment,"[83] logging continues to decimate the forests of British Columbia. "Forty years ago, Vancouver Island's pristine temperate rain forest, one of the few left anywhere in the world, held 4.2 million acres (1.7 million hectares) of trees. By 1990, however, loggers had clear-cut more than half the prime timberland. The rest could be gone in 25 years."[84] It is this double standard which prompts the South to question the seriousness of the Northern commitment to environmental concerns.

The insatiable demand for wood in developed countries, particularly in Japan, is the major incentive which prompts some nations of the South like Malaysia to fell their ancient forests at such an alarming rate. In the weeks and months preceding the Rio Summit, the developing nations were at pains to explain their position on forest utilization to the Western media and to justify their use of this resource because of sheer economic necessity. The South Centre explained its public relations strategy: "The South should launch a global public information campaign to present forcefully the South's position on environment and development and to gain understanding and support among the public in the North for its position. Such a campaign is also necessary to counter the negative image of the South that is often propagated by the media in the North concerning the South's position on global environmental issues."[85] Given the international audience at Rio, the tone became considerably sharper, and some nations were quite strident in their insistence that the right to development had to be made a priority in order to improve the living conditions of their burgeoning populations.

The developing nations argue that their use of resources is not the major ecological problem facing the planet. At the Earth Summit they were not reluctant to point out that according to the United Nations the richest of the world's nations—comprising 20 per cent of a total population in that year of over 5.4 billion people[86]—utilized 83 per cent of its output.[87] Accordingly, in the mind of the nations of the South, it is consumer and business demand in the developed world which determines the type and degree of resource utilization in the countries of Africa, Asia, and Latin America. The example of Indonesia illustrates several facets of the problem. Blessed with vast forest reserves, the country is rapidly logging this resource to earn foreign exchange. Although it has other marketable commodities, it is the wood of its rich forests which elicits foreign consumer demand. Because of Indonesia's financial situation as a developing nation, it has no option but to exploit the available resource and market it. Although its initial ventures were in the export of raw logs, in 1982 Indonesia banned their export as it sought to develop its indigenous plywood industry.[88] In order to secure the necessary degree of dominance in the plywood trade, it hastened its exploitation of the forest resource and secured almost one half of the world's trade in plywood between 1982 and 1987.[89] "In the understandable desire to seize and hold the market, Indonesians have reduced prices much below where they should be, which forces additional productive exploitation on the one hand and encourages profligate consumption practices on the other."[90] The environmentalist Ivan Head complains that Japan imports most of Indonesia's plywood and uses about one third of this material "to fabricate concrete forms for the construction of building foundations. These panels, amounting to 135 million square metres in 1987 alone, are often discarded after being used just once."[91] As Indonesia requires approximately $2 billion per year to service its debt just to Japan, "the linkage between foreign debt and forest depletion can be illustrated dramatically."[92] There are few alternatives available to the nations of the South in their fight to survive, a point that was heavily underscored at the Earth Summit. Indonesia has lost so much topsoil because of its assault on the forests that "the net value of the timber crop has been reduced by approximately 40 percent."[93]

The case of the Philippines is even more serious than that of Indonesia. The rush to development in that nation resulted in the destruction of this vital resource; according to Robert Repetto, an economist with the World

Resources Institute, "failure by the Philippines to collect the full resource value of its timber harvest resulted in the recovery of less than one-sixth of the possible revenue."[94] The Philippines, Brazil, and Indonesia have banned the export of logs,[95] but for the Philippines in particular the measure was taken far too late. Fulgenchio Factoran, the country's Environment Secretary, explained that his nation had "lost the fertility of our soil, the fish-bearing capacity of our rivers and our seas. We've lost our healthy corals, our mangroves, we've lost the opportunity to feed millions of Filipinos through good agricultural practices and we've lost water for irrigation that would have fed the farms. We've lost all of that. We've lost a big future for millions of Filipinos."[96]

Vietnam, "[r]avaged by chemical warfare, collectivized clear-cutting and rampant commercial logging,"[97] has offered monetary inducements to farmers to grow trees. In 1942, the country contained 13 million hectares of forest; by 1982, this had dwindled to 6 million hectares.[98]

The South also argues that the profligate use of resources by the developed nations has brought the world to a state of both environmental and economic crisis. As the Centre for Science and Environment in India explains:

> There is an enormous difference in the economies of developing and industrialised countries in environmental terms. The former continue to depend heavily on the exploitation of their natural capital to meet their current consumption needs and generate the investments needed to build up a stock of human-made capital and a knowledge and skills resource base. The industrialised countries, on the other hand, have already gone through a prolonged phase of natural resource exploitation, both within their own countries and outside, to build up a massive base of human-made capital, knowledge and skills.[99]

The earlier industrialization of the developed world has given it a permanent advantage in terms of assuring its economic supremacy for decades to come. This primacy of Northern priorities is increasingly resented by the South, which suffers both because its resource utilization is based on Northern requirements and because it then has to shoulder Northern criticism for the consequent pollution, soil erosion, deforestation, and climate change which occur. As the Centre for Science and Environment explains: "Today, the reality is that northern governments and institutions can, using their economic and political power, intervene in, say, Bangladesh's development. But no Bangladeshi can intervene in the development processes of Northern economies even if global warming caused largely by Northern emissions may submerge half the country."[100]

The nations which are today collectively labelled the South have for years provided primary materials to the developed world to enable it to progress with its own rapid industrialization. During the colonial era, the economies of Africa and Asia were geared to the production of raw materials, often by the conversion of self-sufficient agrarian systems to ones devoted to more globally marketable agricultural commodities like cotton and tea, which were grown primarily on large plantations. The molding of colonial economies to the requirements of the imperial power set in place a pattern of resource exploitation which continued after decolonization and independence came to Africa and Asia. As the Malaysian Prime Minister commented at Rio, "As colonies we were exploited. Now as independent nations we are to be equally exploited."[101] The economic system was hardly tampered with by cash-strapped governments in the newly emerging nations which desperately craved foreign exchange to buy the manufactured goods they could not yet produce for themselves. Although countries like India embarked on ambitious industrialization plans (and succeeded to some extent), the benefits of this approach have not been sufficient to eradicate poverty in that nation, which already had an estimated 850 million people by the time of the Earth Summit.[102] Although the North is willing to buy raw materials from poor developing nations, it is wary of importing huge quantities of manufactured items, particularly when these are likely to compete with domestic products. John Stackhouse of the *Globe and Mail* (Toronto) commented on the trade barriers India faced when it tried to export its goods to developed nations.[103]

The extent of Northern reliance on raw materials from the South and the impact of this situation on deforestation globally can be illustrated by the example of Canada. In 1988, Canada imported wood and wood products worth $139.6 million and paper and paper products worth $86.8 million from tropical and East Asian nations like Indonesia, Taiwan, Brazil, and Malaysia.[104] "Of much greater significance to deforestation in Canada's merchandise trade accounts are those tropical products grown in circumstances where, for the most part, the forests first have to be cleared. In 1988 the value of these products was: coffee, $432.2 million; citrus, $391.2 million; cocoa and products, $153.8 million; natural rubber, $134.2 million."[105]

Most insidious of all is the fact that "[i]n the Andes, in the Caribbean, and in Southeast Asia, forests are cleared to permit the cultivation of coca, marijuana, and opium poppies for heroin."[106] It appears to the governments of the South that the North continuously sends mixed signals in this great debate about forest protection. The South is confused because on the one hand it is counselled (in a rather patronizing manner, it insists) to conserve its forests, while on the other hand forest products are among its most market-able commodities in the North. Although the biodiversity of the South is perceived as vital to the emerging biotechnological industry of the North, the demand for wood and paper and the enormous debt burden of Southern nations necessitate continued deforestation. Clearly,

> [t]he forest . . . is suffering from multiple attacks from many directions. Many of these attacks combine to strengthen their impact. While farming and ranching eat at the forest from without, logging degrades it from within and opens it up to faster penetration by farmers. . . . As the forest area dwindles, farmers add to forest degradation by hunting, grazing animals and gathering fuelwood.[107]

The South is most resentful of the assumption by developed nations that the remaining forests are a global asset[108] and resource, part of the heritage of mankind and therefore to be cherished and protected. However idealistic such thoughts appear to Northern environmentalists, in the South they are perceived as a new form of imperialism and a threat to national sovereignty. In a television interview given to Stephen Bradshaw of the British Broadcasting Corporation, Prime Minister Dr. Mahathir bin Mohamad of Malaysia, in response to a question about the forest being a common inheritance of mankind, stated very firmly: "No, no. It is not a common inheritance unless you are willing to pay for it. You don't pay for it. You expect us to pay for it. No go!"[109] The Malaysian diplomat Ting Wen-Lian commented: "We are certainly not holding our forests in custody for those who have destroyed their own forests and now try to claim ours as part of the heritage of mankind."[110] The fact that such suggestions come largely from the United States of America, which, like Malaysia, is one of the world's large exporters of timber,[111] adds to the suspicions of the South that the North is really only interested in curbing Southern development, particularly when it is competitive. Another conclusion is that "the North wants to have a direct say in the management of forests in the poor South at next to no cost to themselves."[112] This is again an example of the level of misunderstanding which prevails in both the North and the South, misunderstanding and mistrust colored by history and present economic realities.

The debate can also become quite pointed on occasion. Patricia Adams, writing in the *Globe and Mail*, alleged that in the Malaysian state of Sarawak "almost all timber concessions are owned by Malaysian politicians, their relatives or their companies."[113] Sir Crispin Tickell, former U.K. Ambassador to the United Nations and former advisor to Prime Ministers Margaret

Thatcher and John Major on environmental issues, asserts: "It always sticks in my gullet when we say to someone, 'you must look after the rights of those who live in the forests and act as its trustees,' and we're told not to be colonialists because in fact the big colonialists are the peoples of the poor countries, the elites of the poor countries who simply want to grab . . . land, grab . . . resources and make money . . . all of this in the name of 'development.' "[114] It is certainly true that the profits from forest destruction do not benefit the inhabitants of those regions. Both local entrepreneurs and foreign developers prosper. As United States Vice-President Gore has suggested, "Tropical countries . . . have frequently been encouraged to cut down their rain forests and sell the lumber as a strategy for development, but much of the cash ends up in the hands of a wealthy few (and in bank accounts in industrial countries), leaving the populace even worse off, stripped of their natural resources with little in return."[115] Award-winning environmental author Paul Harrison voices a similar feeling:

> The pillaging of the rainforests is not the result of ignorance or incompetence. In many tropical countries, the rainforests serve as political pork barrel. They have been plundered for quick gains by politicians and their proteges. In most of Latin America, they are sacrificed to the landless as an alternative to land reform. In Asia and Africa they offer an alternative to creating urban jobs or investing in rural areas.[116]

With such assertions foreign environmentalists do little to endear themselves to the leaders of the South who decide the extent to which their governments will support ecologically safe development. The tendency of developing nation governments is to suggest that the North oversimplifies the entire issue and cannot grasp the complexity of the development/environment dichotomy as it exists in the Southern context. The Malaysian Minister of Primary Industries, Dr. Lim Keng Yaik, in a speech given at Kuala Lumpur to mark World Forestry Day, compared the arguments used by environmentalists to those put forth by the Nazis:

> More and more people in the West are falling for such naive, discriminatory and simplistic solutions. This is where the danger lies. The Nazis tried to offer a simplistic solution for the economic and social ills that Germany went through in the 30's. Look what havoc they brought to Germany, Europe and the rest of the world.[117]

With some allowance for rhetorical exaggeration, the very fact of such blunt statements is indicative of the chasm in North-South relations over the development/environment dichotomy. In a kinder, gentler tone, Dr. Lim Keng Yaik underlined the dilemma facing all nations of the South, namely, the problem of "how to protect the environment and have development" and, he continued, to "find an equation that allows for that."[118]

Equations of that type invariably involve money, considerable amounts of it, and financial considerations underlie much of the North-South confrontation over the forests. The most vocal proponents of the Southern position happen to be the Malaysians. As one of that nation's most outspoken diplomats, Ting Wen-Lian, commented: "If developed countries want developing countries to preserve their forests, they should address the poverty, famine and crushing burden of external debt" which impels so much of this deforestation.[119] The Prime Minister of Malaysia, Dr. Mahathir bin Mohamad, spoke for much of the South when he told delegates at the Earth Summit that

> [t]he poor countries have been told to preserve their forests . . . on the off-chance that at some future date something is discovered which might prove useful to humanity. This is the same as telling these poor countries that they must continue to be poor because their forests and other resources are more precious than themselves. . . . Denying them their own resources will impoverish them and retard their development.[120]

Although the South demonstrated a remarkable degree of unity at Rio and in the meetings leading up to the Summit, it is also clear that within Southern nations there are diverse viewpoints on the utilization of the forests. While the indigenous people who have inhabited forested areas for centuries are upset by the upheaval, other inhabitants of those nations find the logging an avenue of escape from abject poverty. The diversity of opinion can be gauged from the remarks of, first, a forest-dwelling villager in Malaysia, who told Stephen Bradshaw of the BBC, "For me, if the leaders who are meeting could stop the logging, that would please me the most."[121] On the other hand, a logger in the state of Sarawak, one of 230,000 employees of logging concerns there, told him: "If we don't cut down the trees there's no money. If we don't work, we won't have enough to eat."[122] The centuries-long way of life of indigenous tribal people is being threatened by the deforestation. These people who have managed to live in harmony with their environment for generations now find their home—the forest—disappearing before their very eyes. Southern leaders are especially sensitive about alleged violations of the human rights of indigenous people, allegations which are usually made by the North. The challenge is to find other sources of employment in the South so that the rush to develop does not result in disruption for ancient ways of life and, ultimately, disaster for the entire planet.

Awareness of the likelihood of such a catastrophe prompted the initiation of the Tropical Forestry Action Plan in 1985 to promote sustainable use of the forests. This plan was the first attempt at international action to alleviate some of the problems of rampant deforestation.[123] That same year the International Timber Agreement came into force, with the objective of encouraging reforestation and forest management.[124] Thailand no longer permits commercial logging of its remaining forests.[125]

The nations of the South have for years demanded no less than a total economic restructuring of the world.[126] This would involve a consolidated and comprehensive shift in trade and the movement of goods to enable the developing nations to derive a fair profit from the sale of their products, along with encouragement from the North to enable more diverse economies to arise in Southern states. In 1984, only 13.9 per cent of global industrial production originated in the South. It is significant that this percentage was lower than Southern production percentages for 1948.[127] "The South's share of world trade fell from about 36 per cent in 1955 to about 25 per cent in 1987. Three-quarters of the investment activities of multinational corporations are in the North; transnational banks have lent 75 per cent of their stock of loans in the North."[128] Commenting on the increase in the developed world of protectionist policies over the preceding decade,[129] the World Bank was reported in June 1992 to have estimated that a 50 per cent "reduction in trade barriers by the European Community, the United States and Japan would raise exports from developing countries by $50 billion U.S. a year. That's roughly equal to the total flow of foreign aid to the poor countries last year. And it would provide critical resources to address environmental and other problems."[130]

A change in attitude, an end to the siege mentality in which elements of the North seek to protect their own standard of living at any environmental cost, would signal a recognition of the need for a more equitable world economic system, a system geared to environmentally safe development and the implementation of basic economic rights for the populations of the developing world. Given the climatic and environmental interdependence of the regions of this planet, it would be naive to assume that the North can somehow remain immune to the problems which plague the South.

The Debt Burden and Development

In a position paper written for the United Nations Conference on Environment and Development, the Centre for Science and Environment, located in New Delhi, maintained:

Since the economic levers of power in the world—aid, trade and debt—lie largely with the North, it is its moral responsibility to provide a lead that gives confidence to the South. The North must indicate its willingness to deal with basic issues that force the South to scrape the earth. And, most importantly, it must stop preaching.[131]

This idea is not restricted to the South. Vice-President Al Gore has suggested that "[a]s the world's leading exemplar of free market economics, the United States has a special obligation to discover effective ways of using the power of market forces to help save the global environment."[132]

The debates and formulations which sprang from the Earth Summit highlighted the most critical problem facing developing nations—the crippling burdens of external debt borne by many of these countries. Although the so-called Third World debt crisis of 1982 was said to be over by the early 1990s, the external debt of developing countries remained at a staggering $1,281 billion by 1991 estimates of the World Bank.[133] Brazil (the world's biggest Third World debtor) owed $98.3 billion, India's debt was $72 billion, and Mexico owed $98.3 billion.[134] At the beginning of the 1990s, interest payments on foreign debt totalled $100 billion annually.[135] "For much of the past decade, developing countries have given more money to the North—through imports, debt repayments and dividends—than they have received."[136] The increase in debt-servicing costs during the 1980s, combined with a decline in exports and a reduction in assistance, brought havoc to many of the poorest of the developing nations.[137] According to Katarina Tomasevski, "there is a net outflow of resources from developing to developed countries."[138] Concern over this issue prompted the United Nations General Assembly to pass a resolution in July 1986 urging that "this net transfer of resources from developing to developed countries has reached such proportions and is increasing at such a pace that concerted action is required on the part of the international community to halt and reverse the process."[139] On 24th February 1992, Tariq Osman Hyder, Director-General of Economic Coordination in the Ministry of Foreign Affairs of Pakistan, confirmed that "in today's recessionary international economic situation, the net transfer of resources between developing and developed countries is in fact negative."[140]

Dr. Helen Caldicott points out that "[t]hird world debt is exacerbating global environmental degradation."[141] She goes on to offer an explanation for the debt crisis in layman's terminology. The enormous quantities of petro-dollars (earned by the oil-exporting nations after the 1973 oil crisis and the consequent rise in the price of that resource) were deposited in a number of

Northern banks. To utilize the money, they lent it to poor developing nations which were short of cash, partly as a result of the sudden upsurge in oil prices. A lending spree ensued, with enormous loans being granted to the Philippines, Chile, Brazil, Argentina, Mexico, and Uruguay, to name only a few. "Lending to Latin America increased from $35 billion in 1973 to $350 billion in 1983."[142] Because of the fluctuation in interest rates on these loans, the cost of servicing these enormous debts rose as the interest rates escalated and remained high in the 1980s. "Each 1 percent increase in the interest rate meant $4 billion more that Latin America had to give American banks in interest payments."[143] Coupled with a global recession which cut Third World exports, the result was that "between 1973 and 1980, Third World debt increased by a factor of four, to $650 billion."[144] The economies of several countries in the developing world have been on a downward spiral since the 1970s, and as yet there appears to be no end in sight for some of them. "The United States in particular is worried that debt is exacerbating political instability in their Latin American 'backyard.' "[145] The United Nations Children's Fund (UNICEF) estimated in the late 1980s that half a million children a year died because of the impact of the debt crisis and the recession, two elements of economic malaise which led to a 10 per cent drop in the standard of living in Latin America and to an even worse drop of 20 per cent in Africa.[146] As the Peruvian economist Javier Iguiniz commented, "I don't like Western solutions to the debt crisis—they kill too many people."[147]

Madagascar's plight illustrates the consequences of the debt crisis. Depending largely on the export of unprocessed agricultural commodities, Madagascar is immediately prey to the fluctuations of world market pricing. Its debt level of $956 million in 1980 had risen to $3,317 million by 1988.[148] "Debt service, which used up less than 4 per cent of export earnings back in 1970, swallowed almost ten times that proportion in 1988. . . . The country . . . had to slash government spending savagely, from 23 per cent of gross national product in 1965, to only 12 per cent in 1988."[149]

In his study *Poverty and the Planet* (1990), Ben Jackson emphasized the fact that in thirty-seven of the poorest countries of the world per capita spending on health had been halved in the previous few years.[150] By 1989, 44 per cent of the Philippine national budget was going to service its foreign debt;[151] the debt-servicing percentage for Malawi was 38 per cent (after rescheduling).[152] Paul Harrison commented that "debt made the developing countries into net donors to the rich countries. Social spending was cut back in Africa and Latin America. Progress in education and health halted and reversed."[153] As Ben Jackson explained:

A debtor—whether an individual, company or country—gets into trouble when the difference between what they earn and the amount they have to repay on the debt (much of it just interest) becomes too great too quickly. This difference is known as the "debt service ratio." In the case of Third World countries, it is the amount of debt they repay each year in comparison with their earnings from trade, since only by exporting can they earn the hard currency, mainly US dollars, in which payments on foreign debts must be made.[154]

Unfortunately, "[m]ost of the profits from commodity sales in the Third World go to retailers, middlemen, and shareholders in the First World. Only 15 per cent of the $200 billion in the annual sales of these commodities in rich countries winds up in the countries that grow them."[155] The result is that some poor countries of the South wind up on a "debt treadmill."[156] According to Ben Jackson,

> The combined effect is to bleed debtor economies dry. The debt crisis has changed the overall inflow of finance from rich countries to poor into a perverse and increasing flow from poor to rich. Third World debtors now transfer over $50 billion a year more to Western banks and governments than they receive in new loans and aid. No other statistics could make a greater mockery of the rich world's idea of itself as a benefactor of the Third World.[157]

So desperate are these poor nations to obtain foreign currency that they are compelled to trade their most marketable products, often at greatly reduced prices and regardless of the environmental destruction they are causing. Natural resources have to be exploited at a frantic pace to meet debt service charges, and in many Southern nations "[e]nvironmental resources are being exhausted just to pay debt."[158] The Third World Network, based in Malaysia, has estimated that twenty-seven of the countries endowed with rain forest resources are burdened with a collective debt to the North of $630 billion.[159]

The economic wish list of the South was articulated by numerous delegates at the Rio Summit, both prior to and during the Conference. There was a general consensus among delegates and economists from the developing world that what the South needed from the North was greater access to its vast consumer markets, debt relief, stabilization or even raising of commodity prices, improved access to international liquidity, and aid for development.[160] The South complained about the fact that so many developed countries used protectionist tariffs on imported manufactures to make Southern goods more expensive and force poorer nations to concentrate on the production of raw materials, which have easier access to Northern markets.[161] The Centre for Science and Environment in India urged the North to

> [g]ive the South a fair deal by reforming the world market system so that it can take into account the ecological costs of producing its commodities and the South will take care of its environment. These costs can be captured only through a series of fiscal and economic instruments as part of a deliberate public policy package.

The Centre then asked: "Are the rich prepared to pay the real costs of what they consume?"[162]

Although for some developing nations the intensity and the brutal consequences of the debt crisis have been eased somewhat, the crisis continues to haunt most nations of the South. Action taken by the Group of Seven (developed countries) resulted in the forgiving of more than $5 billion in official development-assistance loans by the early 1990s.[163] In 1990 a debt-restructuring agreement known as the Brady Plan proposed a package of debt relief estimated to be worth about $25 billion.[164] The aim of the Plan was to convert loans to grants.[165] However, the Plan has been criticized for being "too little, too late, in too few countries;"[166] some nations, like Mexico and the Philippines, simply defaulted on their payments.[167] The United Nations has estimated that a reduction of wealth transfer to the banks of about $10 billion per year could result in some prospect of reaching a modest 2 per cent growth rate.[168] If the debts of all African nations were forgiven, more than $12 billion would be available for development and a better lifestyle for their citizens.[169] Canada authorized the conversion of loans to grants in 1987 and extended this assistance to a number of debt-plagued developing nations, but on condition that they submit to supervision by the International Monetary Fund (IMF).[170] The following year, 1988, the Western Economic Summit "supported a menu of debt relief options that included grace periods, interest rate reductions, rescheduling, and actual debt reductions in a variety of combinations. All this is welcome but is subject to IMF agreements, country by country."[171]

For the nations of the South, submitting to the International Monetary Fund is bitter medicine indeed. To many of these countries, the Fund and its sister organization, the World Bank, are "international lenders of last resort."[172] The Washington-based World Bank is perceived by Southern nations as being run "by and for developed countries—in particular the United States"[173] and as reflecting American interests and policies.[174] As with many

Southern perceptions, this one is somewhat simplistic, if not completely un-justified. In the developing world the entire organization is simply termed the World Bank and derided accordingly, even though it has component institutions like the International Bank for Reconstruction and Development and the International Development Association,[175] as well as the International Finance Corporation[176] and the Global Environmental Facility.[177] The Bank has 159 member countries, with the United States of America owning the largest bloc of shares.[178]

The even more controversial International Monetary Fund was established in 1945 to oversee the world's monetary system and "to stop the financial sneezes of one country giving the whole world trade system a cold."[179] It has had a significant impact on the economic growth of developing countries.[180] Unfortunately, these Northern institutions, whose aim is to assist the developing world, often wind up creating greater problems and exacerbating the North-South confrontation. As Caroline Thomas explains, "The IMF, as guardian of the post-war monetary order, has come to be seen by the developing world as an instrument of Western domination and as a violator of their sovereign rights."[181] Thomas, who lectures at the University of Southampton, continues:

> The root of the problem is that the IMF, which exerts so much control over the economies of Third World states, was never conceived with them in mind. Nor for that matter was the international monetary system, of which it is the cornerstone. Thus Third World states attack the IMF both in general terms and on three specific issues.

First, Thomas points out that the IMF's political philosophy affects its "economic orthodoxy," which neglects the requirements of poor countries. Second, developing nations lack the ability to adequately influence decisions of the Fund. Most unpalatable of all are the stiff conditions imposed on recipients of IMF loans.[182]

Nations of the South compelled to resort to the World Bank or to the International Monetary Fund are required to abide by very tough conditions specified by the two institutions; these are implemented via structural-adjustment programs[183] (in 1991 accounting for $5.9 billion, approximately 26 per cent of the Bank's commitments),[184] which can result in draconian measures to "reduce public deficits, liberalize markets and often devalue . . . currencies. . . . In many cases . . . the immediate results include social service cuts, lower real wages and reduced subsidies for basic goods such as food and oil."[185] In an effort to assist debtor countries in promoting exports, the IMF often urges currency devaluation as a drastic but necessary measure to jump-start the economy. Unfortunately, devaluation does not always have the intended result—poor nations find it harder to pay for necessary imports, and the increased volume of exports can turn into a glut which forces the price down. "Third World countries find themselves producing more and more for less and less."[186] Critics of the Bank assert that it is "undemocratic, unaccountable and unable to support sustainable development."[187]

Although the World Bank's primary activity of approving loans worth an estimated $29 billion a year[188] appeared to be laudable, given the very real need for such funding in the developing nations, it was accused of fostering economic activity at the expense of the environment. For example, the World Bank committed $450 million to a major hydro-irrigation project in India (the Sardar Sarovar Project on the Narmada River) before an independent review suggested that the displacement of approximately 250,000 people by the construction of 165 large and medium-sized dams, 3,000 small dams, and 75,000 kilometers of canals had not been adequately assessed.[189] The Sardar Sarovar Project aimed to create irrigation systems for 1.9 million hectares of land and produce 1,450 megawatts of much-needed electric power.[190] However, international funding for such megaprojects in developing nations has produced a storm of environmental protest in both the developed and developing worlds. The complaints can be quite emotional: "Should you wish to reduce any well-informed environmentalist to tears, try a litany of bank-financed ecological disasters: Sardar Sarovar Dam, India; Pak Mun Dam, Thailand; Kedung Ombo Project, Indonesia; National Livestock Project, Botswana; Polonoreste and Grande Carajas, Brazil; the Tropical Forestry Action Plan, in many places. The list could be lengthened."[191] The problem with such massive schemes is that they destroy as much as they build, and, increasingly, the populations of recipient nations are unwilling to accept the cost of such aid, even if their governments encourage large development projects. Medha Patkar of India, winner of the 1991 Right Livelihood Award for her activism, believes that her countrymen need "a development that is participatory and does not destroy the community or its resources."[192] This idea is now gaining popularity in the developing world, where environmental groups are putting up a stiff resistance to development projects which endanger the environment. As a result of such pressure, Japan withdrew its financial support for the Sardar Sarovar Project.[193]

The involvement of foreign financing has in the past led to a concentration on nonenvironmental considerations in judging the feasibility and success of these projects. Vice-President Al Gore remarks that

when the World Bank, the International Monetary Fund, regional development banks, and national lending authorities decide what kinds of loans
and monetary assistance to give countries around the world, they base their
decisions on how a loan might improve the recipients' economic performance. And for all these institutions, the single most important measure of
progress in economic performance is the movement of GNP. For all practical purposes, GNP treats the rapid and reckless destruction of the environment as a good thing![194]

Commenting on IMF conditions on loans, President Julius Nyerere of Tanzania, commented that the Fund "has an ideology of economic and social development which it is trying to impose on poor countries irrespective of their
own clearly stated policies."[195] Environmentalists in the developed countries
have been very vocal in their opposition to the World Bank and to the IMF.
Helen Caldicott blames the Bank for encouraging debtor nations to fell their
rain forests;[196] Ben Jackson charges that "[u]nder 'adjustment with growth,'
debtors have done plenty of very painful adjusting, but only their debts, not
their economies, have grown."[197]

In the Bank's defense, it ought to be pointed out that its policies have
changed. The Bank has for some years been sensitive to global criticism and
to the environmental activism generated by its support for environmentally
destructive development projects. Although its agenda has been aimed at
alleviating the problem of poverty, the Bank has had some misgivings about
how much the poor are helped by such programs.[198] Now, "after years of
criticism from environmentalists, the bank has got religion and is preaching
it with the fervour of the converted. Its projects are now subject to environmental assessments and it has created and expanded an environment department within itself.[199]

In the year of the Earth Summit, the Bank seemed to emphasize its apparent interest in environmentalism. In its 1992 report on world development,
the Bank urged industrialized nations to contribute to the creation of a
healthier planet. The Bank has attempted to clean up its image by projecting
a greater environmental consciousness and now refuses, for example, to give
financial assistance to logging in primary forest areas.[200] Insisting that
"[w]orld development is consistent with good environmental practices,"[201]
the Bank, following policies enunciated in 1987, is demonstrating a new
awareness of the human and environmental costs of large projects.[202] Its priorities now include provision of clean drinking water, education for women,[203]
and technology transfers to developing countries. The Bank also joined forces
with the United Nations Development Program and the United Nations Envi-

ronment Program to assist in an experiment costing $1.3 billion to fund environmental-protection schemes.[204]

While all this sounds very positive, it has to be remembered that the Southern nations retain their deep suspicion of the World Bank and of the International Monetary Fund. Time alone will tell whether the Bank's new-found eco-sensitivity will strike any responsive chord in the developing countries that receive foreign aid. Indeed, as environmentalism is regarded as a Northern, anti-development notion in many Southern countries, it is possible that the Bank's insistence on environmental assessments for aid projects may be perceived in the South as another facet of a Northern notion which results in an inclination to interfere with the political sovereignty of Southern nations. Eco-sensitivity may be condemned in the South as eco-imperialism. For the North, it will then be a case of being damned if you do and damned if you don't.

Indeed, at a preliminary conference of developing nations in Malaysia, Third World countries formulated their joint initiatives for Rio and were in general agreement with the message of a Malaysian government documentary which urged that "UNCED must not herald the beginning of an era of eco-imperialism."[205] It is entirely possible, given the diverse perspectives brought to the North-South dialogue, that the "introduction of the environment dimension into the development process"[206] will heighten rather than reduce tension. The Southern governments are very conscious of their political independence and, since the Earth Summit, of their collective responsibility to ensure that no Northern nations attempt imperial takeovers, even under the guise of saving the planet. While we in the North may find this attitude quite exasperating, it has to be understood and possibly appreciated if we are to deal with the South in a positive manner. After all, the imperialists of the past ostensibly came to save the souls of the "heathen" and stayed to loot the economies of the colonies. To the Southern mind, there is not that much difference in the present message of saving the planet, particularly when the riches of developing nations are labelled global resources and declared part of the inheritance of mankind.

The Environmental Costs of Development

The idealism of Northern environmentalists can appear to be crass materialism when the message is filtered through the Southern mind, which has its own frame of reference about past exploitation. When, additionally, the nations of the North refuse to take significant steps to curb their own pollution—pollution which accounts for the majority of the environmental problems of this planet—and expect the South to remain poor so that the North can breathe purer air, the chasm between the two sides becomes even wider. International financial institutions like the World Bank walk a thin line as they attempt to cross the North-South divide, advancing aid carefully enough to meet the Northern, donor agenda without appearing to affront the nationalistic sensibilities of the Southern, recipient states. Further, if environmental concerns add to the level of hardship and to the number of stringent financial conditions imposed on developing nations, there might be a repetition in many poor countries of the riots related to IMF austerity measures which have occurred in the past. The 1981 riots in Morocco and unrest the following year in Sudan are but two examples.[207]

Given the negative perceptions of the World Bank, it is not surprising that during the Rio negotiations the developing nations resisted efforts to channel the North's funding of sustainable-development projects through the Bank. As the *Times* (London) commented, "The poorer nations are suspicious of the World Bank, which they regard as a rich countries' institution, and have been demanding a new green fund, which they would control and to which the richer countries would contribute. That has been refused, and when talks on the climate change convention were successfully concluded they accepted the GEF [Global Environment Facility] as the treaty's funding mechanism, at least on an interim basis."[208]

It could also be argued that the South has few alternatives, given its present poverty, but to develop rapidly if its population is not to suffer total destruction. India, which is not by any means one of the most desperate of these poor nations, still has compelling, urgent problems to resolve. To give only one example of the type of crisis such nations face, consider the electricity requirements for that nation, which has a population of nearly one billion.[209] In terms of its power requirements in the future, "[e]ven conservative forecasts put India on a collision course with chaos. Just to keep pace with its swelling population and new-found industrial growth,"[210] the country will have to double its output of electricity.[211] With approximately 16 per cent of the world's people (making it the second most populated nation in the world after China), India consumes only 3 per cent of the planet's energy.[212] By way of contrast, the United States of America, with approximately 5 per cent of the Earth's population, consumes 25 per cent of the world's energy.[213] The average Canadian consumes fifty times as much energy as the average citizen of India.[214] There can be no doubt that without rapid development India is heading down a road to economic disaster. The example of India's electricity

requirements provides dramatic evidence of the dilemma and the crisis facing leaders of many developing nations as they struggle against all types of
obstacles in their attempt to improve the lives of their people.

Some Southern nations proceed rapidly along the path of development and
appear not to heed the perils of the environmental degradation they are creating. China, with about one fifth of the Earth's population,[215] is regarded as
"one of the world's fastest growing polluters."[216] Because China depends on
coal to produce between 70 and 75 per cent of its energy,[217] it consumes
about 900 million tons of coal per year; in 1990, this created about 15
million tons of sulphur dioxide and 20 million tons of coal dust.[218] In the
same year, about "580 million tons of industrial solid waste was dumped,"[219]
and 35.4 billion tons of waste water polluted China's rivers, ports, and
lakes—and this is even more serious a threat when it is realized that only 15
per cent of China's waste water is treated.[220] Chinese pollutants are already
causing acid rain in Japan,[221] and as China increases its energy consumption
(expected to grow sixfold over the next four decades),[222] its carbon emissions—reported in 1992 to be 9 per cent of the world's total[223]—are likely
to become a real health hazard for its population and for those of neighboring
states. Even now, Beijing suffers from so much smog that on occasion it
cannot be seen by orbiting satellites.[224]

In China, approximately 1 person in 74,000 enjoys the luxury of owning a
car.[225] However, "China has been adding over 600,000 vehicles a year in
recent years."[226] When the Chinese decide to trade in their 300 million
bicycles[227] for cars, the consequences in terms of air pollution and energy
consumption will be extremely serious. The sincerity of the Chinese Government's commitment to environmental protection of its own population can be
gauged by the fact that "in a short-sighted approach to saving foreign exchange, China orders imported cars stripped of their catalytic converters."[228]
Even more potential danger to the environment springs from the determination of the developmentally oriented Chinese Government to provide its people with refrigerators, up to an estimated 300 million.[229] If the refrigerators
utilize chlorofluorocarbons, as did older models in the developed countries,
the impact on the delicate and now already damaged ozone layer protecting
the Earth could be very serious.[230] Although the United States has provided
China with aid and advice to produce "environmentally safe refrigerators,"
there are no guarantees that such technology will be preferred.[231]

The situation of China is compounded by the fact that its present geriatric
leadership is propelling a rush to economic development in part to compen-
sate its people for the lack of political freedom and human rights in that

country. China's leaders are more than aware of the fact that the decline of Communism in Eastern Europe puts them very much more on the defensive. Having survived the student demonstrations at Tiananmen Square in 1989 by brutally crushing the demand for democracy, the government hopes to woo its people economically while repressing them politically. It is questionable whether China's attempts to create a free market within a political strait-jacket can succeed. As the present leadership watches totalitarianism wither away throughout the world, it plunges forward with almost reckless zeal into economic development to justify itself to its own population. Environmental concerns are of less significance to China than probably to any other developing nation. Although the Government claims to have spent $2.9 billion on environmental protection between 1986 and 1989,[232] "China's economic revolution isn't likely to slow down out of environmental concerns."[233]

Of even greater concern is China's alleged involvement in the trafficking of cross-border waste. Greenpeace "disclosed that officials in Baltimore were negotiating with authorities in China for permission to dump tens of thousands of tons of municipal solid waste in Tibet. Nothing could be more cynical," suggests Vice-President Al Gore. He continues: "The Tibetan people are powerless to prevent Chinese officials from destroying the ecology of their homeland because of China's armed subjugation of Tibet for the last forty years. But the shipment has not taken place, and the United States has not yet become heavily involved in cross-border waste trafficking."[234]

In the international forums where environmental matters are debated, China points with some justification to the extent of pollution caused by the rich nations and refuses to curtail the exercise of its right to development.[235] The Chinese, in the negotiations leading up to the Earth Summit, campaigned for emphatic developmental rights, with a demand for payment by the North to limit environmental damage in the Southern nations.[236] The Director of China's state administration of the environment, Qu Geping, proposed shortly before the Earth Summit that the developed nations pay at least 20 per cent of the estimated $600 billion required for limitation of environmental damage.[237] China's assertive position found support among Southern nations, who firmly believe that those who created the environmental damage ought to pay to clean up the mess and that environmental concerns ought not to supersede the desperate need for development in the South.

It would be presumptuous for us in the North to decry the Southern economic planners' concentration on immediate necessity. If we had millions of starving and undernourished people and massive unemployment, would we not put environmental concerns on the back burner in the fight to save human lives? Was the reluctance of the Government of the United States to commit to international Conventions on the environment not partly a reflection of its perception that its priorities must rest with the major problem of a recession and the highest unemployment rate in eight years? If we in the North are able to rank our developmental/environmental concerns in this manner, we ought not to resent the South for doing likewise, particularly when the dimensions of their problems are far greater and the need far more compelling than in any Northern nation. "To the North, the issue is environment, but to the poor countries of the South, it's development—their right to develop into prosperous modern economies without interference from the old imperial powers."238

Environmentalists in the North also decry the rush to development in the South as short-sighted and suggest that the pillaging of resources in the forest will ultimately weaken the agrarian lifestyle of the majority of Southern inhabitants. Once trees are felled, topsoil gets washed away, rivers become polluted with land runoff, and rainfall decreases as forests disappear. Although governments in the developing world are more than aware of these consequences, they face the dilemma of coping with immediate crises even if the very solution—quick development—results in long-term problems. Nor can it be suggested that short-sightedness is a particularly Southern affliction. As governmental leaders and environmental activists in the South are quick to point out, the North has always been and continues to be short-sighted in its extravagant use of resources, its construction of pollution-causing industries, and its consumer-oriented lifestyle. The short-sighted policies of the North gave every encouragement to rampant development, destroyed forests in Europe and North America, dramatically changed the world's climate, ripped a hole in the Earth's protective ozone layer, brought death to thousands of rivers and lakes, and even now are killing life in the oceans of the world. Thus does mutual recrimination continue as part of the North-South dialogue.

There are environmentalists in the North who have expressed sympathy with the circumstances facing Southern governments, and the fact that some of these persons are in positions of influence offers some hope that in the future the chasm separating the rich and poor nations may be bridged by mutual understanding. Vice-President Gore has stated that

[r]apid economic improvements represent a life-or-death imperative throughout the Third World. Its people will not be denied that hope, no matter the environmental costs. . . . As a result, that choice must not be

forced upon them. And from their point of view, why should they accept what we, manifestly, will not accept for ourselves? Who is so bold as to say that any developed nation is prepared to abandon industrial and economic growth? Who will proclaim that any wealthy nation will accept serious compromises in comfort levels for the sake of environmental balance?

Vice-President Gore continues: The industrial world must understand that the Third World does not have a choice of whether to develop economically."[239]

In a speech to dignitaries and delegates at the Earth Summit, renowned explorer Jacques Cousteau deplored the "new dictatorship of materialism" and warned that the planet was suffering because of "an interminable succession of absurdities imposed by the myopic logic of short-term thinking."[240] To some extent this "myopia" is understandable as an attempt by some Northern governments, specifically that of the United States, to adjust to an environmental agenda within the frame of reference long dictated by its own much envied lifestyle. American politicians might argue that it would be political suicide for any U.S. president to urge his people towards the massive economic and social sacrifice necessary for rapid global advancement of sustainable development. Politically, it is far safer to appear to be an environmental moderate and effect gradual change in the North, while using one's international influence to pressure Southern nations to hasten their conversion to sustainable development. The Bush Administration made it very clear that "the American life-style is not up for negotiation."[241]

However well this may have played domestically, the remark did nothing to endear George Bush to either the developing world or to environmentalists from the developed world. In view of the fact that the American lifestyle results in a reported 22 to 23 per cent of all the polluting CO_2 produced on this planet,[242] there was some justification for the criticism. The American stance at Rio was perceived as being assertively nationalistic and antagonistic to global concerns. Southern activists argued that if the "lifestyle" was so unassailable, then the obvious conclusion could only be that the developing countries would have to be content forever to remain in poverty in order to sustain and bear the consequences of that extravagant way of living. The Malaysian Prime Minister commented, "The rich talk of the sovereignty of the consumers and their right to their life-styles. . . . they expect the poor peoples of the developing countries to stifle even their minute growth as if it will cost nothing."[243] The South maintained that the consumer-oriented, market-dictated American lifestyle would have to be adjusted if the entire planet is to enjoy some form of sustainable development and to survive. As one Southern think tank asked: "Does anybody seriously believe that the earth can support everybody at the consumption level of a European or a North American?"[244]

Ironically, while condemning the lifestyle of North America and Europe, the governments of most developing countries have at the very least set their sights on some semblance of this lifestyle as a target and aim of development. And although public statements caution about the need to develop in an environmentally safe manner, the sheer pressures of such vast populations generate their own momentum, carrying governments forward in directions which are questionable in the short term and extremely risky in the long term.

Rio and the Right to Development

It is most regrettable that the nations of the world lost the opportunity presented at the Earth Summit and in the months of preparation leading up to it to expand their horizons beyond the confining and restrictive barriers of nationalism, political sovereignty, and economic expediency. Rio needed a wider vision, and the Conference foundered badly on the limited perceptions the world leaders allowed themselves. It was as if they attended a great international gathering clinging tenaciously to limited national agendas, plans of containment which looked inward rather than far-reaching proposals to move the people of the world in a new direction. It is pointless to cast all the blame on President Bush, as indeed the delegates did at Rio. The nations of the South came armed with their own agenda, a plan of action to make up for all the exploitation of the past even if this involved environmental degradation for the future. Instead of using their new unity as representatives of the majority of the world's population, the Southern leadership perceived Rio as a platform for trading insults and for gaining economic leverage against the North. The most vociferous of their advocates, the Malaysians, championed proposals which would assist the South developmentally but resisted efforts to push through a treaty to protect the Earth's remaining forests. That Treaty, backed by the Americans also, partly failed because the United States refused to sign the Biodiversity Convention which gave some benefits to the South for the North's utilization of its vast genetic treasure. The Americans were apprehensive that the Biodiversity Convention would have a detrimental impact on their growing biotechnology industry—which, they feel, ought to be in a position to utilize such diversity found anywhere without having to share the profits with the poor nations which are home to most of the world's remaining species.

An analysis of the Earth Summit demonstrates clearly that both North and South were eager to lay the burden of environmental adjustment on each other. In fairness, it must be stated that not all nations of the North and South are as conscious of national interests as were the Americans and Malaysians at Rio. However, these two nations, one the world's only Superpower and its richest country, the other the most assertive advocate of the development-minded elements of the South, stand as exemplars of the confrontation on which Rio faltered. In their dedication to the primacy of national self-interest over universal values and concerns they were very similar, albeit at opposite poles in the debates and negotiations which occurred. Rio reflected, in the words of one commentator, "a conflict between economics and ecology."[245]

The Conference also demonstrated that the expansion of the protective umbrella of international environmental law requires the parallel creation of a new world economic order in which the gap between rich and poor is narrowed to enable sustainable development to occur in the poorer nations. The South and the North have to find environmentalism mutually beneficial, not only in the long term but in the immediate future of their populations. Ben Jackson has suggested that "environmental destruction in the Third World cannot be prevented without challenging the economic and political pressures that underlie it: from unequal land distribution to international debt and unfair trade."[246]

The creation of law within the boundaries of any democratic nation usually occurs via the process of public consultation, active debate, and, eventually, compromise as legislators indulge in a series of give-and-take deals recognizing the primacy of one interest or another. The resulting legislation may not be perfect (and may indeed be quite flawed by the spirit of compromise which led to its acceptance), but each law marks a step forward in the enlarging sphere of legal protection which has become so fundamental a part of civilized, free societies. The process of formulating international law, although somewhat similar, is much more complicated in that sovereign entities, nation states, undertake to restrict the exercise of their particular sovereignties for the good of all of humanity. The stakes are far higher in international law, and the positions are therefore more rigidly upheld and very reluctantly relinquished. There is far more posturing by world leaders (especially at a summit) because the entire world is watching. Frequently, the public rhetoric is more vehement than the private positions of these personalities. Surrounded by the glow of summitry which plays so well domestically, leaders espouse ideas which they know are probably slightly irresponsible; however, they act out their role before the television cameras and gain either popularity or notoriety, depending on the audience. Soon, the summit meeting is over and the world turns its attention to another matter.

The Earth Summit at Rio foundered in part because leading players were not willing to acknowledge that "[e]conomic security today can become the very basis of ecological security tomorrow."[247] Had there been an enthusiastic commitment by all the participants both to create a new world economic order and to chart a course in sustainable development for the entire planet, the future of mankind would have been brighter. Unfortunately, there was not the courage to seize the moment and effect dramatic change. As with many United Nations sessions, the negotiations became a process of wheeling and dealing which so watered down the resulting Treaties as to make them almost a mockery of the original vision which generated the Earth Summit. The Rio process resulted in five major documents: the effort to protect the world from global warming, commonly referred to as the Climate Treaty; a significant step to protect the Earth's genetic resources, popularly called the Biodiversity Treaty; a statement of general principles on the twin concepts of environment and development, known as the Rio Declaration; a massive blueprint for mankind's future, Agenda 21; and a non–legally binding statement on forestry (reprinted as Appendix 3 to this book). Although the North had hoped for a Forestry Convention, this did not materialize. It would be worthwhile to examine the Rio documents briefly—especially the two main Conventions, which were accepted and which are legally binding—and assess them from the perspective of the South in order to appreciate how Southern priorities fared in the Rio negotiations.

On 9th May 1992, representatives of 143 nations agreed to a convention which would limit the amount of carbon dioxide and other greenhouse gases emitted into the world's atmosphere.[248] This Treaty was signed with much pomp and ceremony by world leaders at the Rio Summit. Labelled both a "historic break-through" and a "sell-out,"[249] the Convention on Climate Change acknowledges "that change in the Earth's climate and its adverse effects are a common concern of humankind."[250] In this Convention, the South won recognition of the fact that "the largest share of historical and current global emissions of greenhouse gases has originated in developed countries,"[251] an important admission by the North. The Convention also concedes that environmental "standards applied by some countries may be inappropriate and of unwarranted economic and social cost to other countries, in particular developing countries."[252] Linking environmental and developmental priorities, the Treaty affirms "that responses to climate change should be coordinated with social and economic development in an integrated manner with a view to avoiding adverse impacts on the latter, taking into full account the legitimate priority needs of developing countries for the achievement of sustained economic growth and the eradication of poverty."[253] This linkage is pursued in Article 3.4: "The Parties have a right to, and should, promote sustainable development. Policies and measures to protect the climate system against human-induced change should be appropriate for the specific conditions of each Party and should be integrated with national development programmes, taking into account that economic development is essential for adopting measures to address climate change."[254]

Developing nations have, in the initial body of principle in this Convention, also sought protection against the imposition of environmental measures that may act to their detriment in the field of international trade.[255] This issue is particularly important for countries like Malaysia, which during the Rio negotiations was active in defeating plans to introduce green labelling to enable consumers to avoid timber from rain forest sources. In defending their stance on this matter, the Malaysians argued that the West would resent it if developing nations attempted to label imports from the North on the basis of their ecological acceptability. "You won't like it," Dr. Lim Keng Yaik, Malaysian Minister of Primary Industries, told BBC correspondent Stephen Bradshaw. To the Malaysian Government, such attempts at green labelling would establish a double standard. Dr. Yaik termed the attempt "hypocrisy to the highest degree."[256]

The Convention on Climate Change calls on States Parties to "stabilize atmospheric concentrations of so-called 'greenhouse' gases at a level that would prevent dangerous human-induced changes in the global climate."[257] The target is to reduce emissions to 1990 levels,[258] hardly a major concession, as this would probably only "reduce carbon dioxide emission by 11 per cent by 2050."[259] The lack of firm timetables in the Convention for implementing such reductions was largely at the insistence of the Government of the United States;[260] that country emits about 22 to 23 per cent of the world's CO_2, the largest percentage of any nation.[261] Sensing a threat to its industrial base, the United States "resisted the idea of any named target whatever"[262] and was accused of having "watered down the summit's central global warming treaty."[263] However, President Bush was not prepared to allow "the extremes of the environmental movement to shut down the US on science that may not be as perfected as we in the US could have it."[264] The attempts by the American President to protect the economic interests of his nation were derided by environmentalists at Rio. Joshua Karliner, representing Greenpeace, commented: "The climate control convention was supposed to be the jewel in the crown. Now it looks more like a rhinestone."[265]

The real gem in the Treaty, from the perspective of the South, was the financial commitment of developed nations to assist developing countries in their efforts to improve the environment. "Industrialized countries are required to help finance and provide technology to developing countries to meet general commitments."[266] The Convention specifies:

> The developed country Parties and other developed Parties . . . shall take all practicable steps to promote, facilitate and finance, as appropriate, the transfer of, or access to, environmentally sound technologies and know-how to other Parties, particularly developing country Parties, to enable them to implement the provisions of the Convention.[267]

This kind of provision caters to a limited extent to the polluter-pays principle[268] espoused by the South before and during the Rio Summit. It also demonstrates a recognition of the fact that the North has the means to save the Earth from further degradation if it can overcome its emphasis on economic self-interest for the larger global interest of humankind. The South believes that it is time now for the North to pay for the costs of its consumption. Although the South shoulders the burden of economic debt, there has now to be a recognition and acceptance of the "environmental debt of the industrialised countries."[269] The South is no longer willing to play the role of humble aid recipient. The mood and tone are now very different and quite assertive: "The South should not be again seen as holding out the begging bowl for 'new and additional resources' or calling for 'technology transfer.' The South should be demanding compensatory measures from the North for errant behaviour as a question of its right over global resources."[270]

Even though the South gained the right to acquire environmental technology, it was compelled to concede on the issue of the handling of the financial provisions of the Convention. The North was anxious to retain the services of the Global Environment Facility, a financing agency controlled by the United Nations and the World Bank.[271] "The Agency's four priority areas are: ozone depletion, global warming, the preservation of biological diversity and international water management. The first three are covered by international treaties that require rich countries to help the poor. The GEF is the conduit for the aid."[272] Because the nations of the North have strong influence within the GEF, they feel more secure in utilizing that institution for the disbursement of the aid to which they have committed themselves in the Climate Change

Convention. The South resisted that approach, perceiving the North-controlled and North-dominated World Bank as "detrimental to the interests of the South"[273] in terms of its autonomy and national sovereignty. It was also pointed out that "the World Bank has no lever on the policies of the countries of the North, which aggravates the imbalance in North-South relations."[274] The eventual compromise instituted a combination of the Global Environment Facility, the United Nations Environment Programme, and the International Bank for Reconstruction and Development as the entity to operate the financial mechanisms of the Convention on an interim basis. Provision was made to restructure the Global Environment Facility to make its membership "universal."[275] In an editorial, the *Times* (London) lamented this realignment of the power structure in the GEF, commenting that the institution would "rapidly become unworkable."[276] Whether this expanded international financial facility will be perceived in the South more positively than the World Bank and its affiliates have been in the past remains to be seen. By linking the right to development and the right to a clean environment, the Climate Change Convention could be considered a victory for Southern moderates, who have now at least obtained international recognition of the interdependence of these two human rights and of the mutual obligation of the rich and poor nations to act together in implementing them.

The widespread global acceptance of the weakened Climate Change Convention was not matched by equal success in the other major formulation produced for signature at Rio. The Biodiversity Convention fell victim to United States resistance and eventual refusal to sign. Although numerous countries have accepted the Treaty and have become Parties to it, the world's leader in biotechnology refused at Rio to accept the terms of the Convention. It was left to President Bill Clinton to reverse the American position with his nation's acceptance of the Biodiversity Convention on 21st April 1993.

The North-South bickering over biodiversity was intense and acrimonious. Having conceded, it felt, far too much in the Climate Convention to gain American adherence, the South dug in its heels and refused to water down the Biodiversity Treaty to meet American requirements. The Southern view was expressed by the environmentalist Vandana Shiva, who argued that

> [m]ost of the biodiversity exists in the South. Two-thirds of it exists in the
> South. It is a Third World resource. By calling it a global resource, by
> calling it a common heritage of mankind, the North is basically preparing
> the ground to assure raw material supply for the emerging biotechnology
> industry which needs this diversity as input. Basically, behind the biodiver-
> sity conflicts is a conflict over who will control this future raw material and

whether that biodiversity will be able to sustain life in the Third World or it will only sustain profits for Northern corporations.[277]

It could be argued that for the South there was far more at stake in this Treaty. It owns the resource and wants to ensure that future exploitation will result in a shared profit between the biotechnological companies, largely in the North (and led by the United States), and the governments of the South, which, in return for a share of the profits, can undertake to conserve the forests and preserve the thousands of species which may in the future provide new medicines, cosmetics, and innumerable other products. Thus far, such genetic riches have been exploited without much thought of compensation to the country of origin.[278] In essence,

> the treaty reflects an attempt to reconcile the goal of preserving species and ecosystems with that of economic development and reduction of poverty. Under the treaty's provisions, nations that ratify it would set up protected areas, promote the preservation of ecosystems and species and restore damaged ecosystems. They would also try to integrate conservation into their economic planning and development, permitting what would be deemed as friendly exploitation of forests for medical resources, food and tourism.[279]

The great achievement of the Convention, according to the Indian delegate Avani Vaish, "is that the value of genetic resources will be really appreciated. Resources were a free commodity, like air and water." Now, he added, the Treaty brings them "under international jurisdiction."[280]

The Treaty affirms "that the conservation of biological diversity is a common concern of humankind," while also reaffirming "that States have sovereign rights over their own biological resources."[281] The South gained recognition, first, of its need for additional financial resources and access to technology,[282] second, of the fact that "economic and social development and poverty eradication are the first and overriding priorities of developing countries,"[283] and, third, of the objective of "fair and equitable sharing of the benefits arising out of the utilization of genetic resources."[284] The Convention recognizes "the sovereign rights of States over their natural resources" and concedes that "the authority to determine access to genetic resources rests with the national governments and is subject to national legislation."[285] The Contracting Parties agree to share "in a fair and equitable way the results of research and development and the benefits arising from the commercial and other utilization of genetic resources with the Contracting Party providing such resources. Such sharing shall be upon mutually agreed terms."[286] The Southern delegates believed that they had adequately addressed American business apprehensions concerning patent rights for biotechnological innovations. Article 16.2 provides for

> [a]ccess to and transfer of technology . . . to developing countries . . . under fair and most favourable terms, including on concessional and preferential terms where mutually agreed. . . . In the case of technology subject to patents and other intellectual property rights, such access and transfer shall be provided on terms which recognize and are consistent with the adequate and effective protection of intellectual property rights.[287]

According to the *Economist*, the United States did not approve of the provisions on intellectual property rights: "The Administration argues . . . that the treaty is hostile to intellectual-property rights."[288] The Americans pointed with some degree of justification to the problems and conflicts which could arise with respect to Article 16.5 of the Convention:

> The Contracting Parties, recognizing that patents and other intellectual property rights may have an influence on the implementation of this Convention, shall cooperate in this regard subject to national legislation and international law in order to ensure that such rights are supportive of and do not run counter to its objectives.[289]

One of the unfortunate aspects of this North-South confrontation is that what may appear as the ultimate in reasonableness to one side can be perceived as totally outrageous by the other. The financial provisions of the Convention on Biological Diversity provide a case in point. Having very little leverage in the rich nation–poor nation debate which underlies all these developmental/environmental negotiations, the South has no option but to focus continually on the few advantages that it does possess. Because most of the biodiversity exists in the South and because its utilization and exploitation will be predominantly carried out by the North, the South sees this as an opportunity to ensure that the North is made to pay as much as possible for use of the "genetic treasure trove."[290] The urgency of the situation has been highlighted by environmentalists around the world. It has been suggested that approximately one hundred species of life are disappearing every day,[291] largely as a result of human activity.

Having for once a coherent and unified agenda to protect its own rights, the South worked hard to ensure that the Convention would not be so weak that future use of the biodiversity would continue as at present with virtually no payment to or even credit given to the country of origin of the species. In a sense, this was an attempt to establish that biological species have a national affiliation which has to be recognized in international law. This was the crucial concept which the Americans were unwilling to recognize, and certainly unwilling to acknowledge with payment to the originating country. The Southern point of view with respect to intellectual property rights was ably expressed by the Centre for Science and Environment in India:

> We have seen that once property rights have been created in favour of companies, the governments of the North plead that there is no way that they can interfere with private sector interest when it comes to issues like a call for technology transfer. Yet there are no such qualms when it comes to demanding free access to the property of the South's farmers and tribals. In the negotiations for a bio-diversity convention, these double standards are writ large and the high sounding plea of the common heritage of humankind is a rhetorical device to disguise the continued exploitation of the poorer countries and their farmers.[292]

Regarding the financial provisions of the Convention, the *Times* (London) was scathing in its condemnation: "A sensible principle that poor countries should be rewarded for protecting species has been turned into a binding obligation on the West to provide a grandiose, multi-course free lunch."[293] Article 20.2 of the Convention states that "[t]he developed country Parties shall provide new and additional financial resources to enable developing country Parties to meet the agreed full incremental costs to them of implementing measures which fulfil the obligations of this Convention."[294] The Treaty strongly favors the position of the South in specifying as well that

> [t]he extent to which developing country Parties will effectively implement their commitments under this Convention will depend on the effective implementation by developed country Parties of their commitments under this Convention related to financial resources and transfer of technology and will take fully into account the fact that economic and social development and eradication of poverty are the first and overriding priorities of the developing country Parties.[295]

Developed nations like Britain and Japan were also concerned about the financial provisions of this Convention.[296] Despite serious misgivings, many nations signed the Biodiversity Treaty because they preferred to participate and work for change within the structure rather than remain isolated like the United States of America and be unable to affect future decisions. The financial measures which were woven into both Conventions signed at Rio reflect an attempt by the South to bring about some form of economic restructuring and greater fairness in the distribution of economic benefits in the world.

The South argues that if the North wants environmental protection, then it must be prepared to pay for it because it is the greatest polluter in the world and because its present messianic fervor for the environment must not be exercised at the cost of Southern development. The South believes that it paid, and paid dearly, for the North's industrialization. It can no longer afford to keep subsidizing the North and refuses now to do so. Hence, the financial measures in the Biological Diversity Convention, measures which were so upsetting to developed nations, appear to the South as nothing more than a reasonable and very moderate beginning in the process of creating a new world economic order in which mutuality of benefit rather than exploitation prevails. Such provisions also reflect the reality of increasing poverty in the South and an unwillingness to continue the downward spiral of life there. Canada's Environment Minister, Jean Charest, commented on the relation between poverty and environmental degradation: "In the past 30 years," he said, "income disparities between the North and South have grown from 20 times to 60 times. This trend is simply not sustainable."[297] Prime Minister Brian Mulroney undertook to ensure that Canada would ratify both the Climate Change Convention and the Biological Diversity Convention.[298] What was being derided by some as a "squiffy little treaty"[299] is obviously considered by others as a significant step forward in a more equitable sharing of the Earth's wealth. Dr. Mostafa K. Tolba, Executive Director of the United Nations Environment Programme, commented that "[t]here will be those who will say the convention is too weak, barely addressing the magnitude of the threat to the global web of life." But, he continued, "the process of international environmental law requires us, for better or for worse, to walk before we can run and to crawl before we walk."[300] One can only agree with the assessment of the *Times* (London) that, "imperfect as the two treaties are, it is important that they are signed. They are markers on the way to more careful, and more equally shared, custodianship of the planet. Even weakened conventions can lead to stronger ones. North and South have here the basis of real bargains."[301]

The visionary who guided the concept of the Earth Summit and who worked strenuously to bring about an international conference to confront and tackle the twin issues of development and environment is Canadian businessman and environmentalist Maurice Strong. The son of a railway worker from Manitoba, Strong made his fortune in oil and has devoted several years to national and international public service, first as chief of Canada's foreign aid program. He was then appointed Secretary-General of the first United Nations environmental conference at Stockholm (1972).[302] Although the Stockholm Conference brought world attention to environmental matters, that meeting "fell short in practice."[303] A "self-made humanist or a practical humanist,"[304] Strong has focused his international efforts on bridging the gap between North and South because "he has seen the environmental crisis in a larger context of rich and poor or developed and developing countries."[305] Strong is firmly committed to the need for all nations to convert their economies to environmentally friendly paths,[306] and his aspirations for Rio were to create the political formulations and the financial mechanisms to encourage and enable both developed and developing nations to achieve that goal. Strong insists that "[w]e have to rise above our . . . differences and forge a new global partnership that will ensure the future of our planet as a secure and hospitable home for all of us, rich and poor."[307] He believes that the Stockholm Conference faltered "because it did not answer the question of how the world's poorest countries would pay for reform."[308] This question was answered, at least to the partial satisfaction of the South, in the main Rio Treaties. Certainly, Maurice Strong's vision "of building a world consensus on making economic development environmentally sustainable"[309] was fulfilled somewhat in the statement of principles known as the Rio Declaration.

In an effort to "set benchmarks for environmental rights and responsibilities,"[310] Canada and other developed nations attempted to create an Earth Charter which would emphasize environmental concerns. The Charter was opposed by many developing countries,[311] which favored a statement emphasizing developmental priorities. The negotiations were intense, even though everyone knew the resulting compromise would not be legally binding. As one Canadian advisor commented, "the visionaries came up against the lawyers and bureaucrats and the lawyers and bureaucrats won."[312] What has emerged is a rather vague series of twenty-seven principles which fulfils to a considerable extent the Southern agenda but which fails to chart with sufficient firmness any innovative directions for the entire planet.

The Rio Declaration[313] affirms the sovereign right of States "to exploit their own resources pursuant to their own environmental and developmental policies" (Principle 2); insists that "[t]he right to development must be fulfilled so as to equitably meet developmental and environmental needs of present and future generations" (Principle 3); acknowledges the significance of environmental protection in the achievement of sustainable development (Principle 4); urges "[a]ll States and all people [to] cooperate in the essential task of eradicating poverty as an indispensable requirement for sustainable development, in order to decrease the disparities in standards of living" (Principle 5); and gives priority to the special needs of developing countries (Principle 6). The Declaration calls on States to "reduce and eliminate unsustainable patterns of production and consumption" (Principle 8). The South also won its point in the acknowledgment by the developed countries of "the responsibility that they bear in the international pursuit of sustainable development in view of the pressures their societies place on the global environment and of the technologies and financial resources they command" (Principle 7). Various activities are recommended, including "exchanges of scientific and technological knowledge" (Principle 9), encouragement of public awareness (Principle 10), passage of environmental legislation at the national level (Principle 11), payment of compensation for victims of environmental damage (Principle 13), and prevention of cross-border movement of harmful substances (Principle 14). With respect to the establishment of environmental standards, the South scored again with the inclusion of this wording: "Standards applied by some countries may be inappropriate and of unwarranted economic and social cost to other countries, in particular developing countries"(Principle 11). Governments in the developing world are also concerned that the North's emphasis on environmentalism may result in restrictions being placed on the importation of Southern products which are environmentally not acceptable. Accordingly, this wording appears in Principle 12: "Trade policy measures for environmental purposes should not constitute a means of arbitrary or unjustifiable discrimination or a disguised restriction on international trade." Clearly, the Declaration caters largely to the requirements and agenda of those Southern governments which came to the Rio process determined to stress the primacy of development over the necessity for environmental protection. To round out the body of principle, there were rather vague concessions made to the significance of women (Principle 20), youth (Principle 21), and indigenous people (Principle 22). As a formulation of proposals for future action, the Rio Declaration fails to meet the test of balance between developmental and environmental concerns. Although it was universally accepted on 11th June 1992 at the Summit, this easy passage at the end was largely because the document is not legally binding.[314]

The massive document Agenda 21, also not legally binding, "became the main forum for North-South wrangling on every topic imaginable, including the spread of deserts, disposal of toxic wastes and protection of women's rights."[315] This "blueprint for global environmental protection"[316] has been accused of being yet one more example of "unsurpassed UN verbosity,"[317] with its inevitable demands for more funding (approximately $125 billion a year from developed countries for implementation).[318] According to Maurice Strong, "[f]inance has the capacity to make or break" the ambitious goals of

Agenda 21.[319] The total annual cost for North and South would be a staggering $625 billion.[320] The problem is that although the document deals with a
number of pertinent and pressing concerns, it does not do so in any dramatic
new context in terms of economic distribution between rich and poor nations.[321] Possibly, that was too daunting a task even to contemplate, but without some restructuring of how wealth moves between nations, there will
probably be no implementation of Agenda 21's grand schemes, no matter
how laudable the ideas may be; and the world will continue its present course
down the path of environmental destruction. During the Rio process, the
developing nations were firm in their insistence that the North must assist
the South in its efforts to implement Agenda 21. As Jamsheed A. Marker,
Pakistan's Representative to the United Nations, commented: "We are saying
we cannot generate the new resources needed to start up Agenda 21 without
new help, so if the North wants us to meet those goals they must treat us
more generously."[322]

In stressing the primacy of Southern self-interest, the developing nations
also combined to scuttle plans for a treaty to protect the world's forests.
Perceiving this as a direct assault on their political sovereignty, some nations
of the South resisted any notion of a Forestry Convention, because, as they
argued, it could be one method of converting their forests into a global, as
against a national, resource. The "vehement opposition of developing countries" arose because, for them, "forests are a major economic asset. These
countries assert that while they believe in protecting forests, they are theirs
to do with as they please . . . they bridle at what they see as an attempt to
abridge their sovereignty by countries that long ago cut down their own trees
for profit but now want to place the main burden of global forest conservation
on countries struggling for economic survival."[323] India's Environment Minister, Kamal Nath, made it clear that his nation would oppose a forest treaty
because forests are no more a global issue than is another vital resource,
oil.[324] India was accused of being "the most intransigent of the developing
countries."[325] The Malaysian Prime Minister, Dr. Mahathir bin Mohamad,
dismissed the idea of a forestry convention: "We know that a forestry convention is about forests which are there. Since most of the countries who are
going to participate do not have any forests, they are going to talk about OUR
forests, not THEIR forests. It's a waste of time for us."[326] The extent of their
resistance may also have been dictated by the fact that the Treaty on Forests
was actively promoted by the Americans, the same Americans who had, according to the South, weakened the Climate Change Treaty, rejected the Biodiversity Convention, and generally done their utmost to stress the primacy

of their country's self-interest during the Rio process. The extent of Southern suspicion can be gauged by the remarks of the Malaysian Ambassador Ting Wen-Lian, who considered "the almost obsessional anxiety to have a forest convention" a reflection of the attempt by developed countries "to appease their public opinion and thus get electoral mileage out of forests."[327] The initial proposal by the United States was for a ban on logging in tropical forests.[328] When developing nations demanded that Northern forests be included, it was clear that the prospects for success were doomed. The United States of America "has strenuously resisted any scrutiny of the logging practices in publicly owned ancient forests in the Pacific Northwest."[329] The North-South rift rendered a binding Treaty on Forests out of the question. In their intense concentration on development, Southern nations are at times inclined to overlook the fact that forest protection is in their own ultimate self-interest as well as that of the North. Once the forests are destroyed the consequences for Southern economies will be very serious indeed. One has only to look at the devastation that forest destruction has brought to the Philippines to get an idea of the likely result which could face other nations in coming decades.

Having failed to gain a legally binding Treaty, the United States and its supporters pushed for the adoption of a nonbinding agreement to be followed by a Treaty at some future date.[330] There was so much hostility to the United States Government at Rio that even its good intentions came under suspicion from the South. Diplomats from developing nations were convinced that the American Government's eagerness to conserve the rain forests sprang from a desire to utilize the resource as a carbon dioxide sink so that "the US will not have to join other developed countries in setting timetables to reduce its own emissions of greenhouse gases."[331]

The South's opinion was explained by the environmentalist Anthony Juniper: "[t]he Americans are keen on a forest convention because they would not have to do much, whereas with strong climate change and biodiversity treaties their commitments would have been considerable."[332] The American President offered $150 million a year for forest conservation.[333] Although some developing countries expressed an interest in receiving the funds,[334] this was not sufficient inducement to convince them to support a Treaty which would be legally binding. According to the *New York Times*, "[A]n American official said that those countries that supported the United States in pushing for a full declaration of principles on forests at the Rio conference would receive special consideration in the distribution of the American aid."[335] But the carrot of funding only resulted in what was titled a *Non–Legally Binding Authoritative Statement of Principles for a Global Consensus on the Management, Conservation and Sustainable Development of All Types of Forests.*[336]

In its preamble, this Forestry Statement acknowledged that "[t]he subject of forests is related to the entire range of environmental and development issues and opportunities, including the right to socio-economic development on a sustainable basis,"[337] a clear concession to the Southern position. The Agreement covers forests in all geographic regions,[338] thereby overcoming Southern objections to any emphasis on tropical forests. The signatories acknowledge that "[f]orests are essential to economic development and the maintenance of all forms of life."[339] The Statement calls for sustainable management of forest resources "to meet the social, economic, ecological, cultural and spiritual human needs of present and future generations. These needs are for forest products and services, such as wood and wood products, water, food, fodder, medicine, fuel, shelter, employment, recreation, habitats for wildlife, landscape diversity, carbon sinks and reservoirs, and for other forest products."[340] It calls for support of the rights of indigenous people and promotion of the participation of women in forest management.[341]

Although the United States balked at the threat it perceived to the interest of its important biotechnology industry in the legally binding Biodiversity Convention, Southern interests were stated very clearly in the Forestry Agreement: "Access to biological resources, including genetic material, shall be with due regard to the sovereign rights of the countries where the forests are located and to the sharing on mutually agreed terms of technology and profits from biotechnology products that are derived from these resources."[342] It could be argued that the Forestry Agreement was probably the most progressive of the Rio documents because of its recognition of the dire necessity to create a new world economic order. The Agreement highlights the importance of "the eradication of poverty and the promotion of food security";[343] the stimulation of "economic and social substitution activities," which would inevitably conserve forests;[344] the provision of "international financial and technical cooperation, including through the private sector, where appropriate";[345] the contribution of financial assistance for "afforestation, reforestation and combating deforestation and forest and land degradation";[346] the transfer of environmental technology "including on concessional and preferential terms";[347] and the removal of tariff barriers and the encouragement of the sale of forest products processed in developing countries.[348] This Agreement even stresses the "importance of redressing external indebtedness, particularly where aggravated by the net transfer of resources to developed countries,"[349] and maintains that "[f]orest conservation and sustainable development policies should be integrated with economic, trade and other relevant policies."[350]

The provisions of the Forestry Agreement presage a new approach in the search for greater economic equity and stability for the developing world. It is unfortunate that the South's emphasis on the sovereignty issue blinded the developing world to the very important concessions that the North was prepared to make in return for forest conservation by the South. In scuttling the Forestry Convention because it was perceived as an attack on national ownership rights, the South may have sacrificed an important tool which could have ensured the survival of the forest for decades to come and the continuation of its benefits for Southern and Northern societies alike. It was obvious as early as April 1992 that the developing world would dismiss the idea of a binding Forestry Convention. At the second Ministerial Conference of Developing Countries on Environment and Development, held between 26th and 29th April 1992 in Kuala Lumpur, the final agreed statement of the participants stressed the national nature of the forest resource and specified that in view of the formulation of the nonbinding Forestry Statement "the negotiation of a legally binding instrument on forests would not be required."[351]

The basic problem facing all nations which participated in the Earth Summit was the question of funding. The North, suffering one of the worst recessions in decades, found itself virtually under siege by the South, which was demanding money in return for exercising environmentally safe development. As one commentator stated, "[O]ne of the real issues facing UNCED is whether money can purchase the Earth."[352] The North also faces additional pressure from the newly democratic states of eastern Europe, which are as demanding of funding as is the South. It is likely to be impossible for the North to provide the poorer nations with the additional $125 billion the latter will need to implement the visionary goals of Agenda 21.[353] If drastic steps are not soon taken to rectify and possibly reverse the huge annual outflow of resources from South to North, the poverty, desperation, and environmental destruction in the South will doubtless become even more serious. Unfortunately, Southern governments refuse to appreciate the fact that tax-paying citizens of Northern nations do not represent a bottomless pool of wealth to be tapped by their governments at will. Most developed nations face staggering deficits of their own and experience constant cutbacks in vital education, health, and welfare programs which form the safety net for the most vulnerable of their citizens.

Although no one denies that people in the North are generally much better off than those in the South, part of the North-South misunderstanding arises because of the inability of the Southern governments to appreciate the dilemma facing Northern governments, which find national and international demands for funding increasing at the very moment when their monetary resources are decreasing. It was a regrettable error by the North not to explain itself more fully during the Rio process; it was equally regrettable that the South was apparently not really willing to adjust to the new reality of a recessionary North which faces serious economic problems of its own.

What the South needed from the Rio process was a new bargain, a new economic restructuring of the world's wealth so that all could benefit. What it got instead was another round of handouts. The result is that the same system which has driven the major part of the world to crushing debt and near bankruptcy will continue into a future which appears rather bleak. Although the Treaties and Agreements formulated during the Rio negotiations were watered down because of the North-South divide, an optimist would argue that we now have a legally binding Treaty on Climate Change and another legally binding Treaty on Biodiversity.

It is still too early to come to any decisive conclusions about the Earth Summit or to determine whether it was a success or a failure. If one perceives UNCED as a vast, global gathering to raise consciousness about environmental and developmental concerns, then the Earth Summit may be deemed a resounding success. If, however, it is studied as we have done, within the context of the North-South confrontation, the reviews can only be tentative and mixed. Although some steps have been taken to extend environmental law in new directions, the formulations are very vague and rather weak. Some participating nations were determined to maintain the Rio momentum, and work to build stronger Conventions in the future and produce treaties which will ensure firm timetables and even firmer deadlines for implementation. UNCED will probably be regarded by historians as a significant step in a process whereby North and South came to realize that the planet is a shared international inheritance, even though it is carved up into nationally controlled spheres. The Earth Summit may also be regarded in future as an important landmark in reinforcing what all of us already know but our governments are slow to acknowledge—that the world is truly interdependent and interrelated, that no one portion can be allowed to deteriorate, because the consequences will ultimately affect us all. To care about the poor, the undernourished, the victims of environmental degradation is no longer just humanitarian concern for others. It is the ultimate form of self-preservation.

The ultimate message of the linkage being perceived globally between the right to development and the right to a healthy environment is very simple and very clear. Either those of us who are fortunate enough to enjoy a decent standard of life act very soon to assure better living conditions for those not as blessed as we are, or we all collapse together, perhaps not immediately but certainly in the very near future. The basic lesson to be learned from Rio may be that soon the rights to development and to a healthy environment will be enjoyed equitably by all of us on the planet or by none of us at all.

Notes

1. For text of these documents see *Everyone's United Nations* (New York: United Nations, 1979), 412–38.

2. David Ott, *Public International Law in the Modern World* (London: Pitman, 1987), 238.

3. *Ibid.*, 239.

4. *Ibid.*, 243.

5. Ranee K. L. Panjabi, Review of James Crawford, ed., *The Rights of Peoples* (Oxford: Oxford University Press, 1988), in *Virginia Journal of International Law*, vol. 30, no. 1 (Fall 1989): 319.

6. James Crawford, ed., *The Rights of Peoples* (Oxford: Oxford University Press, 1988), 41.

7. Panjabi, Review of *The Rights of Peoples*, 320.

8. Ott, *Public International Law*, 244.

9. Crawford, *The Rights of Peoples*, 141–42.

10. Panjabi, Review of *The Rights of Peoples*, 320.

11. Ott, *Public International Law*, 244.

12. Statement by P. V. Narasimha Rao, Prime Minister, India, UNCED, Rio, 12th June 1992.

13. George J. Mitchell, *World on Fire* (New York: Scribner's, 1991), 46.

14. Stephen Bradshaw, BBC News and Current Affairs, report telecast on the Canadian Broadcasting Corporation's News Network, June 1992.

15. Roland Rich, in Crawford, *The Rights of Peoples*, 39.

16. G.A. Res. 41/128, 41 U.N. GAOR Supp. (No. 53) at 186, U.N. Doc. A/41/925 (1986). The vote was 146 in favor, 1 against (United States), 8 abstentions. See Crawford, *The Rights of Peoples*, note 6, 51–52.

17. Ranee K. L. Panjabi, "Describing and Implementing Universal Human Rights," *Texas International Law Journal*, vol. 26, no. 1 (Winter 1991): 197.

18. U.N. Development Programme, *Human Development Report, 1991* (New York: Oxford University Press, 1991), cited in Sandra Postel, "Denial in the Decisive Decade," in *State of the World, 1992* (New York: W. W. Norton, 1992), 4.

19. *Globe and Mail* (Toronto), 29th April 1992, A1:6.

20. World Bank, *World Development Report, 1991* (New York: Oxford University Press, 1991), cited in Postel, "Denial in the Decisive Decade," 4.

21. Dr. Mahathir bin Mohamad, Prime Minister, Malaysia, interview by Stephen Bradshaw, BBC News and Current Affairs, telecast on the Canadian Broadcasting Corporation's News Network, June 1992.

22. Statement by P. V. Narasimha Rao, Prime Minister, India, UNCED, Rio, 12th June 1992.

23. Michael Howard, M.P., interview by Stephen Bradshaw, BBC News and Current Affairs, telecast on the Canadian Broadcasting Corporation's News Network, June 1992.

24. Walter H. Corson, ed., *The Global Ecology Handbook: What You Can Do about the Environmental Crisis* (Boston: Beacon Press, 1990), 44, cited in Helen Caldicott, *If You Love This Planet* (New York: W. W. Norton, 1992), 123.

25. South Centre, "Environment and Development: Towards a Common Strategy for the South in the UNCED Negotiations and Beyond," November 1991.

26. Centre for Science and Environment, "The CSE Statement on Global Environmental Democracy to Be Submitted to the Forthcoming UN Conference on Environment and Development," New Delhi, India.

27. South Centre, "Environment and Development."

28. Centre for Science and Environment, "Statement on Global Environmental Democracy."

29. Bruce Babbitt, *World Monitor*, 30th June 1992, 31:2.

30. Paul Harrison, *The Third Revolution* (London: I. B. Tauris, 1992), 90.

31. *Christian Science Monitor*, 2nd June 1992, 10:2.

32. *Times* (London), 10th June 1992, 12:8.

33. Statement by Dr. Mahathir bin Mohamad, Prime Minister, Malaysia, UNCED, Rio, 13th June 1992.

34. *Times* (London), 2nd June 1992, 10:4.

35. Harrison, *The Third Revolution*, 91.

36. *Ibid.*

37. Mitchell, *World on Fire*, 38.

38. Al Gore, *Earth in the Balance* (Boston: Houghton Mifflin, 1992), 117.

39. Mitchell, *World on Fire*, 37.

40. Gore, *Earth in the Balance*, 115–16.

41. Ben Jackson, *Poverty and the Planet* (London: Penguin, 1990), 29.

42. *New York Times*, 2nd June 1992, A10:6.

43. Gore, *Earth in the Balance*, 106.

44. *Ibid.*

45. *Ibid.*

46. Michael Renner, "Creating Sustainable Jobs in Industrial Countries," *State of the World, 1992* (New York: W. W. Norton, 1992), 150.

47. Gore, *Earth in the Balance*, 116.

48. John C. Ryan, "Conserving Biological Diversity," in *State of the World, 1992* (New York: W. W. Norton, 1992), 10.

49. *Ibid.*, 10–11.

50. Jackson, *Poverty and the Planet*, 26.

51. *Globe and Mail*, 2nd June 1992, B26:4.

52. Augusta Dwyer, *Into the Amazon: The Struggle for the Rain Forest* (Toronto: McClelland-Bantam, 1990), xi.

53. *Time*, 1st June 1992, 22.

54. *Global Outlook 2000* (New York: United Nations, 1990), 85.

55. *Ibid.*

56. FAO Forest Resources Assessment Project, *Second Interim Report on the State of Tropical Forests*, 10th World Forestry Congress, Paris, September 1991, cited in Harrison, *The Third Revolution*, 92.

57. Harrison, *The Third Revolution*, 92.

58. Gore, *Earth in the Balance*, 117.

59. *Ibid.*, 117–18.

60. Alan Durning and Holly Brough, "Reforming the Livestock Economy," in *State of the World, 1992* (New York: W. W. Norton, 1992), 73.

61. *Time*, 20th April 1992, 58:1.

62. *Globe and Mail*, 4th June 1992, A19:2.

63. *Ibid.*, A19:3.

64. *Time*, 20th April 1992, 58:3.

65. Ivan L. Head, *On a Hinge of History* (Toronto: University of Toronto Press, 1991), 102.

66. *Ibid.*

67. *Ibid.*, 125.

68. H. J. Leonard, "Managing Central America's Renewable Resources," *International Environmental Affairs*, 1 (1990): 38–56, cited in Head, *On a Hinge of History*, 125.

69. Head, *On a Hinge of History*, 126.

70. *Newsweek*, 1st June 1992, 23.

71. *Ibid.*

72. *Ibid.*

73. *Ibid.*

74. *New York Times*, 2nd June 1992, A10:6.

75. Head, *On a Hinge of History*, 101.

76. *Globe and Mail*, 6th June 1992, A4:6.

77. *Christian Science Monitor*, 2nd June 1992, 10:1.

78. *Ibid.*

79. *Times* (London), 2nd June 1992, 1:6.

80. *New York Times*, 11th June 1992, A13:1.

81. *Ibid.*, 9th June 1992, A8:3–6.

82. *Ibid.*, 11th June 1992, A13:1.

83. *Globe and Mail*, 6th June 1992, A4:6.

84. *Time*, 1st June 1992.

85. South Centre, "Environment and Development."

86. *Globe and Mail*, 30th May 1992, D8:1.

87. Ibid., 20th May 1992, A2:3.

88. Head, *On a Hinge of History*, 104.

89. *Ibid.*

90. *Ibid.*

91. *Ibid.*

92. *Ibid.*

93. Gore, *Earth in the Balance*, 185.

94. *Globe and Mail*, 2nd June 1992, B26:1.

95. *Global Outlook 2000*, 87.

96. Fulgenchio Factoran, Environment Secretary, Philippines, interview by Stephen Bradshaw, BBC News and Current Affairs, telecast on the Canadian Broadcasting Corporation's News Network, June 1992.

97. *Globe and Mail*, 9th June 1992, A1:2.

98. *Ibid.*, A8:3.

99. Centre for Science and Environment, "Statement on Global Environmental Democracy."

100. *Ibid.*

101. Statement by Dr. Mahathir bin Mohamad, Prime Minister, Malaysia, UNCED, Rio, 13th June 1992.

102. *Globe and Mail*, 18th June 1992, A1:5.

103. *Ibid.*, A11:4–5.

104. Head, *On a Hinge of History*, 105.

105. *Ibid.*

106. *Ibid.*, 106.

107. Harrison, *The Third Revolution*, 108.

108. *New York Times*, 13th June 1992, 4:1.

109. Dr. Mahathir bin Mohamad, Prime Minister, Malaysia, interview by Stephen Bradshaw, BBC News and Current Affairs, telecast on the Canadian Broadcasting Corporation's News Network, June 1992.

110. *New York Times*, 7th June 1992, 20:2.

111. *Ibid.*

112. Statement by Dr. Mahathir bin Mohamad, Prime Minister, Malaysia, UNCED, Rio, 13th June 1992.

113. *Globe and Mail*, 4th June 1992, A19:3.

114. Sir Crispin Tickell, interview by Stephen Bradshaw, BBC News and Current Affairs, telecast on the Canadian Broadcasting Corporation's News Network, June 1992.

115. Gore, *Earth in the Balance*, 312.

116. Harrison, *The Third Revolution*, 98–99.

117. Dr. Lim Keng Yaik, Malaysian Minister of Primary Industries, taped speech included in Stephen Bradshaw's television report on the Earth Summit, BBC News and Current Affairs, telecast on the Canadian Broadcasting Corporation's News Network, June 1992.

118. *Globe and Mail*, 3rd June 1992, B10:4.

119. *Ibid.*, 6th June 1992, A4:6.

120. Statement by Dr. Mahathir bin Mohamad, Prime Minister, Malaysia, UNCED, Rio, 13th June 1992.

121. Interview in Malaysia by Stephen Bradshaw, BBC News and Current Affairs, telecast on the Canadian Broadcasting Corporation's News Network, June 1992.

122. Interview in Sarawak, Malaysia, by Stephen Bradshaw, BBC News and Current Affairs, telecast on the Canadian Broadcasting Corporation's News Network, June 1992.

123. *Global Outlook 2000*, 86–87.

124. *Ibid.*, 87.

125. *Ibid.*

126. South Centre, "Environment and Development."

127. Head, *On a Hinge of History*, 73.

128. *Ibid.*, 59.

129. *Globe and Mail*, 11th June 1992, A17:2.

130. *Ibid.*

131. Centre for Science and Environment, "Statement on Global Environmental Democracy."

132. Gore, *Earth in the Balance*, 347.

133. *Globe and Mail*, 17th August 1992, B1:1–3.

134. *Ibid.*

135. Head, *On a Hinge of History*, 66.

136. *Globe and Mail*, 17th August 1992, B9:3.

137. Head, *On a Hinge of History*, 60.

138. Katarina Tomasevski, *Development Aid and Human Rights* (London: Pinter Publishers, 1989), 6.

139. United Nations General Assembly Resolution 1986/56, 22nd July 1986. See Tomasevski, *Development Aid and Human Rights*, 6.

140. Tariq Osman Hyder, "Joint Statement of the Group of 77 and China to the Fifth Intergovernmental Negotiating Committee for a Framework Convention on Climate Change," 24th February 1992, New York.

141. Caldicott, *If You Love This Planet*, 129.

142. Susan George, *A Fate Worse than Debt*, BBC Documentary, 1990, cited in Caldicott, *If You Love This Planet*, 130.

143. Caldicott, *If You Love This Planet*, 130.

144. *Ibid.*

145. Jackson, *Poverty and the Planet*, 89.

146. UNICEF, *The State of the World's Children*, 1989 and 1990 (Oxford: Oxford University Press, 1989 and 1990), cited in Jackson, *Poverty and the Planet*, 90.

147. Cited in Jackson, *Poverty and the Planet*, 88.

148. Harrison, *The Third Revolution*, 85.

149. *Ibid.*

150. Jackson, *Poverty and the Planet*, 91.

151. Caldicott, *If You Love This Planet*, 132–33.

152. Jackson, *Poverty and the Planet*, 91.

153. Harrison, *The Third Revolution*, 297.

154. Jackson, *Poverty and the Planet*, 93–94.

155. Caldicott, *If You Love This Planet*, 134.

156. Jackson, *Poverty and the Planet*, 94.

157. *Ibid.*

158. *Ibid.*, 97.

159. *Globe and Mail*, 2nd May 1992, A7:3.

160. See South Centre, "Environment and Development."

161. Caldicott, *If You Love This Planet*, 134–35.

162. Centre for Science and Environment, "Statement on Global Environmental Democracy."

163. *Globe and Mail*, 17th August 1992, B9:6.

164. *Ibid.*

165. Head, *On a Hinge of History*, 69.

166. Jackson, *Poverty and the Planet*, 123.

167. *Globe and Mail*, 17th August 1992, B1:2–3 and B9:6.

168. UNCTAD, *Trade and Development Report, 1988* (Geneva, 1988), cited in Jackson, *Poverty and the Planet*, 124.

169. *Globe and Mail*, 29th May 1992, A17:4.

170. Head, *On a Hinge of History*, 69.

171. *Ibid.*

172. *Globe and Mail*, 17th August 1992, B9:1–4.

173. *Ibid.*, 8th June 1992, A8:6.

174. *Ibid.*

175. Canadian International Development Agency, *Annual Report*, 1989–90, 46.

176. *Christian Science Monitor*, 3rd June 1992, 2:2.

177. *Globe and Mail*, 2nd June 1992, B24:1.

178. *Ibid.*, 11th May 1992, A8:1.

179. Jackson, *Poverty and the Planet*, 106.

180. Canadian International Development Agency, *Sharing Our Future* (Hull, Que., 1987), 74.

181. Caroline Thomas, *In Search of Security* (Hemel Hempstead, U.K.: Harvester Wheatsheaf, 1987), 39.

182. *Ibid.*, 40.

183. *Globe and Mail*, 17th August 1992, B9:5–6.

184. *Ibid.*, 29th May 1992, A17:2.

185. *Ibid.*, 17th August 1992, B9:5–6.

186. Jackson, *Poverty and the Planet*, 67.

187. *Globe and Mail*, 29th May 1992, A17:3.

188. *Ibid.*, 11th May 1992, A8:1.

189. *Ibid.*, 19th June 1992, A8:1–2, and 15th June 1992, A0.2.

190. *Ibid.*, 25th April 1992, A7:4.

191. *Ibid.*, 29th May 1992, A17:3.

192. *Ibid.*, 15th June 1992, A8:1.

193. *Ibid.*, 25th April 1992, A7:4.

194. Gore, *Earth in the Balance*, 184–85.

195. Thomas, *In Search of Security*, 54.

196. Caldicott, *If You Love This Planet*, 133.

197. Jackson, *Poverty and the Planet*, 110.

198. Gerald K. Helleiner, "Stabilization, Adjustment and the Poor," in R. M. Bird and S. Horton, eds., *Government Policy and the Poor in Developing Countries* (Toronto: University of Toronto Press, 1989), 35.

199. *Globe and Mail*, 18th May 1992, A7:3.

200. *Ibid.*, 2nd June 1992, B26:1.

201. *Ibid.*, 18th May 1992, A7:3.

202. *Ibid.*, 2nd June 1992, B26:1.

203. *Ibid.*, 18th May 1992, A7:3–6.

204. *Ibid.*, 2nd June 1992, B26:1.

205. Stephen Bradshaw, BBC News and Current Affairs, report telecast on the Canadian Broadcasting Corporation's News Network, June 1992.

206. South Centre, "Environment and Development."

207. Thomas, *In Search of Security*, 55.

208. *Times* (London), 3rd June 1992, 12:2.

209. *Globe and Mail*, 18th June 1992, A1:5.

210. *Ibid.*, 25th April 1992, A7:3.

211. *Ibid.*

212. *Time*, 1st June 1992, 23.

213. *Ibid.*, 22.

214. *Globe and Mail*, 25th April 1992, A7:3.

215. Lester R. Brown, "Launching the Environmental Revolution," in *State of the World, 1992* (New York: W. W. Norton, 1992), 180.

216. *Globe and Mail*, 2nd June 1992, A3:1.

217. *Ibid.*, A3:2; also *Times* (London), 3rd June 1992, 12:8.

218. *Globe and Mail*, 2nd June 1992, A3:2.

219. *Times* (London), 3rd June 1992, 12:8.

220. Harrison, *The Third Revolution*, 197.

221. *Globe and Mail*, 2nd May 1992, A7:1.

222. *Ibid.*

223. *Ibid.*

224. *Newsweek*, 1st June 1992, 31.

225. Caldicott, *If You Love This Planet*, 30.

226. Harrison, *The Third Revolution*, 273.

227. Caldicott, *If You Love This Planet*, 30.

228. *Globe and Mail*, 2nd June 1992, A3:3.

229. *Newsweek*, 1st June 1992, 42.

230. *Ibid.*

231. *Ibid.*

232. *Globe and Mail*, 2nd June 1992, A3:4.

233. *Ibid.*, A3:3.

234. Gore, *Earth in the Balance*, 155.

235. *Globe and Mail*, 2nd June 1992, A3:1, and *Times* (London), 3rd June 1992, 12:7.

236. *Times* (London), 3rd June 1992, 12:8, and *Globe and Mail*, 2nd June 1992, A3:3–4.

237. *Times* (London), 3rd June 1992, 12:8.

238. Stephen Bradshaw, BBC News and Current Affairs, report telecast on the Canadian Broadcasting Corporation's News Network, June 1992.

239. Gore, *Earth in the Balance*, 279.

240. *Times* (London), 6th June 1992, 10:3.

241. *Time*, 1st June 1992, 32:1.

242. *Ibid.*, 22:1, and *Times* (London), 1st June 1992, 12:3.

243. Statement by Dr. Mahathir bin Mohamad, Prime Minister, Malaysia, UNCED, 13th June 1992.

244. Centre for Science and Environment, "Statement on Global Environmental Democracy."

245. *Time*, 1st June 1992, 32:2.

246. Jackson, *Poverty and the Planet*, 30.

247. Centre for Science and Environment, "Statement on Global Environmental Democracy."

248. *Christian Science Monitor*, 2nd June 1992, 10:5.

249. *Times* (London), 1st June 1992, 12:3.

250. United Nations Document, A/AC.237/18 (Part II)/Add.1, 15th May 1992. United Nations Framework Convention on Climate Change. (The Treaty is reprinted as Appendix 4 to this book.)

251. *Ibid.*, Add.1, 2.

252. *Ibid.*

253. *Ibid.*, Add.1, 3.

254. *Ibid.*, Add.1, 5.

255. *Ibid.*, Add.1, 5, Art. 3.5.

256. Dr. Lim Keng Yaik, Malaysian Minister of Primary Industries, interview by Stephen Bradshaw, BBC News and Current Affairs, telecast on the Canadian Broadcasting Corporation's News Network, June 1992.

257. *New York Times*, 13th June 1992, 4:1–2.

258. United Nations Document, A/AC.237/18 (Part II)/Add.1, 7, Art. 4.2 (b).

259. *Globe and Mail*, 3rd June 1992, A6:3.

260. *Ibid.*

261. *Times* (London), 1st June 1992, 12:3, and *Time*, 1st June 1992, 22:1.

262. *Times* (London), 1st June 1992, 12:3.

263. *Ibid.*, 2nd June 1992, 16:6.

264. *Ibid.*

265. *Chronicle-Herald* (Halifax, N.S.), 1st June 1992, A2:3.

266. *New York Times*, 13th June 1992, 4:1–2.

267. United Nations Document, A/AC.237/18 (Part II)/Add.1, 9, Art. 4.5.

268. Centre for Science and Environment, "Statement on Global Environmental Democracy."

269. *Ibid.*

270. *Ibid.*

271. *Globe and Mail*, 8th June 1992, A8:6.

272. *Ibid.*

273. South Centre, "Environment and Development."

274. *Ibid.*

275. United Nations Document, A/AC.237/18 (Part II)/Add.1, 21, Art. 21.3.

276. *Times* (London), 3rd June 1992, 15:2.

277. Vandana Shiva, interview by Stephen Bradshaw, BBC News and Current Affairs, telecast on the Canadian Broadcasting Corporation's News Network, June 1992.

278. *Christian Science Monitor*, 26th May 1992, 3:2.

279. *New York Times*, 6th June 1992, 6:3.

280. *Christian Science Monitor*, 26th May 1992, 3:2.

281. United Nations Document, *Convention on Biological Diversity*, 5th June 1992, Preamble. (The Treaty is reprinted as Appendix 5 to this book.)

282. *Ibid.*, Preamble, 3.

283. *Ibid.*
284. *Ibid.*, Art. 1.
285. *Ibid.*, Art. 15.1.
286. *Ibid.*, Art. 15.7.
287. *Ibid.*, Art. 16.2.
288. *Economist*, 13th June 1992, 94.
289. United Nations Document, *Convention on Biological Diversity*, 10, Art. 16.5.
290. Stephen Bradshaw, BBC News and Current Affairs, report telecast on the Canadian Broadcasting Corporation's News Network, June 1992.
291. *Christian Science Monitor*, 26th May 1992, 3:1.
292. Centre for Science and Environment, "Statement on Global Environmental Democracy."
293. *Times* (London), 3rd June 1992, 15:1.
294. United Nations Document, *Convention on Biological Diversity*, 11, Art. 20.2.
295. *Ibid.*, Art. 20.4.
296. *Economist*, 13th June 1992, 94.
297. *Globe and Mail*, 12th June 1992, A4:6.
298. Statement by Brian Mulroney, Prime Minister, Canada, UNCED, Rio, 12th June 1992.
299. *Economist*, 13th June 1992, 93.
300. *New York Times*, 24th May 1992.
301. *Times* (London), 1st June 1992, 15:2.
302. Ibid., 3rd June 1992, 12:6.
303. *Globe and Mail*, 2nd May 1992, A7:3.
304. *World Monitor*, June 1992, 32.
305. *Ibid.*
306. *Times* (London), 3rd June 1992, 12:7.
307. Maurice Strong, interview by Stephen Bradshaw, BBC News and Current Affairs, telecast on the Canadian Broadcasting Corporation's News Network, June 1992.
308. *Globe and Mail*, 2nd May 1992, A7:3.
309. *New York Times*, 4th June 1992, A10:2.
310. *Globe and Mail*, 2nd June 1992, 1:4.
311. *Ibid.*
312. *Ibid.*, 2nd June 1992, A3:4.
313. United Nations Document, A/CONF.151/5/Rev.1, *The Rio Declaration on Environment and Development*, 13th June 1992 (reprinted as Appendix 2 to this book).
314. *Times* (London), 12th June 1992, 1:7.
315. *Time*, 1st June 1992, 29:1.
316. *New York Times*, 14th June 1992, 6:6.
317. *Times* (London), 1st June 1992, 15:1.
318. *Ibid.*, 9th June 1992, 27:1.
319. *Globe and Mail*, 2nd May 1992, A7:1.
320. *Ibid.*
321. *Ibid.*, 15th June 1992, A8:5.
322. *New York Times*, 14th June 1992, 10:1.
323. *Ibid.*, 7th June 1992, L20:1–2.

324. *Ibid.*, 9th June 1992, A8:3.

325. *Globe and Mail*, 10th June 1992, A6:4.

326. Dr. Mahathir bin Mohamad, Prime Minister, Malaysia, interview by Stephen Bradshaw, BBC News and Current Affairs, telecast on the Canadian Broadcasting Corporation's News Network, June 1992.

327. *New York Times*, 7th June 1992, L20:1–4.

328. *Time*, 1st June 1992, 25:2.

329. *Ibid.*

330. *New York Times*, 9th June 1992, A8:3.

331. *Irish Times*, 11th June 1992, 7:2.

332. *Times*, 2nd June 1992, 10:2–3.

333. *Ibid.*, 1:5.

334. *New York Times*, 9th June 1992, A8:4.

335. *Ibid.*, A8:4–5.

336. United Nations Document, A/CONF.151/6/Rev.1, 13th June 1992 (reprinted as Appendix 3 to this book).

337. *Ibid.*, Preamble (a).

338. *Ibid.*, Preamble (e).

339. *Ibid.*, Preamble (g).

340. *Ibid.*, Art. 2(b).

341. *Ibid.*, Arts. 5(a) and 5(b).

342. *Ibid.*, Art. 8(g).

343. *Ibid.*, Art. 7(a).

344. *Ibid.*, Art. 7(b).

345. *Ibid.*, Art. 8(c).

346. *Ibid.*, Art. 10.

347. *Ibid.*, Art. 11.

348. *Ibid.*, Art. 13(b).

349. *Ibid.*, Art. 9(a).

350. *Ibid.*, Art. 13(d).

351. Kuala Lumpur Declaration on Environment and Development, SMCED/MC/DOC.2, 26th–29th April 1992.

352. *Irish Times*, 9th June 1992, 8:1.

353. *Ibid.*, 8:2.

CHAPTER FOUR

Can International Law Improve the Climate? An Analysis of the United Nations Framework Convention on Climate Change Signed at the Rio Earth Summit in 1992

When delegates from 178 nations[1] gathered at Rio de Janeiro in June 1992 to attend the United Nations Conference on Environment and Development (UNCED), they were well aware that "[t]he road to environmental hell is paved with good intentions."[2] Such awareness did not, unfortunately, generate the collective will needed to create a body of effective international law to deal with the serious environmental crises now facing all the inhabitants of this planet. Although there was no shortage of rhetoric at Rio, the formulations which will forever be associated with that mega-conference fall short of what was universally expected and what is now needed to remedy the pollution and atmospheric decline plaguing almost every part of the world. Associated closely with the Rio Summit are two international treaties, the Convention on Biological Diversity[3] and the Framework Convention on Climate Change.[4]

This chapter will attempt an analysis of the Climate Change Convention and assess its significance within the global context of international problems which it attempted to resolve. Although the chapter will explore some controversial issues surrounding the Convention, such as the position of the United States and the European initiative, it will not deal with the lengthy negotiating process which led up to the Convention. Rather, the emphasis will be on international opinion about the Climate Change Convention as it stands. A brief summary of the entire Convention, with appropriate endnote citations, will precede the detailed analysis, where, for the reader's convenience, the number of the relevant article will be given in parentheses immediately fol-

Reprinted with kind permission from an original version published by the *North Carolina Journal of International Law & Commercial Regulation*.

lowing quotation of the provision (in italics). The clauses of the Treaty will not be analyzed seriatim but under specific subheadings which reflect the major aspects of global politics that dominated the Convention and which continue to bedevil all attempts at environmental amelioration. Occasionally, it will be necessary to repeat the wording of a provision because of its relevance to various topics being discussed. An attempt will be made to give the reader a fairly comprehensive view of the significant aspects of the Climate Change Convention and to determine whether the best of intentions will make any difference to the fate of the planet.

It is hoped the analysis will clarify and explore the Treaty within a global context which is fraught with complex political and economic problems, problems which come to the fore whenever the world sets out to take action to change the way human beings interact with their planet. National self-interest, North-South economic divisions, and a reluctance to take decisive measures to improve the environment all had an impact on the final convention on climate which was signed at Rio. There was no shortage of opinion on its pros and cons; as one commentator explained in *Newsweek*:

> The . . . treaty has inspired wild verbiage. It's anything from a last-ditch attempt to save a dying planet to a cynical plot to impose a socialist industrial order in the guise of climate protection. Global warming can engender such polar positions because the actuality of the subject is so pleasingly nebulous. . . . The greenhouse effect is a blank slate onto which partisans can project whatever they wish to behold.[5]

The Problem of Global Warming

The author Ivan Head comments that "[i]t is unlikely that any other single environmental issue contains a potential hazard to human well-being as great as climate change."[6] This single environmental issue has the potential to affect the political, economic, and social systems of every nation on Earth, and it will take nothing short of a miracle to reverse the apparent damage which experts are now describing with a sense of inevitable doom.

Basically, global warming involves damage to the atmosphere surrounding this planet and nourishing its millions of life forms in a beneficent environment which, until man interfered, had sustained all manner of plant and animal life for centuries. The atmosphere consists of "a mass of gases that surrounds the earth and is bound to it by the force of gravity. The composition of the atmosphere consists overwhelmingly of nitrogen (about 78 per cent) and oxygen (about 21 per cent) that coexist in a constant, fixed proportion.

The balance of 1 per cent is a mixture of several gases—carbon dioxide (CO_2), water vapour, argon, neon, helium, and methane (CH_4) among them."[7] In this very delicate, balanced system each component serves a vital function. Nitrogen and oxygen "are primary contributors to plant and animal life: nitrogen passes from the air into the soil where bacteria transform it into nitrates to be taken up as nourishment by plants; oxygen is our primary source of energy, responsible for the respiration of living organisms and combustion."[8] It is man's interference with this carefully balanced mechanism which has apparently resulted in climatic change.

The complexity of the global warming problem is exacerbated by the fact that the science surrounding it is as murky as the politics now engulfing the issue. As Stephen Strauss suggests, "The science on this has the clarity of a mudpie."[9] International lawyers who turn to scientists for precise answers before formulating legal solutions find themselves facing uncertainty, contradictions, and differing conclusions—all of which render the task of creating international law very difficult.

Global warming and consequent man-made climate change have been linked to the acceleration of the greenhouse effect—the result of approximately two centuries of industrialization. In nonscientific layman's terms, there is a natural greenhouse effect:

> short-wave solar radiation passes through a clear atmosphere relatively unimpeded; but long-wave radiation emitted by the warm surface of the Earth doesn't have such an easy time of it—some of it is absorbed and then reemitted by a number of trace gases in the cooler atmosphere above—since the outgoing long-wave radiation must, on average, balance the incoming short-wave radiation, both the atmosphere and the Earth's surface will be much warmer than they would be without these so-called "greenhouse gases."[10]

As Andrew Revkin explains, the greenhouse effect occurs because "[w]ater vapor, carbon dioxide, and a few other atmospheric gases act like the glass panes of a greenhouse, allowing sunlight in to warm the planet but preventing heat from escaping."[11] The greenhouse gases "trap the reflected energy of the sun as it radiates outward from the earth's surface. The mix of these greenhouse gases at any given time determines what proportion of the radiation is contained and held close to the Earth. This in turn determines the temperature of the surface itself."[12]

Industrialization has contributed more greenhouse gases, specifically carbon dioxide, and has added more dimensions to the problem through the extensive use of chlorofluorocarbons (CFCs),[13] which also deplete the ozone layer. Other gases responsible for the problem include methane and "emissions from automobiles, coal-burning industries and garbage dumps."[14] As Senator George Mitchell explains, "Beginning with the industrial revolution . . . man began burning fossil fuels—petroleum, coal, oil, and natural gas—at an unprecedented rate and throwing their residues, the greenhouse gases they create, into the atmosphere by the millions of tons annually. . . . The unavoidable result has been what scientists call the 'greenhouse warming' of the planet."[15] The greenhouse gases, with the addition of man-made chemical compounds, "are altering two of the atmosphere's primary functions: trapping the heat from the sun, and blocking some of the sun's harmful radiation."[16]

The problem is that no one knows how much of these man-made emissions the earth's systems can absorb safely before the entire atmosphere overloads, with horrifying consequences for the future of all the species which now inhabit the Earth. Concern about the effect of these emissions has grown, even though scientists are still scrambling to keep up with the popular demand for certainty and solutions. It is now widely believed that "human activity, mainly the combustion of fossil fuels, is causing the concentrations of these heat-trapping gases to increase at an unprecedented rate."[17]

Although scientists agree on the existence of the greenhouse effect, both natural and man-made, there is uncertainty regarding the impact of this phenomenon, the speed with which it operates, and its ultimate consequences worldwide. According to Stephen Strauss, "[d]etermining how much or how little any given greenhouse gas contributes to temperature change has proved to be nightmarish."[18] Lacking scientific certainty, the consensus is to come down on the side of caution and act now to prevent catastrophe later. "Scientists generally agree that it has been getting warmer over the last hundred years."[19] The environmental author Andrew Revkin asserts that "[t]he planet is warming. There has been some criticism from skeptics who say that temperature records are inaccurate. But other data strongly support the idea that things are heating up. . . . The winter snow pack covering the northern hemisphere has retreated markedly over the past few decades."[20] Human activity and economic development have both had a dramatic impact on the Earth's land, its oceans, and now its atmosphere. The human propensity to create and use masses of chemicals is partly to blame for the man-made greenhouse effect which is causing so much concern now. As Vice-President Al Gore stated, "The chemical revolution has burst upon the world with awesome speed."[21] In view of the fact that world chemical production now doubles every seven or eight years,[22] the atmospheric consequences of this massive utilization of chemicals are bound to become even more serious in coming years. The United Nations Environment Programme estimates that the human species has discovered or created over seven million chemicals, and commonly uses approximately eighty thousand of these.[23] The world is only now beginning to discover the consequences to land, water, and air from this addiction to chemicals. The existence of waste products derived from chemical utilization is an issue which has caused universal concern. Essentially, the global warming problem is simply the result of human developmental activity, and in that sense the parameters of the problem are similar to the consequences of deforestation, overuse of land and consequent desertification, depletion of scarce freshwater resources, pillaging of fish stocks from oceans, and so on.

Man is now proving to be the most dangerous species ever created, and the entire planet is reaping the results of what its dominant species has sown. Having polluted the oceans, killed freshwater lakes, deluged the Earth with acid rain, and destroyed the soil with contaminants, we should not be surprised that the atmosphere is now also suffering from human depredation. The challenge is now to live human lives within the context of our chemically oriented civilization without destroying the environment in the process. Time alone will tell whether human beings will be able to cope. That the atmosphere is no longer coping as well as it once did is becoming more and more obvious with each passing year; for example, those of us who live in northern climes like Canada find our winters becoming milder and our brief summers warmer. Whatever benefits this situation may have for people who have endured the trials of Canadian winters are outweighed by the realization of the threat that global warming poses for all of us on the planet.

As Senator George Mitchell comments, "Once greenhouse gases get into the atmosphere, they are like the man who came to dinner; they stay. They can hang there for decades or centuries."[24] The dimensions of the atmospheric danger can be gauged by considering just one of the culprit gases, carbon dioxide (CO_2) "the gas with the highest public profile."[25] Carbon dioxide, termed "the world environment's leading menace,"[26] poses a serious threat because it remains in the atmosphere for a long time and because it is still the gas primarily responsible for climatic change.[27] Approximately 55 per cent of present global warming has been attributed to carbon dioxide.[28] The burning of fossil fuel has resulted in the emission of about 185 billion tons of CO_2 into the atmosphere since 1860.[29] "Half the CO_2 added to the atmosphere in all of human history has been emitted in just the past 30 years."[30] Approximately 75 per cent of total CO_2 emissions originate in industrialized countries.[31] Vice-President Gore concludes that it does not seem reasonable "or even ethical . . . to assume that it is probably all right to keep driving up CO_2 levels. In fact, it is almost certainly not all right."[32] If, as has been stated by Gregg Easterbrook, human actions emit 7 billion tons of carbon into the air annually,[33] there is reason for alarm about the consequences of this tampering with the atmospheric balance. In Vice-President Gore's view,

> the artificial global warming we are causing threatens far more than a few degrees added to average temperatures: it threatens to destroy the climate equilibrium we have known for the entire history of human civilization. As the climate pattern begins to change, so too do the movements of the wind and rain, the floods and droughts, the grasslands and deserts, the insects and weeds, the feasts and famines, the seasons of peace and war.[34]

There is a definite need for more scientific research, a point that was emphasized by G. O. P. Obasi, Secretary-General of the World Meteorological Organization (WMO). He told delegates at UNCED that "[w]ith the advent of satellites and computers, our understanding of the global climate system, including knowledge of its interactions and processes, has increased dramatically over the past few years." Nevertheless, according to Mr. Obasi, "WMO is among the first to recognize that new types of information are needed and much more research will have to take place, so as to answer urgent questions now facing us, such as the timing, magnitude and regional pattern of the expected climate change."[35]

The Impact of Climate Change

The extent of international dedication to resolving the problem of global warming and climate change is dependent on the degree of seriousness with which humanity considers the threat of this phenomenon. As Andrew Revkin states: "Global warming presents a critical test of two uniquely human attributes: reason and foresight. It is up to all of us to seek out the facts and decide on a course of action."[36] If we agree that there is a majority consensus among scientists about the severity of the problem,[37] and if "levels of carbon dioxide continue to rise at the current rate, there is a significant chance that disruptive climate shifts will occur within the lifetimes of children born today."[38] Given this scenario, the threat is imminent and the need for environmentally effective action urgent.

The United Nations Environment Programme predicts that over the next five to ten decades the trapping of additional heat because of human-made gases could "result in a severe decline in productivity in some regions, shifts in climate zones toward the poles, rising ocean levels and extensive flooding, and accelerated animal and bird extinctions."[39] To this depressing prediction, Andrew Revkin adds the possibility of "wars over shrinking water supplies."[40] For a resource-rich developed nation like Canada, the impact on fisheries, prairie wheat production, forests, and fruit orchards could be serious.[41] The decade of the 1980s produced five of the ten hottest summers recorded in Canada.[42]

For the United States of America, economist William Cline of the Institute for International Economics projects that global warming of 2.5 degrees could lower American economic output by 1 per cent by the year 2050. This may not seem like much, but the cost could be at least sixty billion dollars annually by the middle of the next century.[43] "The $60 billion-a-year cost to the U.S. economy includes $18-billion in agricultural losses, $11-billion for extra air conditioning, $7-billion for water and $7-billion in land loss and dike building as oceans rise."[44]

A study commissioned by the United States Environmental Protection Agency and carried out jointly by Oxford University and New York's Goddard Institute for Space Studies concluded in its 1992 report that global warming would result in poorer crop yields in developing countries, "adding up to 360 million people to those at risk from hunger in the next century."[45] The devastation caused by global warming is likely to have its worst impact on those areas which are economically underdeveloped and therefore least likely to be able to cope with this added economic problem. As Martin Parry, head of Oxford's environmental change department, concluded about the threat to agriculture, "The balance of likelihood is on the negative side."[46]

Although the predictions are frightening, they are not exact, because the number of variables which have to be considered leave even the most complex of computers projecting a variety of scenarios depending on the factors they have to weigh. This only adds to popular confusion, misconception, and a dangerous degree of complacency about the problem. As the world's first Climate Conference occurred in 1979,[47] there is now an established, lengthy history of international action, research, and consequent studies on the subject. However, implementing effective remedies for the problems which affect our climate continues to be elusive. In 1990, the Intergovernmental Panel on Climate Change (IPCC) wrote a report on the scientific aspects of climate change. The report was the result of a joint effort by the United Nations Environment Programme and the World Meteorological Organization to gather the knowledge of the world's experts on this complex issue.[48] The IPCC was established in 1988 to assess scientific knowledge about the climate.[49] The team of 170 scientists[50] from a number of countries "based the 1990 projections on a variety of scenarios for future emissions of carbon dioxide (CO_2) and other greenhouse gases."[51] The IPCC report concluded "that the future warming rate could speed up considerably, with Earth's mean temperature climbing about 1.12 degrees Celsius by 2025 and 2.7 degrees by 2100."[52] Given present trends of reliance on coal, current levels of deforestation until this resource is depleted, and modest checks on carbon dioxide emissions, IPCC projections estimate a sea level rise of twenty centimeters by 2030 and of sixty-five centimeters by 2100.[53] Conceding that Earth's temperature has risen by about half a degree Celsius over the past hundred years and that the level of the oceans has risen by ten to twenty centimeters, the IPCC also stated that these developments are "of the same magnitude as natural climate variability."[54]

The IPCC projections of global warming have been revised, specifically by the work of climatologists Tom Wigley and Sarah Raper of the University of East Anglia in England. The figures have been pushed downward to a "best guess" now of 2.5 degrees warming and a sea level rise of forty-eight centimeters. Wigley and Raper caution against optimism because of their research, insisting that "the warming corresponds to a rate roughly five times that observed over the past century and the sea level rise is at a rate roughly four times that estimated for the past century."[55] All of this leaves the nonscientist layman, particularly the international lawyer, in considerable doubt and confusion. Henry Hengeveld of the Atmospheric Environment Service in Toronto explains that "[t]he evidence of scientific facts is a bit of a myth."[56] A number of climatologists feel that "there is no doubt that humans are perturbing the climate system. But computer simulations can't yet tell exactly how that system will adjust. They can't deal well with clouds and moisture or with the effect of the oceans."[57] Henry Hengeveld describes the scientific process which leads to all these contradictory conclusions: "What happens is that we have pieces of facts which converge to support conclusions, but these happen to be built on subjective evaluations. Personal expectations do influence how we interpret the evidence before us."[58]

The International Political Response to Global Warming

The contradictions in science have been amply reflected in the positions taken by politicians on the issue of global warming and its likely consequences. The lack of scientific certainty has bedeviled attempts to deal conclusively with the problem of climate change. "Indeed, the greenhouse effect has become such a source of dispute and speculation that it serves as a sterling example of the way political decision making and popular understanding can lose their way in the scientific thicket."[59] It is important to keep this fact in mind when analyzing the Climate Change Convention signed at the Earth Summit. As the *Economist* points out:

> Although there are good grounds for believing that the accumulation of greenhouse gases in the atmosphere will warm the planet, there is much less certainty about the pattern which that warming will take, let alone its impact. Scientists are unsure about the extent to which warming will be offset or reinforced, or about the speed with which it will occur. They are unsure how far expanding oceans will raise sea levels, or whether and where droughts and rainfall will increase. . . . [I]t is as yet impossible to predict which countries may gain, which lose, and how.[60]

Considering the extent of technical confusion and uncertainty, the global reaction to the problem has been both surprising and encouraging. If the measures are weak and lack teeth, that is a situation which can be rectified as awareness of the serious nature of global warming grows. Indeed, some commentators are amazed that the world's politicians have actually managed to produce an international convention concerning a matter which is scientifically still riddled with contradictions, however strong the underlying message of alarm sounded by many climate experts. "That the world's nations have already agreed on a treaty on climate change . . . is extraordinary. It is, after all, less than a decade since global warming was first discussed outside laboratories."[61] The United Nations has, for once, not been dilatory in debating and increasing international awareness about the possible dangers of climate change. It is encouraging to observe the United Nations responding to avert catastrophe rather than waiting around to react to it once it has occured. The significance of the Climate Change Convention goes far beyond the problem of global warming. It indicates a new, refreshing trend in global political cooperation—a trend which may err on the side of caution and hopefully save the environment of Spaceship Earth for future generations. Given the diversity of scientific conclusions, the divergence of international sentiment, and the opinions of various vested interests in both the developing and developed worlds, the creation of an initial Convention is a substantial achievement.

It would be worthwhile briefly to examine a few of the international measures taken with respect to climate change. Given its length constraints, this chapter cannot discuss all the international formulations which preceded the Climate Change Convention of 1992. Accordingly, the two-year period preceding UNCED is all that will be considered here. (The reader is reminded that the United Nations General Assembly had passed resolutions on the climate issue in previous years.) It is hoped that a brief overview of the plethora of international activity just prior to UNCED will establish both the degree of concern which propelled the formulation of the Convention and the global significance of the issue. This background will also set the stage for the analysis of the Convention and explain why the Treaty could not be stronger in the obligations it imposed on signatory nations.

By 27th January 1989, the General Assembly had already adopted its resolution for the Protection of Global Climate for Present and Future Generations of Mankind, which recognized that "climate change is a common concern of mankind" and directed United Nations officials to prepare for a "possible future international convention on climate."[62] In the following year, the General Assembly reiterated its call for negotiations on a framework convention on climate, declared it a matter of urgency, and proposed that the convention include concrete commitments "in the light of priorities that may be authoritatively identified on the basis of sound scientific knowledge."[63] In March 1990, the General Assembly declared "[p]rotection of the atmosphere by combating climate change" an environmental issue of major concern.[64]

Regional activity resulted in the development of bodies of consensus on the significant issues related to climate change in particular areas. The Kenyans hosted a conference in May 1990 to explore the possible "impacts of global climatic change on the ecosystems, economies and infrastructures of African Countries."[65] The consequent Nairobi Declaration emphasized the need for action to reduce greenhouse gas emissions; called for economic diversification and for the promotion of afforestation and reforestation; supported research on climate change and its impact on coastal areas, agriculture, and livestock; and, significantly, endorsed the substantial reduction of greenhouse emissions by the year 2005.[66]

The European Community was also active in this area. Representatives met at Luxembourg on 29th October 1990 to adopt a common declaration on global warming, reflecting a "common stand . . . to lead the international effort to curb greenhouse gases."[67] The twenty-four Community ministers agreed to stabilize CO_2 emissions at Community level by the year 2000.[68] A few days later, on 5th November 1990, the Environmental Ministers of the

European Free Trade Area (Austria, Finland, Iceland, Norway, Sweden, and Switzerland) endorsed the freezing of CO_2 emissions by the year 2000.[69]

At its forty-fifth session, which convened in 1990, the General Assembly received the Report of the Secretary-General on the Protection of Global Climate for Present and Future Generations of Mankind.[70] On 21st December 1990, the General Assembly adopted another Resolution on climate, establishing a broad-based negotiation process for the creation of the convention and specifying that the climate convention should be completed before the Earth Summit in June 1992 so that it could be opened for signature during that conference.[71]

The General Assembly also took account of the contents of the Ministerial Declaration adopted at the Second World Climate Conference, held at Geneva between 29th October and 7th November 1990.[72] That conference, organized by a group of United Nations agencies, concluded that "[c]limate issues . . . are increasingly pivotal in determining future environmental and economic well-being. . . . If the increase of greenhouse gas concentrations is not limited, the predicted climate change would place stresses on natural and social systems unprecedented in the past 10,000 years."[73] The 747 participants representing 120 countries[74] stated that a clear scientific consensus had "emerged on estimates of the range of global warming which can be expected during the twenty-first century." They suggested that, "notwithstanding scientific and economic uncertainties, nations should now take steps towards reducing sources and increasing sinks of greenhouse gases through national and regional actions, and negotiation of a global convention on climate change and related legal instruments." The Conference cautioned that "[t]he remaining uncertainties must not be the basis for deferring societal responses to these risks."[75] The Conference Statement also stressed that "[c]limate change may compound existing serious problems of the global mismatch between resources, population and consumption. In many cases the impacts will be felt most severely in regions already under stress, mainly in developing countries."[76]

The process for the creation of the Framework Convention on Climate Change began with the initial session held in the United States of America between 4th and 14th February 1991. That process, with all its bitterness, intensity of viewpoints, and somewhat frustrating outcome, is not the subject matter of this chapter. Suffice it to say that the analysis to follow will inevitably have to explore some aspects of the divergence which arose, particularly with respect to the controversial position taken by the Bush Administration of the United States.

No less controversial, from the international perspective, was the Tlatelolco Platform on Environment and Development adopted by Latin American and Caribbean governments meeting between 1st and 7th March 1991. This declaration endorsed what might be called the standard developing nation stance, emphasizing the responsibility of developed nations for greenhouse gas emissions and the vulnerability of Latin American and Caribbean states to the consequences of climate change. The members of the Economic Commission for Latin America and the Caribbean (ECLAC) insisted that "their participation in any agreed solution should be consonant with the economic and technical resources available to the developing countries."[77]

The Declaration adopted by developing nations meeting in Beijing on 19th June 1991 was even more blunt in asserting the claims of the South. The ministers of forty-one developing nations voiced their concern over the "accelerating degradation of the global environment" and insisted that "it is the developed countries which are mainly responsible for excessive emissions of greenhouse gases, historically and currently, and it is these developed countries which must take immediate action to stabilise and reduce such emissions." More alarming was the strident assertion that "[d]eveloping countries cannot be expected to accept any obligations in the near future." More predictably, these nations insisted that the convention being negotiated "must include . . . firm commitments by developed countries towards the transfer of technology to developing countries, the establishment of a separate funding mechanism, and the development of the economically viable, new and renewable energy sources as well as sustainable agricultural practices, which constitute an important step to deal with the major cause of climate change. In addition, the developing countries must be provided with the full scientific, technical and financial cooperation necessary to cope with the adverse impacts of climate change."[78] The tone of developing nations had moderated somewhat with respect to the climate change problem by 29th April 1992, when the ministers of fifty-five developing countries issued the Kuala Lumpur Declaration following a three-day meeting in Malaysia. In this Declaration, signatories urged the "developed countries to undertake meaningful and specific commitments on the stabilization and reduction of emissions of carbon dioxide and other greenhouse gases."[79]

Earlier, on 19th December 1991, the United Nations General Assembly passed a resolution in an attempt to give a boost to the negotiating process for a convention on climate change, urging its speedy and successful completion.[80] Meanwhile, as the deadline for the opening of UNCED approached, there was a flurry of activity in other sectors. Specifically, the Governing

Council of the United Nations Environment Programme (UNEP) called for all states to support implementation of the World Climate Programme and for the creation of a Global Climate Observing System to monitor the weather (World Weather Watch), the atmosphere (Global Atmosphere Watch), and the oceans (Global Ocean Observing System).[81] The preparatory committees of UNCED were concerned as well with the issue of climate change generally and with its specific aspects, such as transboundary air pollution, energy transition, and energy supply.[82] All these proposals will, if implemented successfully and meaningfully, have an impact on the process of reversing the negative consequences of climate change.

An excellent factual summary of the activity preceding the adoption of the Convention is contained in the Report of the Chairman (Jean Ripert of France) of the Intergovernmental Negotiating Committee for a Framework Convention on Climate Change (1st June 1992).[83]

Clearly, the issue of climate change received international attention, at least at the level of formulating resolutions and issuing declarations relevant to rectification of the problem. It is unfortunate that the level of active implementation does not yet match the range of enthusiastic verbal activity in the field of environmentalism. However, the plethora of resolutions, declarations, and formulations does bring the matter to the forefront of international attention. With such heightened awareness, we can hope that, some day really strong measures will be taken by all the nations of the world. As Jean Ripert emphasized, "countries are now agreed that, for present and future generations of mankind, something must be done to protect the global climate from anthropogenic change: the Convention is a good beginning."[84]

A Brief Summary of the Convention

In order to analyze the Framework Convention on Climate Change, it would be useful briefly to summarize its main provisions so that the reader has a point of general reference to comprehend the specific topical discussion which follows this section.

The Convention defines climate change as signifying "*a change of climate which is attributed directly or indirectly to human activity that alters the composition of the global atmosphere and which is in addition to natural climate variability observed over comparable time periods.*"[85] The Convention acknowledges that "*change in the Earth's climate and its adverse effects are a common concern of humankind*";[86] that "*the largest share of historical and current global emissions of greenhouse gases has originated in developed coun-*

tries";[87] and that developing countries have special difficulties because their *"economies are particularly dependent on fossil fuel production, use and exportation,"*[88] and therefore will require *"access to resources . . . to achieve sustainable social and economic development."*[89] The basic objective of the Convention is rather vaguely stated as *"stabilization of greenhouse gas concentrations in the atmosphere at a level that would prevent dangerous anthropogenic interference with the climate system. Such a level should be achieved within a time frame sufficient to allow ecosystems to adapt naturally to climate change, to ensure that food production is not threatened and to enable economic development to proceed in a sustainable manner."*[90] The Convention concerns itself with greenhouse gases not controlled by the Montreal Protocol.[91]

The Convention adopts the precautionary approach: *"The Parties should take precautionary measures to anticipate, prevent or minimize the causes of climate change and mitigate its adverse effects. Where there are threats of serious or irreversible damage, lack of full scientific certainty should not be used as a reason for postponing such measures."*[92]

Acknowledging the existence of *"common but differentiated responsibilities,"*[93] developed nations are to *"take the lead in combating climate change and the adverse effects thereof."*[94] The Convention accordingly endorses the precautionary principle;[95] declares that sustainable development is a right;[96] and cautions against environmental measures becoming *"a means of arbitrary or unjustifiable discrimination or a disguised restriction on international trade."*[97]

Signatory nations enter into a series of commitments to develop national inventories of emissions;[98] implement national programs to mitigate climate change;[99] cooperate in the process of technology transfer;[100] develop plans for conservation of sinks and reservoirs of greenhouse gases, coastal zone management, water resources, and agriculture;[101] and participate in education programs to increase public awareness about climate change.[102] Encouragement is given to the idea of developing international programs *"aimed at defining, conducting, assessing and financing research, data collection and systematic observation."*[103]

Provision is also made for periodic reporting on the progress of parties which are mainly developed countries and those former Communist states now termed economies in transition.[104] These two groups of nations have special responsibility to take the lead in implementing measures to limit emissions, *"recognizing that the return by the end of the present decade to earlier levels of anthropogenic emissions of carbon dioxide and other green-*

house gases . . . would contribute" to mitigating the effects of climatic change.[105] The language is so ambiguous as to be almost meaningless. The target is to return to 1990 levels of CO_2 emissions,[106] but no firm deadline is specified. It is important to note that some of the Convention's clauses are subject to periodic review and amendment.[107]

The participation of developing countries is conditional: *"The extent to which developing country Parties will effectively implement their commitments under the Convention will depend on the effective implementation by developed country Parties of their commitments under the Convention related to financial resources and transfer of technology and will take fully into account that economic and social development and poverty eradication are the first and overriding priorities of the developing country Parties."*[108]

The Convention establishes a Conference of the Parties,[109] to promote information exchange,[110] assess the implementation of the Convention by the Parties,[111] make recommendations pursuant to the goal of implementation,[112] and *"mobilize financial resources."*[113] A Secretariat is also established,[114] to perform the usual administrative functions of such bodies and to *"facilitate assistance to the Parties, particularly developing country Parties, on request, in the compilation and communication of information required in accordance with the provisions of the Convention."*[115] The Convention also establishes a *"subsidiary body for scientific and technological advice"*[116] to assess the *"state of scientific knowledge relating to climate change and its effects,"*[117] act as a scientific resource for the Conference of the Parties,[118] advise on the important issue of technology transfer,[119] and provide advice *"on ways and means of supporting endogenous capacity-building in developing countries."*[120] The Convention also creates a subsidiary body (open to participation by all signatories) for implementation to assist the Conference of the Parties.[121]

The financial provisions of the Climate Change Treaty call for *"equitable and balanced representation of all Parties within a transparent system of governance."*[122] Article 11 (Financial Mechanism) states, somewhat nebulously: *"A mechanism for the provision of financial resources on a grant or concessional basis, including for the transfer of technology, is hereby defined. It shall function under the guidance of and be accountable to the Conference of the Parties, which shall decide on its policies, programme priorities and eligibility criteria related to this Convention. Its operation shall be entrusted to one or more existing international entities."*[123] On an interim basis, the Global Environment Facility of the United Nations Development Programme, the United Nations Environment Programme, and the International Bank for Reconstruction and Development constitute the international entity allotted the task of operating the financial mechanism.[124] The same article specifies that *"the Global Environment Facility should be appropriately restructured and its membership made universal."*[125]

As regards the settlement of disputes between Parties, provisions are made for *"negotiation or any other peaceful means"* chosen by the disputants.[126] Parties may also resort to the International Court of Justice and/or binding arbitration.[127] Parties are not allowed to make reservations to the Convention.[128] The Convention may be amended either by consensus or by a three-fourths majority of the Parties present and voting.[129]

The U.S. Government and the Climate Change Convention

What was excluded from the Climate Change Convention was as important as what was ultimately accepted. The most controversial issue related to the desire of the majority of negotiating nations to impose firm time deadlines for the reduction of greenhouse gases and the successful resistance of one nation to that goal. It was the Bush Administration which drew the most condemnation, nationally and internationally, for its environmental positions in the series of negotiations leading up to Rio. The American Government took a firm stand against definite deadlines and ultimately prevailed.

During both the Earth Summit and the preceding media blitz, it was widely reported that the United States emits the largest share of greenhouse gases of any nation, approximately 17.6 per cent of the total.[130] With a mere 5 per cent of the world's population, the United States consumes 25 per cent of its energy and emits 22 per cent of all CO_2 produced.[131] Thus one nation, by emitting almost a quarter of the carbon dioxide produced each year, contributes as much to global warming as all the developing countries combined—and there are over a hundred such nations now.[132]

Although the U.S. Environmental Protection Agency (EPA) revealed that about a million tons of dangerous chemicals were emitted into the air in 1987,[133] the reaction of the Government of the United States was cautious. Equally cautious was President Bush's stand on the global warming issue, a vital environmental cause in which, regrettably, the United States lost the initiative and ultimately gained only opprobrium from the rest of the world. As the journalist Christopher Young commented, "American leadership of a new world order has been stopped cold on the environmental front."[134] Leading the Senate delegation to the Rio Earth Summit, Al Gore was critical of President Bush for failing to play "a leadership role on global environmental preservation."[135] The *Christian Science Monitor*, in an editorial, contrasted the positions of Republicans and Democrats in the 1992 election campaign, the latter party insisting that the United States "must become a leader in the fight against global warming, agreeing to limit carbon dioxide emissions to 1990 levels by the year 2000," while the Republicans stressed development compatible with conservation.[136]

The dilemma confronting the Bush Administration was explained by a former Undersecretary of State, David Newsom. He indicated that because the United States emits the most gases damaging to climate and because of its wealth, "Washington will be expected to take a leading role in measures to reverse environmental decline and contribute to the related costs." Newsom suggested that there was an "ideological resistance in a conservative administration to any acceptance of scientific evidence of global warming."[137] John Knauss, Head of the American Delegation to the Second World Climate Conference in Geneva in 1990, made it very clear that his country "was not prepared to bind itself to any target percentages for reductions which we are not able to guarantee to fulfil."[138]

President Bush apparently believed that there was "insufficient scientific data on which to base rational policy decisions" on global warming,[139] and the American delegation to the Convention's negotiations process resisted all efforts to establish firm time deadlines for the reduction of emissions. When the opposition of European and other nations became very strong, the American President made his attendance at the Summit conditional on global acquiescence in the American position. His spokesman announced that President Bush would "not attend . . . unless the rest of the world backs down on limiting earth-warming carbon dioxide gases."[140] Michael McCarthy reported in the *Times* (London) that "Mr. Bush had held back from announcing his attendance until the international treaty on global warming . . . had been negotiated to the satisfaction of the US. At American insistence the treaty as it stands now contains no legally binding commitments to control emissions of gases thought to be causing the greenhouse effect such as carbon dioxide, of which the US is the world's biggest producer."[141]

According to a Reuters report published in the *Globe and Mail*, one important reason for the reluctance of the President to accept a strong treaty was his conviction that measures aimed at precise reduction of greenhouse gases would threaten the American economy and jobs[142]—a vital factor for a President going into an election during an economic recession. As he confronted the vital environmental issues posed at Rio, the American President was also aware that unemployment in his country had reached 7.5 per cent, its worst point in about eight years.[143] Speaking to a group of business executives, President Bush expressed his concern that the Rio Summit might limit his nation "to a course of action that could dramatically impede long-term economic growth in this country."[144] The United States Department of Energy concluded that emission-reduction measures would have an adverse economic impact because of the nation's reliance on coal and oil.[145] Hence, facing the development/environment dichotomy which was at the very heart of UNCED, the leader of the world's lone superpower settled very clearly for development, even at the expense of environment.

This presidential position was unfortunate in terms of environmentalism, but it was consistent with the conservative ethic which deplores the proliferation of governmental regulation on business. Peter Stothard of the *Times* (London) commented that "[o]pposition to timetables and targets at Rio became . . . a mantra for the Republican right."[146]

This attitude toward environmental regulation is not, however, exclusively a matter of ideology. There is the issue of cost to be considered as well. Robert Crandall of the Brookings Institution in Washington suggested that "the full cost of environmental policy will soon exceed defense spending" (which totalled almost three hundred billion dollars a year in 1992).[147] The Environmental Protection Agency has estimated that pollution control would cost the nation more than one and a half trillion dollars in the 1990s.[148]

To counter the avalanche of hostile international publicity, President Bush (who eventually did attend the Summit, to become its most controversial participant) proposed to delegates at UNCED that participating countries meet the following year "to lay out our national plans for meeting the specific commitments in the Framework Convention. Let us join in translating the words spoken here into concrete action to protect the planet."[149] Having done its utmost to weaken the Treaty, it now appeared that the United States would take the lead in "speedier implementation" of its provisions.[150] Alden Meyer, Director of the Program on Climate Change and Energy of the Union of Concerned Scientists, suggested, "the United States is probably likely to exceed the European goals, even though we won't commit to them."[151] Clara Germani, writing in the *Christian Science Monitor*, reported that the United States also "pledged $75 million in aid to developing countries to help them curb the emissions believed to cause climate change."[152] The contradictions in the American position were apparent to many observers of the negotiation process that led to the Convention. As the *Times* (London) explained in an editorial, "America's Clean Air Act will probably bring its CO_2 emissions within the targets President Bush has declined to endorse."[153] The *Christian*

Science Monitor also believed that "the US might have little problem meeting the proposed standard of holding CO_2 emissions to 1990 levels."[154] Given these facts, the position of the American President would in retrospect appear to have been unnecessarily risky, risky both to the global leadership expected of the United States and to his re-election as President.

The Head of the United States Delegation to the Earth Summit, William K. Reilly, Chief Administrator of the Environmental Protection Agency, tried to put the best face on the public relations disaster his country faced at Rio. He emphasized his nation's strong support for the Convention and informed delegates that "[t]he United States has taken the lead in developing an action plan for controlling greenhouse gases that is detailed and effective. Most of these actions are already underway; others are awaiting legislative approval."[155]

Another American attempt to win back international support was by aggressively lobbying for a forestry agreement. The protection of forests, especially tropical forests, has been a significant U.S. initiative. Though the Americans secured a nonbinding agreement on forestry, they failed to get the developing nations, which still enjoy the luxury of vast forest resources, to accept legally binding commitments to conserve their forests. Indeed, seen in the context of American resistance to strengthening the Climate Change Convention, the forest initiative was cynically perceived by delegates from the developing world. The White House was accused of attempting "to win agreement on the preservation of the tropical rainforests as carbon dioxide 'sinks' so that the US will not have to join other developed countries in setting timetables to reduce its own emissions of greenhouse gases."[156] President Bush was also accused of being "willing to reduce carbon emissions into the atmosphere from third-world forests but not from American smokestacks and tailpipes."[157]

In a very real sense the President was confronted by a situation which defied solution. Facing re-election in a recessionary economy, he believed that environmental issues had to accommodate one priority, namely, that the American lifestyle would not be negotiable.[158] He also had to accommodate the diverse elements within the Republican Party which emphasize the primacy of business and free-market systems. Commenting on his obvious plight, the *Globe and Mail* suggested that "Bush is a perfect presidential name, considering all the hedging he has to do."[159] Political necessity dictated that the man who had portrayed himself as the environmental president would conclude his career being termed the "dead weight of the environmental world order."[160]

The European Initiative to Strengthen the Climate Change Convention

The extent of dissatisfaction with the United States position can be gauged by the fact that a number of European countries attempted some last-ditch efforts to sponsor a declaration which would strengthen the commitments contained in the Framework Convention on Climate Change. The mainspring of the movement was the Austrian Government, specifically its Environment Minister, Ruth Feldgrill-Zankel; other supporting countries included Switzerland and Liechtenstein. After signing the Climate Change Convention, these three nations issued a joint declaration which stated that they would "continue to implement the measures that are necessary at least to stabilize, as a first step, their CO_2 emissions by the year 2000 at the level of 1990, and to reduce thereafter their emissions of CO_2 and other greenhouse gases . . . on the basis of national policies and strategies, taking into account the best available scientific, technical and economic evidence."[161] This document then urged "other countries, in particular industrialized countries, to . . . achieve the earliest and most effective operation of the Convention."[162]

The European initiative garnered international publicity for the cause of halting climate change and highlighted the fundamental differences between the United States of America and the great majority of nations. That, in this instance, a minority of one had prevailed testifies to the very real strength of the United States in international negotiations. According to the *New York Times*, Austria was firmly supported by Switzerland and the Netherlands, all three nations having tried very hard but unsuccessfully to negotiate a stronger treaty.[163] Richard Mott of the World Wildlife Fund commented that the three European states were highly frustrated and were "trying to put the limelight on the countries that have weakened the treaty."[164]

The Austrians denied that their efforts were aimed against the United States.[165] As one Canadian environmentalist argued, "Nobody is isolating the Americans—they have isolated themselves."[166] However, according to the *Edmonton Journal*, the Canadian Government sensed a real danger in supporting the Austrian declaration because of a fear of isolating the Americans.[167] It was apparent that Ruth Feldgrill-Zankel did not approve of the American President's position on global warming. She explained that she could not "understand why the United States sees an opposition between the environment and the economy."[168] It was also reported by James Brooke in the *New York Times* that Germany and other European nations were separately seeking support "for the rapid adoption of a treaty protocol to impose targets and timetables."[169] William Reilly, Head of the United States Delegation, said that his country would block the German initiative,[170] and the American Government was alleged to have sent a "threatening" letter to the Government of Austria.[171]

Eventually, the European Community attempted (without too much success) to smooth ruffled American feathers by rejecting the Austrian initiative and adopting instead "a uniform statement setting targets and timetables for reduction of carbon-dioxide emissions to 1990 levels by the end of the century."[172]

The North-South Conflict and the Climate Change Convention

It is virtually impossible to analyze any international convention today without considering the North-South confrontation and examining how the particular treaty fits into that context. The relationship between developed and developing nations has become the most important global issue since the end of the Cold War. It affects every issue of global significance and has come in the 1990s to dominate international politics. The basic aspects of this confrontation can be examined only briefly before we proceed to fit the Climate Change Convention into this situation.

The volatile relationship between the nations which are developed and relatively prosperous (collectively labelled the North) and those that are developing and relatively poor (collectively labelled the South) has been going on for decades. It is just one more unfortunate legacy of the colonial past, a system of economic inequity in which the rich resources of the colonies were utilized to develop the economies of the colonial power, often to the detriment of the colonized populations. Independent, diverse, mainly agrarian societies in the South were quite suddenly made part of a complex global market system which assigned to them the roles of hewers of wood and drawers of water. As providers of raw materials, Africa, Asia, the Middle East, and Latin America produced the primary commodities which enabled Europe and North America to industrialize and thereby develop a standard of living that gave the majority of the North's citizens a level of personal comfort unheard of in any era in history. Conversely, this high standard of living was created at the expense of those who not only provided the raw materials to keep the factories running but also acted as vast markets to absorb the continuing flow of industrial products. The beneficiaries were in political control of those who provided the benefits. And so the system worked, until the sheer economic injustice of the situation, along with political, nationalist awakening in ancient societies like India, generated massive revolts against foreign rule.

Eventually, the nationalistic fervor, grounded in economic discontent and the low standard of living of the majority, brought political independence to most of the former colonies. The rapid process of decolonization following the end of the Second World War would change the political map of the world. Unfortunately, in the economic realm it remained largely business as usual.

The former colonial powers were willing to give huge donations in foreign aid to their onetime colonies, but the world market system was not prepared to adjust to accommodate a new set of industrialized nations. The newly independent countries found that they were still expected to remain exporters of primary products; they had little or no hope of breaking into Northern markets, which were protected to ensure the stability of domestic industries. Although some countries in Southeast Asia have made inroads into the developed world with manufactured goods, the majority of developing nations still find that it is their raw materials which sell in the North. Data collected by the International Monetary Fund (IMF) show that "between 1982 and 1989, the terms of trade of the developing countries as a whole fell by 20 per cent. During the same period, the terms of trade of the industrialized countries improved by 16 per cent."[173] Although world trade has grown by approximately 600 per cent since 1955, the South has enjoyed an increase of only 12 per cent.[174] Indeed, "in comparative terms, the developing countries' share of world trade dropped from about one-third of the total to one-quarter of the total."[175] Consequently, many governments in the South have realized that unless there is a drastic shift in the way the economic pie is sliced their vast populations are destined for generations of poverty. Political independence has not brought the economic miracles of development promised by the nationalistic revolutionaries who formed the first postcolonial governments. Some countries, like India and China, have created diverse economies, but they have yet to meet the needs of their enormous populations for such bare necessities as adequate food, shelter, clothing, and a basic education.

It is important that the policy makers, and particularly the environmentalists, in the developed world realize the amount of frustration and despair which motivates the often harsh rhetoric flowing from the South. Having few cards to play in the international game, these countries are determined to use every tactic and exercise every option they have to reverse the poverty which has been their lot for decades. Regrettably, one casualty of the North-South economic conflict has been the cause of global environmentalism. Having developed to a point which gives a majority of their citizens a comfortable standard of living, most industrialized societies are now expressing concern about the price the planet has paid for this development. The air over most countries is now befouled, the oceans are polluted, the lakes and ponds are dying, the atmosphere is warming to levels which may be dangerous in the coming century, and the quality of environmental life is threatened globally. Hardly anyone disputes the responsibility of the industrialized North for the creation and continuation of the environmental crises which plague this planet today. However, accepting responsibility and doing something about it are two different matters. We have already seen how the position of the United States of America in the climate change negotiations precluded the formulation of a strong treaty. When developed nations urge the developing world to exercise environmental caution when industrializing and not emulate the North in its rapid, untrammelled rush to develop, the South inevitably sees this as yet one more attempt by the North to keep its people in perpetual poverty. Basically, the South is now telling the North that if the latter wants to indulge in this environmental concern and expects the South to follow suit, then the North (which created most of the environmental problems) must be willing to pay for the South's participation in this clean-up activity.

A new bargain is being forged now because the North is, in a very real sense, at the mercy of the South for the first time in history. If the South proceeds to destroy its forests and industrialize at the same frantic pace as did the North, Earth's fragile ecosystems will probably not be able to sustain the consequences. Hence, the developing world has found a card—environmentalism—and is playing it to demand less economic inequity and a fairer share of the economic wealth of this planet. Whether a new economic order is likely to emerge from the mutual vulnerability and interdependence of these two great blocs of nations remains to be seen. Till then, every United Nations resolution of global scope and every international treaty is formulated with this confrontation in mind. The analysis of every treaty must consider whether the North gained or the South, or whether the agreement reflects a balance between these two economic groups.

The North-South issue was fundamental throughout the process leading up to the signing of the Climate Change Convention at Rio. Sun Lin, Head of the Chinese Delegation to the Intergovernmental Negotiating Process, voiced the realization of most participating nations: "Owing especially to the enormous disparities between developed and developing countries in levels of economic development, and also due to the great difference amongst various countries in geographical environment and natural endowment, the negotiations on this Framework Convention will be all the more difficult."[176]

The chief delegate from India was more blunt in expressing the Southern position once substantive negotiations began. His statement is indicative both of the extent of North-South tension prevalent in the entire process and of the refusal of developing nations to bear the burden of environmental measures. As he explained to delegates: "If per capita emissions of all countries had been on the same levels as that of the developing countries, the world would not today have faced the threat of global warming. It follows therefore, that developed countries with high per capita emission levels of greenhouse gases are responsible for incremental global warming."[177] Although this chapter cannot explore the entire negotiating process, it is important to gain some insight into the viewpoints which sparked the polarized positions which plagued all the formulations associated with the Rio Summit. The Indian delegate insisted that the "principle of equity should be the touchstone for judging any proposal. Those responsible for environmental degradation should also be responsible for taking corrective measures."[178] Clara Germani of the *Christian Science Monitor* explained the developing nations' position: "Third-world diplomats don't want to be held to emissions standards that even industrialized nations won't meet. Moreover, those standards could limit needed economic growth in developing countries unless they receive financial assistance to help develop alternative energies. The US's $75 million was not enough for many third-world participants."[179] The amount of American aid pledged scarcely matches the $75 billion increase that the World Bank estimated that developing countries would require annually to fund environmental projects by the end of the 1990s.[180] In contrast, John Major, Prime Minister of the United Kingdom, was reported by Robin Oakley of the *Times* (London) to have pledged "an extra £100 million as Britain's contribution to Third World programmes on climate change funded by the Global Environment Facility."[181]

With respect to this aspect of the North-South debate, the Convention reflects acknowledgment of the Southern position:

Noting *that the largest share of historical and current global emissions of greenhouse gases has originated in developed countries* . . . (Introduction)

and further,

Recognizing also *the need for developed countries to take immediate action in a flexible manner on the basis of clear priorities, as a first step towards comprehensive response strategies at the global, national and, where agreed, regional levels that take into account all greenhouse gases, with due consideration of their relative contributions to the enhancement of the greenhouse effect.* (Introduction)

Although the North was willing to acknowledge responsibility for having created most of the global warming problem, the agenda of the South required more concessions from the developed world. The Convention reflects the primacy of that agenda.

The South and the Climate Change Convention

The lack of unity among developed nations during the Rio process and in the discussions leading to the creation of the Climate Change and Biodiversity Conventions would have a significant impact on the final product which emerged from these involved and complex, often tortuous negotiations. With the benefit of hindsight one could suggest that a more progressive stance by the United States of America might have made for a stronger treaty, one with firm commitments for both developed and developing nations to implement. As it is, the North was fragmented, disunited, and unable, because of the American position, to exert any serious pressure on the South to undertake the type of economic sacrifice for environmentalism which alone will ease the rate of global warming. This is important because measures to alleviate global warming have to be taken internationally if the planet is to succeed in curbing this problem.

If reduction occurs only in the developed world, and if the South proceeds at its current pace of industrialization and forest depletion, Southern emissions would outpace the positive achievements of the North and the planet would be in worse shape. To give only one example, China, which is rapidly becoming one of the world's most determined polluters, has been described in *Time* as the "world's fourth biggest contributor to global warming," following the United States, the former Soviet Union, and Brazil.[182] According to *Newsweek*, Benxi, a Chinese city near the border with Korea, spews out over eighty-seven million cubic meters of gases each year from its two hundred or more factories.[183] *Time* reported that this was happening despite the fact that "[s]ince 1978, the city's particle discharge has been cut in half by the installation of equipment designed to suck pollutants . . . from factory smoke."[184] According to a report in the *Times* (London), Deng Nan, China's Deputy Minister of Science, Technology and the Environment (and incidentally, daughter of Chinese leader Deng Xiaoping), admitted that "[t]he country's overall ecological and environmental situation continues to deteriorate."[185] According to John Stackhouse and Jan Wong, reporting in the *Globe and Mail*, China already emits approximately nine per cent of the world's CO$_2$,[186] with its per capita carbon dioxide emissions averaging between one and five tons in 1989.[187]

At UNCED, Chinese Premier Li Peng informed delegates that "[i]n the past decade or more . . . China's GNP has more than doubled. Yet the quality of China's environment has remained basically stable, or even improved in some areas. Our environment and development strategy suited to China's conditions has proved to be successful."[188] The rest of the world is no longer as sanguine as the Chinese Government appears to be about its environmental situation. "As China is increasingly identified as one of the main contributors to ozone depletion and global warming, its environmental problems are becoming the world's problems and are being intensely scrutinized at home and abroad."[189] As the Chinese example demonstrates, the world cannot afford the luxury of African, Asian, and Latin American development proceeding at its present pace without paying a very heavy environmental price in the not-too-distant future.

On the other hand, it does not appear fair to the nations of the South to restrict their development and consign their enormous populations to ever-increasing poverty. In a very real sense, developing nations have no option but to raise their citizens' material standards as rapidly as possible. Without development and a basic minimum standard of living, these nations face the prospect of economic turmoil and possibly even political revolution. The dilemma of China is more severe than that of most other developing nations. Functioning as a Communist society with a political and administrative system now considered outmoded and outdated in most of the world, the Chinese Government's only hope of clinging to totalitarian power is to give its population as much economic development as possible, regardless of the environmental cost. The solution, as China sees it, is to make the developed world pay for the clean-up. "Beijing estimates the cleanup will cost developing countries more than $600-billion. It has called on developed countries to pay more than $125-billion of that."[190]

It could be argued, with some justifiable cynicism, that the South's agenda in the process leading up to the Earth Summit boiled down to acquiring as much money as possible from the North for environmental projects. It could also be argued that the Convention on Climate Change reflected and catered to the South's needs more than to the overall cause of reducing greenhouse gas emissions globally. The Convention is replete with obligations to the developing world to be undertaken by the rich countries, with special provisions to consider the specific needs of the least developed (Article 4.9, for instance). The Introduction includes the idea of *"common but differentiated responsibilities"* as between developed and developing nations, thereby setting the tone for the imposition of different obligations. Bearing in mind that the developed countries have created most of the global warming problem, this would appear to be quite fair.

The Convention also addresses the financial apprehensions of poorer nations by recognizing that

> *environmental standards, management objectives and priorities should reflect the environmental and developmental context to which they apply, and that standards applied by some countries may be inappropriate and of unwarranted economic and social cost to other countries, in particular developing countries.* (Introduction)

and also affirms that

> *the responses to climate change should be coordinated with social and economic development in an integrated manner with a view to avoiding adverse impacts on the latter, taking into full account the legitimate priority needs of developing countries for the achievement of sustained economic growth and the eradication of poverty.* (Introduction)

The requirements of developing nations are given clear recognition and are expressed:

> *all countries, especially developing countries, need access to resources required to achieve sustainable social and economic development and . . . in order for developing countries to progress towards that goal, their energy consumption will need to grow taking into account the possibilities for achieving greater energy efficiency and for controlling greenhouse gas emissions in general, including through the application of new technologies on terms which make such an application economically and socially beneficial.* (Introduction)

The Convention has a section on Principles which includes mention of the specific concerns of developing countries which must guide the Parties in implementing the Convention:

> *The specific needs and special circumstances of developing country Parties, especially those that are particularly vulnerable to the adverse effects of climate change, and of those Parties, especially developing country Parties, that would have to bear a disproportionate or abnormal burden under the Convention, should be given full consideration.* (Article 3.2)

The UNCED Conference was an attempt to merge two main considerations, environment and development, making the latter possible but with safeguards which would protect the environment for generations to come. UNCED also demonstrated the growing global awareness that unchecked development is destroying the planet and that this trend cannot continue if human beings and the millions of other species which share this planet are to survive and thrive in the next century.

However, a main focus of the South's agenda in the pre-UNCED process and in the different international fora for negotiating such conventions has been to emphasize the primacy of development—a primacy based on the fact that the poor nations have no alternatives left. Ecologically safe development was promoted as the favored choice, but if the poor could not afford that, then any development was preferable to no development. Hence, one could expect to see a reflection of that viewpoint in the Climate Change Treaty. It came in the form of this acknowledgement:

The Parties have a right to, and should, promote sustainable development. Policies and measures to protect the climate system against human-induced change should be appropriate for the specific conditions of each Party and should be integrated with national development programmes, taking into account that economic development is essential for adopting measures to address climate change. (Article 3.4)

One way of encouraging sustainable development in the South was by taking steps to reverse the serious economic inequities which consign the majority of the world's population to a life of grinding poverty. Twenty-three per cent of the world's people enjoy 85 per cent of its income, and this level of economic injustice means that over one billion people have to "survive on less than $1 a day."[191] The reaction in the Climate Change Convention was to urge Parties to address the root cause of the economic injustice which prevails:

The Parties should cooperate to promote a supportive and open international economic system that would lead to sustainable economic growth and development in all Parties, particularly developing country Parties, thus enabling them better to address the problems of climate change. Measures taken to combat climate change, including unilateral ones, should not constitute a means of arbitrary or unjustifiable discrimination or a disguised restriction on international trade. (Article 3.5)

Michel Camdessus, Managing Director of the International Monetary Fund (IMF), informed delegates at UNCED that the IMF supports more open trading and explained that this policy "is consistent with promotion of a better environment. More open trade regimes can be associated with lower levels of pollution intensity of production, because they encourage investment in more modern and less polluting technology. Moreover," he said, "trade liberalization is generally associated with increased efficiency, and with higher growth and incomes and can facilitate a shift to production technologies that use less inputs and so conserve natural resources."[192]

Realization that performance of the obligations of the Convention could be a severe financial burden for the poor nations prompted the developed countries to recognize their responsibility to help developing nations create national inventories of greenhouse gas emissions and formulate plans for dealing with the problem (Articles 4.3 and 12.1). Emphasizing the importance of *"adequacy and predictability in the flow of funds,"* the Convention also provides developed country funding for technology transfers to help developing countries implement their obligations under the Convention (Article 4.3). Further commitments call for *"the transfer of, or access to, environmentally sound technologies and know-how,"* particularly to developing country Parties (Article 4.5).

It is significant that the participation of developing countries in this Convention is made conditional *"on the effective implementation by developed country Parties of their commitments under the Convention related to financial resources and transfer of technology and will take fully into account that economic and social development and poverty eradication are the first and overriding priorities of the developing country Parties"* (Article 4.7). This provision does almost as much damage to the hopes for rapid implementation of the Convention as did the American rejection of timetables for reduction of emissions. In a blunt speech at UNCED, Dato' Abdullah Haji Ahmad Badawi, Foreign Minister of Malaysia, explained: "Already overburdened by severe economic and social pressures, the South should not be made to bear the brunt of further sacrifices. . . . After all the North . . . is . . . responsible for the bulk of greenhouse gas emissions. . . . The North must therefore help the South to accelerate its development by providing the necessary flow of resources and technology. *Without such action, the developed countries do not have the moral authority to tell the rest of the world to save resources and to stop pollution."*[193] The retreat on both fronts—to the Americans on emissions deadlines and to the South on this issue of conditionality—considerably weakened the Convention.

Developing countries are to be assisted and protected not just from the adverse effects of climate change but also from the *"impact of implementation of response measures"* (Article 4.8). Article 5 (Research and Systematic Observation) takes *"into account the particular concerns and needs of developing countries"* and encourages cooperation to improve *"their endogenous capacities and capabilities"* (Article 5[c]). The North has also committed itself to training experts, *"in particular for developing countries"* (Article 6[b][ii]).

On the administrative front, the Secretariat of the Convention is specifically directed to *"facilitate assistance to . . . developing country Parties,"* among its many duties (Article 8.2[c]). The subsidiary body for Scientific and Technological Advice established by the Convention is directed to *"provide advice . . . on ways and means of supporting endogenous capacity-building in developing countries"* (Article 9.2[d]). Developing countries will also be assisted with technical and financial support in the task of compiling and communicating information according to the terms of the Convention and to enable them to identify their technical and financial requirements in this regard (Article 12.7). By conceding the necessity for a *"transparent system of governance"* in the financial mechanism of the Convention, the North appears to have listened to the South's insistence that the mechanism have *"an equitable and balanced representation of all Parties"* (Article 11.2). The call to restructure the Global Environment Facility to make its membership universal (Article 21.3) also responds to a Southern perception of unfairness in financial mechanisms of this type. The Convention specifies that developing countries may *"avail themselves of financial resources related to the implementation of the Convention through bilateral, regional and other multilateral channels"* (Article 11.5). There is also provision for developing nations to

> *propose projects for financing, including specific technologies, materials, equipment, techniques or practices that would be needed to implement such projects, along with, if possible, an estimate of all incremental costs, of the reductions of emissions and increments of removals of greenhouse gases, as well as an estimate of the consequent benefits.* (Article 12.4)

It is apparent from the provisions of this Convention that the North has gone far to meet the demands of the South for a more equitable world economic order. The phraseology adopted in the Convention strongly endorses the concerns of the South. Hopefully, it will also help to ease the fears which developing nations had that any such treaty, with unfair provisions, could "seriously retard their oil consumption, electricity production, irrigated rice farming and livestock programmes" and that a climate convention would "set limits on almost every sphere of economic activity."[194] It remains to be seen, however, whether the performance of the North will match the promises it has made to developing nations. It is clear that the text is the product of a delicate compromise and that the entire treaty would probably have benefitted if the Americans had lent their massive weight to a stronger Convention. However, this Treaty is only a first step in a long process which will, hopefully, alleviate the consequences of global warming. To have created a document which brings such diverse nations with so many different interests into the framework plan is in itself a major achievement.

The Obligations of All Signatory Nations

One positive aspect of this rather weak Convention is that it promotes a great deal of activity, internationally, regionally, and nationally, to fulfil its requirements. This very fact makes it obvious that the Climate Change Convention cannot be overlooked or shelved; states are required to undertake certain activities within specified time frames, and developed countries, in particular, have a heavy bureaucratic burden imposed on them. States are urged to *"enact effective environmental legislation"* (Introduction), a step which is doubtless of primary importance to the whole concept of global improvement of the environment. Parties are also asked to *"promote sustainable development"* (Article 3.4).

More precisely, signatory nations have to undertake the following specific tasks:

[1.] *Develop, periodically update, publish and make available to the Conference of the Parties . . . national inventories of anthropogenic emissions;* (Article 4.1[a])

[2.] *Formulate, implement, publish and regularly update national and, where appropriate, regional programmes containing measures to mitigate climate change . . . and measures to facilitate adequate adaptation to climate change;* (Article 4.1[b])

[3.] *Promote and cooperate in the development, application and diffusion, including transfer, of technologies, practices and processes that control, reduce or prevent anthropogenic emissions . . . in all relevant sectors, including the energy, transport, industry, agriculture, forestry and waste management sectors;* (Article 4.1[c])

[4.] *Promote sustainable management, and . . . the conservation and enhancement . . . of sinks . . . including biomass, forests and oceans as well as other terrestrial, coastal and marine ecosystems;* (Article 4.1[d])

[5.] *Develop and elaborate appropriate . . . plans for coastal zone management, water resources and agriculture;* (Article 4.1[e])

[6.] *Employ . . . impact assessments, formulated and determined nationally with a view to minimizing adverse effects on the economy, on public health and on the quality of the environment, of projects or measures undertaken by them to mitigate or adapt to climate change;* (Article 4.1[f])

[7.] *Promote and cooperate in scientific, technological, technical, socio-economic and other research, systematic observation and development of data archives related to the climate system;* (Article 4.1[g])

[8.] *Promote and cooperate in the full, open and prompt exchange of relevant scientific, technological, technical, socio-economic and legal information related to the climate system and climate change, and to the economic and social consequences of various response strategies;* (Article 4.1[h])

[9.] *Promote and cooperate in education, training and public awareness related to climate change and encourage the widest participation in this process.* (Article 4.1[i])

Additionally, all Parties to the Convention have to report to the Conference of the Parties regarding the creation of national inventories of anthropogenic emissions (Article 12.1[a]) and steps taken to implement the Convention (Article 12.1[b]).

The Parties are also required to

[s]upport and further develop, as appropriate, international and intergovernmental programmes and networks or organizations aimed at defining, conducting, assessing and financing research, data collection and systematic observation, taking into account the need to minimize duplication of effort; (Article 5[a])

and to

[s]upport international and intergovernmental efforts to strengthen systematic observation and national scientific and technical research capacities and capabilities . . . and to promote access to, and the exchange of, data and analyses . . . obtained from areas beyond national jurisdiction. (Article 5[b])

As part of the process of popularizing the cause of reducing emissions, Parties agree, *"within their respective capacities,"* to promote and facilitate at the national and regional level

(i) *the development and implementation of educational and public awareness programmes on climate change and its effects;*

(ii) *public access to information on climate change and its effects;*

(iii) public participation in addressing climate change and its effects and developing adequate responses; and

(iv) training of scientific, technical and managerial personnel. (Article 6[a])

Internationally, similar efforts are aimed at

(i)   the development and exchange of educational and public awareness material on climate change and its effects; and

(ii) the development and implementation of education and training programmes, including the strengthening of national institutions and the exchange or secondment of personnel to train experts in this field, in particular for developing countries. (Article 6[b])

The Obligations of Specified Parties to the Climate Change Convention

The Convention imposes obligations which are quite weighty on all its signatories. However, certain nations, particularly developed countries, have to carry a proportionately heavier burden of the tasks involved in performance of its clauses. Essentially, the obligations imposed are commensurate with the capacities. In this sense, the Convention strikes a fair balance between Parties and recognizes the economic differences between rich and poor nations. However, the President of the Commission of European Communities thought that the obligations should have been more specific. In his speech at UNCED, Jacques Delors stated that "the European Community would have preferred the Convention on climate change to establish more precise commitments and objectives, especially for the industrialized countries."[195]

It is likely that the duties examined in the previous section may become quite onerous for some of the least developed countries. Time alone will tell how effectively the developing nations will be able to implement their responsibilities and whether they will perform or renege on their commitments. Because of an awareness of the likely economic problems associated with carrying out their duties, they have made their participation conditional on funding. The relevant clause caters heavily to the viewpoint of the South:

The extent to which developing country Parties will effectively implement their commitments under the Convention will depend on the effective implementation by developed country Parties of their commitments under the Convention related to financial resources and transfer of technology and will take fully into account that economic and social development and poverty eradication are the first and overriding priorities of the developing country Parties. (Article 4.7)

Although this provision puts responsibility on the developed countries to make it possible for the poor nations to perform their obligations under the Convention, in principle such a clause does a disservice to the cause of international environmentalism. It is nebulous and provides an escape route for any developing nation which may seek to shirk its environmental responsibilities in favor of development of its economy. Had the North not been so fragmented during the negotiating process, a firmer set of commitments for the South might have been possible. Howard Mann believes that the position of the South "becomes a cause for particular concern from an environmental perspective, when one realizes the extent to which energy-related harmful emissions, if they are wholly unaddressed, have the capacity to grow in many developing countries in order to meet the very development objectives that are being placed as a precondition to taking environmental measures."[196]

The Convention divides the non-developing nations of the world on the basis of two annexes attached to the end of the Treaty.[197] Annex II consists of the developed nations, mainly in Europe and North America. Annex I includes both these countries and the former Communist states of Eastern Europe, now termed economies in transition. Nations listed in the annexes are expected to fulfil all the obligations already discussed and take on a few more. Acknowledging the important concept of *"common but differentiated responsibilities"* (Introduction) and noting that *"environmental standards, management objectives and priorities should reflect the environmental and developmental context to which they apply"* (Introduction), the Convention expresses recognition of the

> *need for developed countries to take immediate action in a flexible manner on the basis of clear priorities, as a first step towards comprehensive response strategies at the global, national and, where agreed, regional levels that take into account all greenhouse gases, with due consideration of their relative contributions to the enhancement of the greenhouse effect.* (Introduction)

Developed countries are expected to *"take the lead in combating climate change and the adverse effects thereof"* (Article 3.1). To demonstrate that they are heading the fight against global warming, the Parties listed in Annex I[198] agree to

> *adopt national policies and take corresponding measures on the mitigation of climate change, by limiting . . . anthropogenic emissions of greenhouse gases and protecting and enhancing . . . greenhouse gas sinks and reservoirs.* (Article 4.2[a])

The obligations are unfortunately so vague as to be almost without substance. The Convention goes on to imply that the *"return by the end of the present decade to earlier levels of anthropogenic emissions of carbon dioxide and other greenhouse gases"* (Article 4.2[a]) would be consistent with its objectives. In the same article, the Convention proposes *"the aim of returning individually or jointly to . . . 1990 levels* [of] *anthropogenic emissions of carbon dioxide and other greenhouse gases"* (Article 4.2[b]). Reuters News Agency said of these provisions, "Dubbed 'constructive ambiguities' by negotiating committee Chairman Jean Ripert of France, the two paragraphs were necessary to get the United States to agree."[199] Michael McCarthy of the *Times* (London) reported that the provisions are "couched in convoluted language . . . a guideline, rather than a legal commitment."[200]

These constructively ambiguous obligations require nations specified in Annex I to report within six months of the entry into force of the Convention and periodically thereafter to the Conference of the Parties on their performance of these duties (Article 4.2[b]). There is a requirement of Annex I nations for detailed descriptions of policies and measures adopted to implement the means for reducing greenhouse gases and for a specific estimate of their impact (Article 12.2[a]) and [b]). The nations listed in Annex I have also to

> coordinate as appropriate with other such Parties, relevant economic and administrative instruments developed to achieve the objective of the Convention; (Article 4.2[e][i])

and

> identify and periodically review . . . policies and practices which encourage activities that lead to greater levels of anthropogenic emissions of greenhouse gases. (Article 4.2[e][ii])

Despite these provisions for review, there was serious disappointment because, in the words of the Environment Minister of Kenya, "no specific targets were agreed upon."[201] Dr. Kofi N. Awoonor, Ambassador of Ghana to the United Nations, voiced the apprehensions of many delegates when he said that "without any stipulated targets for controlling emissions, we fear the Convention may not produce the desired effect of arresting the process of current global warming."[202] The Italian Minister for Environment, Giorgio Ruffolo, believed that the absence of precise targets and timetables impaired the Convention.[203] The Prime Minister of Malaysia, Dr. Mahathir bin Mohamad, was more blunt, stating that the U.S. position "rendered the agreement inequitable and meaningless."[204]

The developed countries listed in Annex II undertake to assist developing countries financially to fulfil the obligations to prepare national inventories and environmental plans to implement the provisions of the Convention (Article 4.3). Additionally, developed countries will *"provide such financial resources, including . . . the transfer of technology, needed by the developing country Parties to meet the agreed full incremental costs of implementing"* the various obligations undertaken by all Parties as enumerated above (Article 4.3). Technology transfer to the developing world forms a significant aspect of the South's hope for environmentally sustainable development and is accepted as a commitment placed on the developed nations:

> *In this process, the developed country Parties shall support the development and enhancement of endogenous capacities and technologies of developing country Parties.* (Article 4.5)

Because of the economic problems being faced by the former Soviet Union and its Eastern European neighbors as they convert from a Communist to a free-market economy, the Convention on Climate Change allows them *"a certain degree of flexibility"* in performing their obligations with respect to reducing emissions, even with respect to the constructively ambiguous goals of returning to the 1990 levels specified for developed countries (Article 4.6). Victor I. Danilov-Danilian, Minister for Ecology and Natural Resources of the Russian Federation, explained the problems facing his country: "The state of environment in Russia is alarming. Natural systems have already been seriously damaged by annual per capita emissions into the air of 130 kg. of pollutants, spoils, dumping places for wastes, covering dozens of thousands of square kilometres, polluted waters of rivers and lakes. Only the sheer size of the territory of Russia still saves its nature from total collapse."[205]

Annex II countries are also required to provide detailed descriptions of measures taken to assist developing country efforts to deal with climate change (Article 12.3).

Though the Convention lacks teeth, it has promoted and generated considerable activity on the scientific, academic, and bureaucratic levels. One hopes that this activity will increase public awareness about the problem of emissions, facilitate popular approval of active implementation of the Convention's objectives, and keep this subject at the forefront of environmental concerns.

National Sovereignty Considerations in the Convention

It has become almost a ritual now for international law instruments to pay formulary homage to the concept of nationalism by acknowledging the sovereignty of nations. This type of provision appears frequently in international declarations and conventions and has become part of the usually accepted baggage which governments bring with them to international negotiation processes. The Climate Change Convention was no exception:

> *States have, in accordance with the Charter of the United Nations and the principles of international law, the sovereign right to exploit their own resources pursuant to their own environmental and developmental policies, and the responsibility to ensure that activities within their jurisdiction or control do not cause damage to the environment of other States or of areas beyond the limits of national jurisdiction.* (Introduction)

and this reiteration of the importance of sovereign rights:

> Reaffirming *the principle of sovereignty of States in international cooperation to address climate change.* (Introduction)

In view of the fact that the Parties to the Convention are nation states, the reiteration of sovereign rights may appear to be somewhat redundant and unnecessary. After all, it is as nations that they will participate in the efforts to curb greenhouse emissions. It is national efforts which will implement or fail to implement the provisions of the Convention.

The first provision quoted above balances the concept of sovereign rights with the idea of national responsibility. The second appears to cater to the apprehensions of countries which have become independent since the end of the Second World War and which still view the former imperial powers—the North—with considerable suspicion in any international forum. However, as we have seen, it is not just the recently independent states which cling to the trappings of nationalism in the international arena. The United States of America under George Bush adopted a very nationalistic stance both at the Rio Summit and during the negotiations which preceded it, including the meetings which produced the Climate Change and Biodiversity Conventions.

The problem with nationalism, from an environmental perspective, is that it is the single greatest obstacle to rapid, effective change. Frequently, environmental problems are global problems. Their solution demands a wider, universalist outlook which transcends the narrow framework of exclusivity that national considerations impose on governments. Political leaders come to environmental conferences clutching their sovereign rights like security blankets. They fight to ensure that the end product of the global discussion is most pleasing to, or, at the least, least unpleasing to, their particular national interest. This is all very well, but when over 150 countries are playing this type of game the consequences can only be vacuous, nebulous pronouncements which are so devoid of real meaning as to become the despair of environmentalists everywhere. This was to some extent the ultimate fate of the Climate Change Convention. This is why many of its provisions tend to be general rather than specific, loose rather than precise, and indeterminate rather than definite. John Major, Prime Minister of the United Kingdom, tried to take a positive approach to the Convention when he said:

> [W]hat we are seeing here is something quite unique: over 100 nations, 100 Heads of Government, 170-odd nations actually here seeking to reach agreement. That was never going to be easy but what it actually has demonstrated is two things: firstly the importance of the occasion; and secondly the fact that so many nations are prepared to come together, often swallowing their own domestic national interests, to try and reach an agreement on matters of importance to the environment.[206]

Unfortunately, the problems of the environment are not merely global but also long-term. The careers of most politicians are not lengthy enough to allow them to lead their nations in the sustained commitments of money and effort required. Hence, while environmental problems are widespread and often, as with greenhouse gases, all encompassing, the solutions can be applied only in a patchy and piecemeal fashion. One can only hope that future trends will be increasingly in the direction of regional and international endeavors, with less emphasis on the primacy of country-by-country performance, however significant that may be.

The Impact of Global Warming on Low-Lying Areas

One ironic and tragic aspect of the crisis of greenhouse gas emissions is that those parts of the world least responsible for creating the global warming problem will be the first to suffer its horrifying consequences. As the rise in ocean levels is scientifically associated with global warming, all low-lying coastal and island regions are likely to be endangered by it. There are indications that the problem is already being faced by a number of states. One of the positive public relations achievements of the UNCED process was to alert international attention to the plight of the inhabitants of these areas. For the people of these regions, remedial measures are compellingly and urgently needed. Unfortunately, most of these areas lie in the developing world; many of them, desperately poor, can command only a limited portion of the world's environmental attention. Their situation now is that of a disaster waiting to happen. When it does, on a large enough scale, the world may pay attention to the fact that global warming is not a matter which concerns just international lawyers, scientists, and politicians. The greenhouse gases we emit into the atmosphere will ultimately first destroy the most vulnerable physically (because they are poor) and politically (because they lack the clout to pressure the world's polluting nations to rectify the situation).

The likely scenario for island nations was graphically described at UNCED by Resio S. Moses, Secretary of the Department of External Affairs for the governments of Micronesia:

> Virtually all oceanic islands and low-lying coastal areas stand to suffer first consequences of human-induced climate change. Total inundation caused by sea-level rise is, of course, the ultimately terminal event for islands, but long before that, living conditions on the islands will have become unbearable due to increased storm activity, destruction of reefs, land and beach erosion and disappearance of fresh water and foodstocks. Options for mitigating or adapting to these effects are quite limited for the low-lying islands and atolls. The ultimate defensive measure, namely relocation of the population, may save lives at the time, but at a tragic cost. History shows that relocation means the end of the cultures involved.[207]

Moses called for "drastic reductions in current levels of carbon dioxide emissions by the developed countries."[208] Emission reduction and provision for additional resources by industrialized countries were the demands of the Governor-General of Papua New Guinea, who explained that his nation, consisting of over six hundred islands, has vast areas of wetlands which would be severely affected by sea level rise and by cyclones.[209]

Teatao Teannaki, President of the Republic of Kiribati, informed delegates at the Earth Summit that his country "consists of 33 low and flat coral atolls, surrounded by a vast area of ocean. The ocean is encroaching on land, as land retreats." He continued, "This is our share of the cost of industrialization and economic development. It is clearly unproportional to our negligible share, if any, in causing global environmental problems." The President asked, "Should we continue to bear this cost until we are wiped out?"[210] Kinza Clodumar, Minister of Finance for the Republic of Nauru, in the central Pacific, eloquently outlined the impact of environmental catastrophe when he spoke to UNCED: "We are . . . small, vulnerable island states, entrenched on the front lines of the ecological crisis. In the event of global ecological collapse, we will be the first to go, but we will not be the last. . . . It is said that no man is an island unto himself, but given our shared ecological fate, we propose that all countries would do well to consider themselves as an island."[211] The President of the Republic of the Marshall Islands, Amata Kabua, was equally apprehensive about the fate of his island nation, which has an average elevation of only two meters above sea level. As he explained, "any significant sea level rise will be catastrophic to the Marshall Islands, the homeland of the Marshallese people for thousands of years."[212] Tom Kijiner, the country's Minister for Foreign Affairs, dramatically reminded delegates that the world "faced an intricate ecological time bomb" and warned that "sea level rise could annihilate the Marshall Islands as effectively as a nuclear bomb."[213] The urgency of the situation was emphasized by Tofilau Eti Alesana, Prime Minister of Western Samoa, who informed delegates about the severity of recent tropical cyclones and cautioned: "For us, it is not just a question of reaching or maintaining sustainable levels of development. It is not just a question of hunger and poverty. For us, it is a question of pure survival."[214] This sentiment was echoed by delegates from a variety of nations like Jamaica, Sri Lanka, Cyprus, and Bangladesh. The Bangladeshi Foreign Minister reminded UNCED that global warming and sea level rise "would lead to a reduction of an already minimal land-person ratio and an increased pressure on natural resources."[215]

The delegate from Vanuatu did not hesitate to lay blame for the situation facing Pacific island countries. He pointed out that "Vanuatu has been compelled to assume a significant portion of the hidden costs of the conspicuous consumption of people in other regions whose standard of living—although not necessarily their quality of life—is considerably higher than that of our own people."[216]

The vulnerability of these nations was not ignored by the United Nations, largely because of the action taken by these states to influence international opinion. In November 1989 the small states held a conference in the Maldives on the issue of sea level rise and produced the Male Declaration on Global Warming and Sea Level Rise.[217] This statement highlighted their specific predicament; called for an international response from the industrialized countries, which have a "moral obligation" to initiate remedial action; and proposed negotiations for a framework convention on climate change.[218] The United Nations General Assembly recognized the significance of the issue, endorsed it in a resolution of December 1989,[219] and recommended that this matter be considered during discussions for the formulation of a climate change convention.[220] Research on and monitoring of the impact of global warming and sea level rise on coastal zones was proposed by the Second World Climate Conference, which met in Geneva between 29th October and 7th November 1990.[221] In its Ministerial Declaration, the Conference proposed a number of measures to deal with this problem. "Such response strategies include phasing out the production and use of CFCs, efficiency improvements and conservation in energy supply and use, appropriate measures in the transport sector, sustainable forest management, afforestation schemes, developing contingency plans for dealing with climate related emergencies, proper land use planning, adequate coastal zone management, review of intensive agricultural practices and the use of safe and cleaner energy sources."[222]

The Governing Council of the United Nations Environment Programme also considered global warming and sea level rise during its sixteenth session. The council highlighted its concerns by acknowledging that "the First Assessment Report of the Intergovernmental Panel on Climate Change, adopted in August 1990," had predicted that sea level rises "between three to ten centimetres a decade under the business-as-usual emissions scenario" could be expected and "that, even if greenhouse gas emissions were reduced, there would still be a continuing need to address sea-level rise."[223] Accordingly, UNEP urged governments to address the issue of "vulnerability to sea-level rise."[224]

The Climate Change Convention responds to this flurry of international activity by making the United Nations General Assembly Resolution of 1989 on sea level rise a point of reference (Introduction). The Convention goes on to recognize the vulnerability to climate change of various areas, including *"low-lying and other small island countries"* (Introduction), and reiterates the needs of those areas *"that are particularly vulnerable to the adverse effects of climate change"* (Article 3.2). The commitments entered into by Parties include the development of plans for coastal zone management (Article 4.1[e]); Annex II states are asked to *"assist the developing country Parties that are particularly vulnerable to the adverse effects of climate change in meeting costs of adaptation to those adverse effects"* (Article 4.4). Further provisions call for funding, insurance, and technology transfer, with specific reference to vulnerable regions, including small island countries and low-lying coastal areas (Article 4.8 [a] and [b]). As many of these island states are barely developed, the Parties *"shall take full account of the specific needs and special situations of the least developed countries in their actions with regard to funding and transfer of technology"* (Article 4.9).

Although the obligations to assist these vulnerable areas are quite vague, the issue of their particular predicament was expressed and reiterated in an international convention. Although global warming will have an impact on all countries, its adverse effects will unfortunately be felt first by those least equipped to cope with them. The plight of these small, poor, low-lying island and coastal nations simply cannot be ignored; nor can their ancient cultures and civilizations be allowed to be swept away some day by oceans which have risen because we in the developed world want to continue to enjoy our comfortable but environmentally destructive lifestyle.

Consideration of Other Vulnerable Areas

It is important to emphasize that there are a number of areas in the world which are vulnerable to environmental degradation. It is impossible in this chapter to consider them all; however, the United Nations and the international community have been involved in producing the usual plethora of verbal and written pronouncements on these problems. The consideration of low-lying island and coastal states is only one example of the type of concern which has been generated worldwide. While acknowledging the plight of the insular states, the Convention on Climate Change also paid some attention to the catastrophe of desertification, particularly in Africa (Introduction, Article 4.1[e] and Article 4.8[e]). Mountain ecosystems were also considered (Introduction and Article 4.8[g]), along with marine ecosystems (Introduction), regions prone to floods and drought (Introduction, Article 4.1[e] and Article 4.8[e]), and forested areas (Article 4.8[c]). All the problems of so many regions cannot yet be blamed on global warming. However, since the Climate Change Convention is a consensus-driven document a number of items were probably included as a form of international acknowledgment of concern, particularly as the science on climate change may well verify its linkage with all these various environmental problems which have already begun to affect the lifestyle of millions of people.

The Efforts of the Oil-Producing Countries

It has been suggested, with some justification, that the "chance that the climate treaty will significantly change the world's output of fossil fuels over the next century is extremely slender."[225] Dr. Hans Blix, Director-General of the International Atomic Energy Agency, reminded delegates at UNCED that approximately 90 per cent of the planet's commercial energy is derived from fossil fuels.[226] John Wakeham, Energy Secretary for the United Kingdom, deemed the reliance on fossil fuels for 80 percent of the world's energy requirements "the most fundamental problem facing civilisation."[227]

The oil-producing nations, many of them wealthy if not yet highly industrialized, mounted an effective and vociferous campaign against resort to alternative energy sources and indeed even against the concept of energy efficiency. Because of the linkage between the burning of fossil fuels, greenhouse gas emissions, and global warming, and because a large part of the developed world relies heavily on oil as an energy source, the emphasis on curbing emissions had to consider this connection and deal with it by formulating some firm principles for implementation. Helga Steeg, Executive Director of the International Atomic Energy Agency, suggested that removing subsidies on fossil fuels worldwide "would achieve dramatic reductions in CO_2 emissions."[228] The issue is primary and was of crucial importance for such UNCED documents as Agenda 21 and the Climate Change Convention.

The battle to defend fossil fuels was fought on all fronts. The Arab oil-producing countries saw a serious threat to their one great resource; "[d]uring the preparatory negotiations for the Earth Summit, Arab countries, acting on behalf of the Organization of Petroleum Exporting Countries, protested against what they regarded as . . . overemphasis on energy efficiency and fossil-fuel reduction."[229]

Rashid Abdullah al-Noaimi, Minister of Foreign Affairs of the United Arab Emirates, emphasized that "the results of a good deal of scientific research confirm that oil and natural gas contribute less to carbon emissions and the resultant air pollution than other sources of energy such as coal," and he cautioned against "focusing on a single source."[230] The oil producers' determined campaign was headed by the Saudi Arabian delegation, which sought removal of references "to policies promoting energy efficiency and to the need for research into alternatives to fossil fuels. While fossil fuels are the major source of gases that cause global warming, their sale is the main source of income for oil-producing countries, such as Saudi Arabia."[231] According to James Rusk, reporting in the *Globe and Mail*, the Saudis also proposed that if fuel consumption was curbed, they would have to be compensated by the developed countries for lost sales.[232]

The case for the oil producers was made eloquently and somewhat less confrontationally by Dr. Subroto, Secretary-General of the Organization of

Petroleum Exporting Countries. He explained the position of OPEC to delegates at UNCED:

> We in OPEC, although producers and exporters of a fossil fuel which has been the target of certain interests because of carbon dioxide emissions, share the general concern over the apparent deterioration of the air we breathe and the water we drink, to name only the two most important life-sustaining elements. Indeed, as developing countries, we have solid grounds for being even more concerned than the nations of the North, from whose industrialization process in the last 250 years or so, the whole problem of environmental degradation stems.[233]

Dr. Subroto went on to explain that oil "accounts for more than 90 per cent of total export earnings" for developing country oil producers and exporters. "These revenues," he told the delegates at UNCED, "are our major source of foreign exchange, capital formation and for promoting growth and development. . . . In short, if oil is in trouble, the economies of the oil exporters are in danger."[234] Dr. Subroto then urged UNCED against adopting any drastic measures "that would penalize oil producers before there is substantive evidence to show that these measures are right and necessary."[235] The OPEC agenda was to ensure that any measures adopted by the world community would be "compatible with continued economic growth in both developed and developing countries."[236] Some encouragement was given to the exporters' position by Ali Hassan Mwinyi, President of Tanzania, who argued that "[r]eduction of fossil fuel consumption in developing countries would not be a viable option in the short-term. Such a move would only mean larger numbers of users of wood fuel and other biomass fuel, with obvious implications for deforestation and soil degradation."[237] It was apparent that the campaign to influence world opinion on the issue of fossil fuel had succeeded. Dr. Subroto expressed his satisfaction with the acknowledgment of the concerns of oil exporters during the negotiating process, "reflected in the text of . . . [the] Convention," and hoped "that its implementation will be carried out in the same spirit."[238]

The Climate Change Convention recognizes the concerns of the oil producers not only by what is stated in its provisions but, ironically, also by what is excluded. In its Introduction, the Convention notes

> *the special difficulties of those countries, especially developing countries, whose economies are particularly dependent on fossil fuel production, use and exportation, as a consequence of action taken on limiting greenhouse gas emissions,* (Introduction)

and concedes that

> *[m]easures taken to combat climate change, including unilateral ones, should not constitute a means of arbitrary or unjustifiable discrimination or a disguised restriction on international trade.* (Article 3.5)

This point of detriment to international trade is of considerable importance to oil producers whose revenues depend on the export of one commodity. The Convention also asks Parties to *"give full consideration to what actions are necessary"* (Article 4.8) to meet the specific needs and concerns of *"[c]ountries whose economies are highly dependent on income generated from the production, processing and export, and/or consumption of fossil fuels and associated energy-intensive products"* (Article 4.8[h]).

Such needs could arise from *"the adverse effects of climate change and/or the impact of the implementation of response measures"* (Article 4.8). The provision is reiterated:

> *The Parties shall . . . take into consideration in the implementation of the commitments of the Convention the situation of Parties, particularly developing country Parties, with economies that are vulnerable to the adverse effects of the implementation of measures to respond to climate change. This applies notably to Parties with economies that are highly dependent on income generated from the production, processing and export, and/or consumption of fossil fuels and associated energy-intensive products and/or the use of fossil fuels for which such Parties have serious difficulties in switching to alternatives.* (Article 4.10)

The Convention also specifies that *"in order for developing countries to progress towards that goal* [sustainable development], *their energy consumption will need to grow taking into account the possibilities for achieving greater energy efficiency"* (Introduction). There is no doubt that the Convention bends over backward to accommodate the demands of the oil producers. It could be argued that their lobbying, along with the intense pressure of the United States of America against firm timetables, had the greatest impact on the provisions which were finally accepted by all members. The result, from the environmental perspective, is a weak, toothless Convention with no clear measures encouraging reduction of fossil fuel consumption, no clarion call to all nations to switch from oil to other, less-polluting sources of energy. The inference is that reducing CO_2 emissions will inevitably lead to some reduction in the use of fossil fuels, but the wording of the Climate Change Convention reflects the political reality of a major campaign by one group—the oil producers—which succeeded in influencing all Parties to endorse its vested interest at the expense of the environmental cause which ought to have been at the forefront of international concern. Ironically, the oil-producing nations were not satisfied with the Convention. For Kuwait, Saudi Arabia, the United Arab Emirates, Iran, and Iraq, "the Convention puts too much emphasis on CO_2 as being one of the causes of the deterioration of the atmosphere and the climate."[239] On a more positive note, Olof Johansson, Minister of the Environment for Sweden, argued that the very existence of the Convention lent urgency to the development of new energy sources "in order to lessen dependence on fossil fuels."[240]

The Financial Mechanism of the Climate Change Convention

One of the biggest challenges facing Northern governments in their quest to reduce greenhouse gas emissions is the factor of cost. As many developed countries are only now crawling out of a deep recession, the problem of being able to afford environmental solutions is very real and compelling. Because there was a perception that environmental remedies could lead to higher unemployment, the Administration of President George Bush was wary of involving the United States in financial commitments which could endanger the American economy. Unfortunately, the mass of evidence pointing to the fact that environmentalism could easily be encouraged to become the big job-creation program of the future was apparently overlooked.

A related problem springs from the firm determination of the developing countries to make their participation in greenhouse gas reduction conditional on monetary and scientific assistance from the developed nations. Hence, the Climate Change Convention involves two enormous financial commitments for the North, in funding both Northern and Southern participation in measures to reduce global warming. The problem of affordability is compounded by the nationalistic sensitivities of developing nations. Many of them want the assistance without any appearance of political or even environmental strings if these appear to infringe on their sovereign rights. The cause of alleviating the problem of global warming has to take all these factors into consideration and yet come up with the funding to deal with the issue so that the whole world can benefit. As James Rusk of the *Globe and Mail* commented, "One of the toughest fights at the Earth Summit is about who is going to pay to save the Earth."[241]

Another complicating factor is the need for development in the South— rapid development if the populations are to be given a bare-minimum standard of living and some hope of upward mobility, some opportunity to acquire consumer goods, some kind of chance to educate their children and basically improve their lives. Human civilization has been built on certain premises, one of which is the notion that each generation seeks to improve conditions for the next. This instinct is deeply rooted in human nature and could well be the mainspring of much that is progressive and decent in civilization. However, in recent years, because of burgeoning world population and declining resources, this instinct has encountered obstacles in every country on this planet. Citizens of both developed and developing nations have found themselves consistently hindered in their search for economic and social betterment. It could be argued that Bill Clinton was elected president in 1992 largely because people wanted to change and reverse this negative trend which was becoming so pervasive, even in the world's only superpower. If Americans have sensed the urgent need to change the downward spiral of their lives, how much more desperate must be the feelings of the millions of inhabitants of developing nations who are so much poorer and who live so precariously on the brink of life-threatening destitution and poverty. In promoting the cause of environmentalism, one has to consider the very real and compelling human interests of millions and weigh these carefully against the equally vital need to protect our planet—our only home—from man-made degradation. The balance will be difficult to achieve, and the alternatives are grim, no matter what one does. These problems are compounded by the mutual misunderstandings which prevail between North and South, similar in form to those which prevailed between East and West until the end of the Cold War. There is the same confrontational rhetoric, the same defensiveness. Nations do not talk to each other in comprehension of each other's dilemmas but mouth cliches about each other, cliches which oversimplify issues that are complex.

During the process before UNCED, including the negotiations leading up to the Convention on Climate Change, it was expected that the financial problems of both North and South would become a significant topic of discussion internationally and within the governments of participating nations. As Amy Kaslow reported in the *Christian Science Monitor*, "Many developing countries argue that they need help from the developed world to finance environmental protection. And some say that environmental protection should take a back seat to their plans for economic development, as it has in developed countries until recently."[242] However, there are those who argue that hurling money at the South will not solve the problem. Patrick McCully suggests in the *Ecologist* that a huge increase in aid to fight global warming "ignores the fact that aid has left a legacy of neo-colonialism, debt dependency, corruption and failure. The history of the transfer of western technologies to the Third World has been similarly dismal. Emphasis on the need for transfers of money and machinery to the Third World obscures the urgent need for radical changes in First World consumption patterns and global economic and political structures."[243] Patricia Adams, in the *Globe and Mail*, emphasizes another aspect of this problem: "Third World governments want money, and to get it they are prepared to hold hostage their people and the environment upon which their people depend. The Western governments—reeling from often-justified criticism of their own environmental records—want to buy the silence of their critics. But throwing money at the problem only promises to compound the damage."[244] Whether one believes that the North owes the South because the North has created the environmental degradation and ought now to pay to clean it up, or whether one believes that the South is eco-blackmailing the North, it is obvious that the financial aspect of international environmental treaties is inevitably both complex and fraught with potential conflict. An understanding of this background and the fact of polarized opinion is essential when analyzing the financial provisions of the Climate Change Convention.

The Financial Mechanism is contained in Article 11 of the Convention, which defines *"a mechanism for the provision of financial resources on a grant or concessional basis, including for the transfer of technology"* (Article 11.1), which *"shall function under the guidance of and be accountable to the Conference of the Parties, which shall decide on its policies, programme priorities and eligibility criteria related to this Convention. Its operation shall be entrusted to one or more existing international entities"* (Article 11.1). The international entity, on an interim basis, is named the Global Environment Facility (GEF) (Article 21.3).

The GEF, created in November 1990 for an initial three-year term,[245] is managed by the United Nations Development Programme, the United Nations Environment Programme,[246] and the International Bank for Reconstruction and Development (a component of the organization which is termed the World Bank).[247] The GEF was created as a program of loans and grants "to help developing countries deal with environmental problems."[248] Its main functions are to deal with ozone depletion, global warming, water management on an international level, and conservation in the area of biological diversity.[249]

Both before and during the Rio Summit, the GEF was vigorously criticized in both the North and the South for representing the interests of donor countries rather than the needs of recipient nations. Greenpeace was reported in the *Irish Times* (Dublin) as having alleged that the GEF was "being used to 'greenwash' much larger bank projects which have a detrimental impact on the environment of developing countries."[250] Michael Gucovsky of the United Nations Development Programme responded that the GEF was being "driven by the people who are the beneficiaries of its projects." Colin MacKenzie reported in the *Globe and Mail* that the World Bank insisted that development is "consistent with good environmental practices."[251] Yet its detractors were many and they were very vocal. On the subject of global warming, Susan George, Associate Director of the Transnational Institute in Amsterdam, warned that "[b]ank lending in the energy sector would overwhelm anything the GEF might contribute to reducing greenhouse-gas emissions. The great bulk of the bank's loans will continue to be business as usual, although the GEF may tack a few environmental tails on some quite vicious dogs."[252]

The developed countries, which assumed the role of major aid donors in the Climate Change Convention, felt more comfortable with the GEF—they control its parent organization, the World Bank. Speaking for the European Community, Carlos Borrego, Portugal's Environment Minister, explained that an appropriately adapted GEF "should play a leading role as the multilateral funding mechanism."[253] The South had fears about the World Bank in part because at the time of UNCED the United States of America, of the 159 member countries, controlled the largest bloc of shares in the World Bank.[254] For Northern donors, this fact ensured that the funds disbursed to the GEF would be spent for the benefit of the recipient populations. Yet this very idea irked some developing nation governments, which viewed it as presumptuous and alleged that the Bank's projects reflect the ideological priorities of the United States.[255]

Nations of the South argued strenuously for more equitable representation in these international financial institutions. Alhaji Sir Dawda Kairaba Jawara, President of the Republic of the Gambia, articulated developing country sentiment in calling for a restructuring of the GEF, with expansion of its financial base to "allow financing of more diverse projects."[256] The consensus which finally emerged was probably encouraged by the forty-third meeting of the International Monetary Fund and the World Bank Development Committee in Washington, D.C., in April 1992.[257] Participating member nations called for "a reformed GEF to serve as the leading mechanism for new and additional UNCED funding."[258]

The Climate Change Convention concedes that "*the Global Environment Facility should be appropriately restructured and its membership made univer-*

sal" (Article 21.3) and that the *"financial mechanism shall have an equitable and balanced representation of all Parties within a transparent system of governance"* (Article 11.2). There are provisions to reconsider and review the funding mechanism and to review the amounts required for undertaking the tasks specified in the Convention (Articles 11.3 and 11.4). "The Conference of the Parties will determine the policies, program priorities and eligibility criteria relating to the provision of . . . resources."[259]

The Americans would have preferred much tighter control of environmental funding, with the World Bank having a decisive voice in GEF projects. But in this matter, as James Rusk reported in the *Globe and Mail,* "the tide seem[ed] to be running somewhat against the United States."[260] The attitude of the South with respect to the entire gamut of environmental issues was to prefer a "greenfund that would provide aid in the name of the environment, or an increase in funds for current development programs."[261] Mostafizur Rahman, Foreign Minister of Bangladesh, explained the green-fund concept:

> It is our belief that the new and additional finances would be best administered by a separate Green Fund which should be used to implement the activities approved by this Conference. The Fund should be democratically governed with equal voice for all members in setting priorities, identifying projects, and taking decisions on disbursements. While the Global Environment Fund (GEF) can be an appropriate mechanism to fund global programmes, it cannot address national problems, for which the separate Fund will be essential. [262]

The South also wanted to ensure that the financing for the Climate Change Convention would be in addition to "existing flows of Official Development Assistance. Most developed countries, including Japan, U.S., Germany, Canada and Australia, were not prepared to provide this assurance and, consequently, efforts to define 'new and additional' were set aside."[263]

The Climate Change Convention, in conceding to neither extreme, represents a balance based on international consensus. Given the environmental consciousness which has now swept all nations and such institutions as the World Bank, this balance may just be quite effective in combating global warming.

Conclusion

> Over the next half century, the damage caused by global warming may be quite modest. Over the much longer term, it may be greater, although not

(at least for rich countries) catastrophic. The world can react to this prospect in two ways: it can take action to slow down climate change, or it can
wait until it happens and then adapt. The balance between these courses
will depend on how the costs of action compare with the costs of wait-and-
see.[264]

The Framework Convention on Climate Change takes a middle course between the two alternatives presented above. It takes a number of initial steps
to study, categorize, and reduce greenhouse emissions, yet it does so tenta-
tively, without the firm commitments or guidelines which alone could have a
dramatic impact on the problem of global warming. Although the result is not
"business as usual," the measures outlined in this chapter will not dramatically reverse the harmful effects of climate change unless individual states
take firm steps to curb emissions with specific targets and deadlines. The
European interest in following this direction is very encouraging. Soon after
the Earth Summit, the people of the United States of America elected a
dedicated environmentalist, Al Gore, as their Vice-President. At the time,
there was reason to believe that this was likely to result in a change of heart
by the world's leading producer of CO_2 emissions. On 22nd April 1993,
President Bill Clinton committed his nation "to reducing our emissions of
greenhouse gases to their 1990 levels by the year 2000."[265] Nicholas Lenssen
of the World Watch Institute suggested that "[i]t would be very easy to . . .
reduce emissions below 1990 levels with strong leadership on the issue in
Washington."[266] Maurice Strong, Secretary-General of UNCED, expressed
his expectation that President Clinton and Vice-President Gore would "raise
sustainable development to a higher level in international politics."[267] However, even with the encouragement of Washington, "a long-term, complex
problem such as global warming will not be solved overnight. What is important initially is to establish a meaningful process for addressing the issue."[268]
As Thorbjorn Berntsen, Norway's Environment Minister, explained, implementational measures "must cover all climate gases, address sources and
sinks and all economic sectors."[269]

The Secretary-General of the United Nations tried to view the Convention
in a positive light. While admitting that "[t]he initial level of commitment is
not as high as many would have wished," Boutros Boutros-Ghali suggested
that "a low level of threshold should maximize participation—which is one
condition for effectiveness. And the process of policy review should improve
commitments over time."[270]

If implemented by the developed nations, such ideas as the carbon tax could have an impact on the problem, although the economic consequences would have to be balanced carefully against the intended environmental benefits. Flavio Cotti, Head of the Swiss Delegation to UNCED, emphasized that "[i]n order to avoid distortions affecting economic competitivity it is of fundamental importance to introduce such measures simultaneously in all industrialised countries, or at least in a considerable number of them."[271] Nations like Sweden have long supported the carbon tax idea as part of an integrated process of environmental clean-up;[272] the Netherlands has supported the introduction of a CO_2 tax in the European Community.[273] The carbon tax proposal has its supporters and detractors, and it is too early to tell whether the idea will become popular enough to be accepted by the business communities and populations of North America. Helga Steeg, Executive Director of the International Atomic Energy Agency, has suggested that "[e]nergy markets work best when they are competitive and where prices reflect costs. Internalizing environmental costs, through economic instruments such as taxes, is the best way to use the market's strengths." But, she concedes, "we must be able to estimate these costs. For many environmental problems, and especially for climate change, we are far from knowing all of the costs."[274]

The level of activity on the issue of combating global warming did not diminish with the signing of the Convention on Climate Change. In his report of 1st June 1992 (just prior to UNCED), Chairman Jean Ripert of the Intergovernmental Negotiating Committee emphasized that "what is essential now is for States to keep up the momentum of the global partnership of nations created in the negotiation of the Convention."[275] Representing the European Community, Carlos Borrego, Environment Minister of Portugal, confirmed the Community's interest in contributing to the preparation of Protocols covering specific issues, "especially the limitation of CO_2 emissions."[276]

Meanwhile, some European states made initial announcements concerning their specific plans. The United Kingdom committed itself to reducing emissions of carbon dioxide and of other greenhouse gases to 1990 levels by the year 2000, "provided others do so as well."[277] Germany set its target at a 20 to 30 per cent reduction in CO_2 emissions by the year 2005;[278] Denmark, at a 20 per cent reduction by 2005 (compared to 1988 levels);[279] and the Netherlands, at a reduction of 3 to 5 per cent in 2000 (compared to 1990 levels).[280] In a public statement, the European Community confirmed its "target to reduce CO_2 emissions to 1990 levels by the year 2000."[281]

Richard E. Benedick, formerly of the United States Department of State, commented that the Earth Summit "should not be judged by the immediate results but by the process it sets in motion."[282] Michael Howard, Secretary of State for Environment of the United Kingdom, optimistically declared that the Convention would "not be the international community's last word on this subject."[283] Chancellor Helmut Kohl of Germany invited delegates to his country to attend the first follow-up conference on climate change.[284]

This chapter began with the suggestion that "[t]he road to environmental hell is paved with good intentions."[285] We have analyzed in considerable detail a document which is replete with good intentions, a Convention which in some ways charts a new course in environmental considerations. Despite its many imperfections, its significance as a marker in international environmental law cannot be underestimated and ought not to be underrated. Having created the legal document, we have now to see whether the provisions will result in progressive action and implementation or remain within the realm of good intentions. Eiour Gounason, Iceland's Minister for Environment, made the point eloquently: "Our efforts will not be measured by the number of pages that pour out of this conference, not by the words but by our deeds."[286] Although we are not absolutely certain that global warming could result in environmental hell, can we afford simply to stop at "good intentions" and take a chance with the future of our children and of our planet?

Notes

1. *New York Times*, 14th June 1992, 10:2.

2. Statement by Domingo L. Siazon, Jr., Director-General of UNIDO, UNCED, Rio, 4th June 1992.

3. For text see United Nations Document, Convention on Biological Diversity, 5th June 1992, reprinted as Appendix 5 to this book.

4. United Nations Document A/AC.237/18 (Part II)/Add.1, reprinted as Appendix 5 to this book.

5. *Newsweek*, 1st June 1992, 33:1.

6. Ivan L. Head, *On a Hinge of History* (Toronto: University of Toronto Press, 1991), 91.

7. *Ibid.*, 92.

8. *Ibid.*

9. *Globe and Mail*, 3rd June 1992, A6:1.

10. Thomas H. Shillington, *Our Changing Climate: Building the Global Response* (Environment Canada: Prepared for Climate Change Convention Negotiations Office, Atmospheric Environment Services, 1991), 3.

11. *Christian Science Monitor*, 30th June 1992, 19:1.

12. Head, *On a Hinge of History*, 92.

13. Shillington, *Our Changing Climate*, 3.

14. *Globe and Mail*, 7th May 1992, 1:4–5.

15. George Mitchell, *World on Fire* (New York: Scribner's, 1991), 225–26.

16. Head, *On a Hinge of History*, 92.

17. *Christian Science Monitor*, 27th February 1992, 19:1.

18. *Globe and Mail*, 4th June 1992, A6:5.

19. Washington Post, "Global Warming May Be Just Lot of Hot Air," *Vancouver (B.C.) Sun*, 1st June 1992, A10:4.

20. *Christian Science Monitor*, 30th June 1992, 19:2–3.

21. Al Gore, *Earth in the Balance* (Boston: Houghton Mifflin, 1992), 147.

22. *Ibid.*, 148.

23. *Ibid.*

24. Mitchell, *World on Fire*, 67.

25. *Globe and Mail*, 4th June 1992, A6:5–6.

26. Mitchell, *World on Fire*, 49.

27. *Development and the Environment: The French Approach* (Government of France, Department of the Environment, 1991), 20.

28. Shillington, *Our Changing Climate*, 5.

29. Mitchell, *World on Fire*, 63.

30. Shillington, *Our Changing Climate*, 5.

31. United Nations Document, General Assembly, A/45/696/Add.1, (1990), 5.

32. Gore, *Earth in the Balance*, 96.

33. *Newsweek*, 1st June 1992, 32:1.

34. Gore, *Earth in the Balance*, 98.

35. Statement by G. O. P. Obasi, Secretary-General, World Meteorological Organization, UNCED, Rio, 11th June 1992.

36. *Christian Science Monitor*, 30th June 1992, 19:4.

37. *Globe and Mail*, 7th May 1992, A2:3.

38. *Christian Science Monitor*, 30th June 1992, 19:3.

39. *Ibid.*, 2nd June 1992, 11:2.

40. *Ibid.*, 30th June 1992, 19:3–4.

41. Shillington, *Our Changing Climate*, 8.

42. *Ibid.*, 6.

43. *Globe and Mail*, 7th May 1992, 1:2, and 3rd June 1992, B10:3.

44. *Ibid.*, 7th May 1992, A2:3.

45. *Times* (London), 20th May 1992, 7:1–2.

46. *Ibid.*, 7:2.

47. "United Nations Activities," *Environmental Policy and Law*, vol. 20, nos. 1/2 (March 1990): 3.

48. Washington Post, "Global Warming," A10:4.

49. "United Nations Activities," *Environmental Policy and Law*, 3.

50. *Christian Science Monitor*, 9th June 1992, 8:1.

51. *Ibid.*

52. Washington Post, "Global Warming," A10:4.

53. Shillington, *Our Changing Climate*, 6.

54. Washington Post, "Global Warming," A10:4.

55. *Christian Science Monitor*, 9th June 1992, 8:1.

56. *Globe and Mail*, 3rd June 1992, A6:1–2.

57. *Christian Science Monitor*, 9th June 1992, 8:3.

58. *Globe and Mail*, 3rd June 1992, A6:2.

59. *Ibid.*, A6:1.

60. *Economist*, 30th May 1992, 18:2.

61. *Ibid.*

62. United Nations Document, General Assembly Resolution A/RES/43/53, 27th January 1989.

63. U.N. Document, General Assembly Resolution A/RES/44/207, 15th March 1990.

64. U.N. Document, General Assembly Resolution A/RES/44/228, 22nd March 1990.

65. "Climate Change: Nairobi Declaration," *Environmental Policy and Law*, vol. 20, no. 6 (December 1990): 199:2–3.

66. "International Conference on Global Warming and Climate Change: African Perspectives," Nairobi, Kenya, 2nd–4th May 1990. For text, see *Environmental Policy and Law*, vol. 20, no. 6 (December 1990): 234.

67. "E.C.: Agreement on CO_2 Emissions," *Environmental Policy and Law*, vol. 20, no. 6, (December 1990): 207.

68. *Ibid.*

69. "EFTA: Freeze on Carbon Dioxide Emissions," *Environmental Policy and Law*, vol. 20, no. 6 (December 1990): 206.

70. United Nations Document, General Assembly, A/45/696, 2nd November 1990.

71. United Nations Document, General Assembly, A/RES/45/212, 17th January 1991.

72. United Nations Document, General Assembly Resolution A/RES/45/212, 17th January 1991.

73. United Nations Document, A/45/696/Add.1, 3, 8th November 1990.

74. United Nations Document, A/45/696/Add.1, 8th November 1990, Annex II.

75. United Nations Document, General Assembly A/45/696/Add.1, 8th November 1990, 3.

76. Conference Statement, Second World Climate Conference, Geneva, 29th October–2nd November 1990. For text, see *Environmental Policy and Law*, vol. 20, no. 6 (December 1990): 226.

77. "Tlatelolco Platform on Environment and Development," *Environmental Policy and Law*, vol. 21, nos. 3/4 (July 1991): 181–82.

78. "Developing Countries: Preparing for UNCED—Beijing Declaration," *Environmental Policy and Law*, vol. 21, nos. 5/6 (December 1991): 267–68.

79. Kuala Lumpur Declaration on Environment and Development, Second Ministerial Conference of Developing Countries on Environment and Development, 26th–29th April 1992, Kuala Lumpur, Malaysia.

80. United Nations Document, General Assembly Resolution A/RES/46/169, 18th February 1992.

81. UNEP: Governing Council Decisions, 16th Session, *Environmental Policy and Law*, vol. 21, nos. 3/4 (July 1991): 173–74.

82. "UNCED: Decisions Relating to Climate Change," *Environmental Policy and Law*, vol. 21, nos. 5/6 (December 1991): 242–46.

83. U.N. Document A/CONF.151/8. For text, see Nicholas Robinson, ed., *Agenda 21 and the UNCED Proceedings* (New York: Oceana Publications, 1992), vol. 3, 1715.

84. U.N. Document, A/CONF.151/8, 1st June 1992, Report of the Chairman of the Intergovernmental Negotiating Committee for a Framework Convention on Climate Change. Text in Robinson, *Agenda 21 and the UNCED Proceedings*, 1719.

85. United Nations Document, Framework Convention on Climate Change, A/AC.237/18 (Part II)/Add.1, Art. 1.2.

86. *Ibid.*

87. *Ibid.*

88. *Ibid.*

89. *Ibid.*

90. *Ibid.*, Article 2, Objective.

91. *Ibid.*, Article 4.2 (a) and (b).

92. *Ibid.*, Article 3.3.

93. *Ibid.*, Article 3.1.

94. *Ibid.*

95. *Ibid.*, Article 3.3.

96. *Ibid.*, Article 3.4.

97. *Ibid.*, Article 3.5.

98. *Ibid.*, Article 4.1(a).

99. *Ibid.*, Article 4.1 (b).

100. *Ibid.*, Article 4.1(c).

101. *Ibid.*, Article 4.1(d) and (e).

102. *Ibid.*, Article 4.1(i).

103. *Ibid.*, Article 5(a).

104. *Ibid.*, Article 4.2(b).

105. *Ibid.*, Article 4.2(a).

106. *Ibid.*, Article 4.2(b).

107. *Ibid.*, Article 4.2(d).

108. *Ibid.*, Article 4.7.

109. *Ibid.*, Article 7.1.

110. *Ibid.*, Article 7.2(b).

111. *Ibid.*, Article 7.2(e).

112. *Ibid.*, Article 7.2(g).

113. *Ibid.*, Article 7.2(h).

114. *Ibid.*, Article 8.1.

115. *Ibid.*, Article 8.2(c).

116. *Ibid.*, Article 9.1.

117. *Ibid.*, Article 9.2(a).

118. *Ibid.*, Article 9.2(e).

119. *Ibid.*, Article 9.2(c).

120. *Ibid.*, Article 9.2(d).

121. *Ibid.*, Article 10.1.

122. *Ibid.*, Article 11.2.

123. *Ibid.*, Article 11.1.

124. *Ibid.*, Article 21.3.

125. *Ibid.*

126. *Ibid.*, Article 14.1.

127. *Ibid.*, Article 14.2(a) and (b).

128. *Ibid.*, Article 24.

129. *Ibid.*, Article 15.3.

130. Head, *On a Hinge of History*, 93.

131. *Time*, 1st June 1992, 22:1.

132. *Christian Science Monitor*, 9th April 1992, 11.

133. "President Challenged on Environment," *Environmental Policy and Law*, vol. 19, nos. 3/4 (July 1989): 115.

134. *Edmonton Journal*, 12th June 1992, A17:6.

135. *Globe and Mail*, 13th June 1992, A4:3–5.

136. *Christian Science Monitor*, 22nd September 1992, 20:2.

137. *Ibid.*, 8th April 1992, 18:3–4.

138. *Environmental Policy and Law*, vol. 20, no. 6 (December 1990): 198.

139. *Ibid.*, vol. 19, nos. 3/4 (July 1989): 115.

140. *Globe and Mail*, 30th March 1992, A8:3.

141. *Times* (London), 13th May 1992, 1:5–6.

142. *Globe and Mail*, 30th March 1992, A8:3.

143. *Ibid.*, 6th June 1992, B2:2.

144. *Christian Science Monitor*, 23rd April 1992, 8:2.

145. *Ibid.*, 9th April 1992, 11.

146. *Times* (London), Life and Times Section, 1:3.

147. *Christian Science Monitor*, 2nd June 1992, 7:1.

148. *Ibid.*, 7:5.

149. Statement by George Bush, President, United States of America, UNCED, Rio, 12th June 1992.

150. *New York Times*, 13th June 1992, 1.

151. *Ibid.*, 4:3.

152. *Christian Science Monitor*, 2nd March 1992, 9:4.

153. *Times* (London), 11th June 1992, 17.

154. *Christian Science Monitor*, 19th May 1992, 20.

155. Statement by William K. Reilly, Chief Administrator of the U.S. Environmental Protection Agency, UNCED, Rio, 3rd June 1992.

156. *Earth Summit Times*, cited in *Irish Times*, 11th June 1992, 7:2.

157. *Christian Science Monitor*, 3rd June 1992, 9:4.

158. *Time*, 1st June 1992, 32:1.

159. *Globe and Mail*, 10th June 1992, 1:1.

160. *Christian Science Monitor*, 27th February 1992, 19:4.

161. Statement and Declaration by Austria, Liechtenstein, and Switzerland.

162. *Ibid.*

163. *New York Times*, 5th June 1992, A6:6.

164. *Ibid.*

165. *Edmonton Journal*, 5th June 1992, 1:3.

166. *Ibid.*, 1:2.

167. *Ibid.*, 1:2–3.

168. *Globe and Mail*, 5th June 1992, A4:6.

169. *New York Times*, 8th June 1992, A5:4–5.

170. *Ibid.*, A5:5.

171. *Ibid.*, A5:4.

172. *Ibid.*, 10th June 1992, A8:4.

173. IMF, *World Economic Outlook*, May 1990, cited in Head, *On a Hinge of History*, 46.

174. Head, *On a Hinge of History*, 44–45.

175. *Ibid.*, 45.

176. Statement by Sun Lin, Head of the Chinese Delegation to the First Session of the Intergovernmental Negotiating Committee for a Framework Convention on Climate Change, Washington, D.C., 4th–14th February 1991.

177. Statement by leader of the Indian delegation to the substantive negotiations session of the Intergovernmental Negotiating Committee for a Framework Convention on Climate Change, June 1991.

178. *Ibid.*

179. *Christian Science Monitor*, 2nd March 1992, 9:4.

180. *Globe and Mail*, 23rd May 1992, 7:3.

181. *Times* (London), 13th June 1992, 1:2.

182. *Time*, 29th April 1991, 44:2.

183. *Newsweek*, 1st June 1992, 31:3.

184. *Time*, 29th April 1991, 45–46.

185. *Times* (London), 3rd June 1992, 12:7–8.

186. *Globe and Mail*, 2nd May 1992, A7:1.

187. *Ibid.*, 2nd June 1992, A3:22.

188. Statement by Li Peng, Premier, People's Republic of China, UNCED, Rio, 12th June 1992.

189. *Time*, 29th April 1991, 47.

190. *Globe and Mail*, 2nd June 1992, A3:4.

191. U.N. Development Programme, *Human Development Report, 1991* (New York: Oxford University Press, 1991), and World Bank, *World Development Report, 1991* (New York: Oxford University Press, 1991), cited in Sandra Postel, "Denial in the Decisive Decade," in *State of the World, 1992* (New York: W. W. Norton, 1992), 4.

192. Statement by Michel Camdessus, Managing Director, International Monetary Fund, UNCED, Rio, 8th June 1992.

193. Statement by Dato' Abdullah Haji Ahmad Badawi, Foreign Minister, Malaysia, UNCED, Rio, 10th June 1992.

194. Centre for Science and Environment, "Environmental Colonialism to the Fore" (New Delhi: CSE, January 1991).

195. Statement by Jacques Delors, President, Commission of the European Communities, UNCED, Rio, 13th June 1992.

196. Howard Mann, "The Rio Declaration," in American Society of International Law, *Proceedings of the 86th Annual Meeting* (Washington, D.C., 1st–4th April 1992), 410.

197. Annex I: Australia, Austria, Belarus, Belgium, Bulgaria, Canada, Czechoslovakia, Denmark, Estonia, European Community, Finland, France, Germany, Greece, Hungary, Iceland, Ireland, Italy, Japan, Latvia, Lithuania, Luxembourg, Netherlands, New Zealand, Norway, Poland, Portugal, Romania, Russian Federation, Spain, Sweden, Switzerland,

Turkey, Ukraine, United Kingdom of Great Britain and Northern Ireland, and United States of America.

Annex II: Australia, Austria, Belgium, Canada, Denmark, European Community, Finland, France, Germany, Greece, Iceland, Ireland, Italy, Japan, Luxembourg, Netherlands, New Zealand, Norway, Portugal, Spain, Sweden, Switzerland, Turkey, United Kingdom of Great Britain and Northern Ireland, and United States of America.

198. *Ibid.*

199. *Globe and Mail*, 11th May 1992, A8:6.

200. *Times* (London), 1st June 1992, 12:3.

201. Statement by Philip Leakey, Minister for Environment, Kenya, UNCED, Rio, 12th June 1992.

202. Statement by Dr. Kofi N. Awoonor, Ambassador of Ghana, UNCED, Rio, 5th June 1992.

203. Statement by Giorgio Ruffolo, Minister for Environment, Italy, UNCED, Rio, 4th June 1992.

204. Statement by Dr. Mahathir bin Mohamad, Prime Minister, Malaysia, UNCED, Rio, 13th June 1992.

205. Statement by Victor I. Danilov-Danilian, Minister for Ecology and Natural Resources, Russian Federation, UNCED, Rio, 8th June 1992.

206. Transcript of interview given by Prime Minister John Major to Paul Reynolds, BBC Radio, 11th June 1992, en route from Bogota to Rio, in Policy Background 11/92, 11th June 1992, British Information Services, Ottawa.

207. Statement by Resio S. Moses, Secretary of the Department of External Affairs, of the Federated States of Micronesia, UNCED, Rio, 10th June 1992.

208. *Ibid.*

209. Statement by Sir Wiwa Korowi, Governor-General, Papua New Guinea, UNCED, Rio, 13th June 1992.

210. Statement by Teatao Teannaki, President, Republic of Kiribati, UNCED, Rio, 13th June 1992.

211. Statement by Kinza Clodumar, Minister of Finance, Republic of Nauru, UNCED, Rio, 9th June 1992.

212. Statement by Amata Kabua, President, Republic of the Marshall Islands, UNCED, Rio, 13th June 1992.

213. Statement by Tom Kijiner, Minister for Foreign Affairs, Republic of the Marshall Islands, UNCED, Rio, 4th June 1992.

214. Statement by Tofilau Eti Alesana, Prime Minister, Western Samoa, UNCED, Rio, 13th June 1992.

215. Statement by Mostafizur Rahman, Foreign Minister, Bangladesh, UNCED, Rio, 10th June 1992. See also statements by P. J. Patterson, Prime Minister, Jamaica, Vincent Perera, Minister of Environment, Sri Lanka, and Andreas Gavrielides, Minister of Agriculture, Republic of Cyprus, UNCED, Rio, 10th June 1992.

216. Statement by Robert F. Van Lierop, Permanent Representative to the United Nations from the Republic of Vanuatu, UNCED, Rio, 8th June 1992.

217. For text, see "Global Warming and Sea Level Rise," *Environmental Policy and Law*, vol. 20, nos. 1/2 (March 1990): 58–59.

218. *Ibid.*

219. United Nations General Assembly Resolution, Possible Adverse Effects of Sea-Level Rise on Islands and Coastal Areas, Particularly Low-Lying Coastal Areas, 22nd December 1989.

220. *Ibid.*

221. "Second World Climate Conference," *Environmental Policy and Law*, vol. 20, no. 6 (December 1990): 228.

222. Second World Climate Conference, Geneva, 29th October to 7th November 1990, Ministerial Declaration; for text, see *Environmental Policy and Law*, vol. 20, no. 6 (December 1990): 220.

223. UNEP Sixteenth Session Governing Council, *Environmental Policy and Law*, vol. 21, nos. 3/4 (July 1991): 174.

224. *Ibid.*

225. *Economist*, 30th May 1992, 24.

226. Statement by Dr. Hans Blix, Director-General, International Atomic Energy Agency, UNCED, Rio, 4th June 1992.

227. *Times* (London), 6th November 1991, 27:6.

228. Statement by Helga Steeg, Executive Director, International Atomic Energy Agency, UNCED, Rio, 4th June 1992.

229. *Globe and Mail*, 5th June 1992, A4:6.

230. Statement by Rashid Abdullah al-Noaimi, Minister of Foreign Affairs, United Arab Emirates, UNCED, Rio, 9th June 1992.

231. *Globe and Mail*, 10th June 1992, A1:3.

232. *Ibid.*

233. Statement by Dr. Subroto, Secretary-General of OPEC, UNCED, Rio, 9th June 1992.

234. *Ibid.*

235. *Ibid.*

236. *Ibid.*

237. Statement by Ali Hassan Mwinyi, President, United Republic of Tanzania, UNCED, Rio, 13th June 1992.

238. Statement by Dr. Subroto, Secretary-General of OPEC, UNCED, Rio, 9th June 1992.

239. "Convention on Climate Change Signed," *Environmental Policy and Law*, vol. 22, no. 4 (August 1992): 207.

240. Statement by Olof Johansson, Minister of the Environment and Natural Resources, Sweden, UNCED, Rio, 8th June 1992.

241. *Globe and Mail*, 8th June 1992, A8:6.

242. *Christian Science Monitor*, 3rd June 1992, 2.

243. "The Case against Climate Aid," *Ecologist*, vol. 21, no. 6 (November/December 1991): 244.

244. *Globe and Mail*, 4th June 1992, A19:5.

245. "Global Environmental Facility," *Environmental Policy and Law*, vol. 22, no. 4 (August 1992): 218.

246. *Globe and Mail*, 29th May 1992, A17:2.

247. Canadian International Development Agency, *Annual Report, 1989–90*, 46.

248. *Christian Science Monitor*, 18th May 1992, 7:4.

249. *Globe and Mail*, 8th June 1992, A8:6.

250. *Irish Times*, 1st June 1992, 7:1.

251. *Globe and Mail*, 18th May 1992, A7:3.

252. *Ibid.*, 29th May 1992, A17:3.

253. Statement by Carlos Borrego, Minister of Environment, Portugal, UNCED, Rio, 3rd June 1992.

254. *Globe and Mail*, 11th May 1992, A8:1.

255. *Ibid.*, 8th June 1992, A8:6.

256. Statement by Alhaji Sir Dawda Kairaba Jawara, President, Republic of the Gambia, UNCED, Rio, 12th June 1992.

257. "IMF and World Bank Development Committee Meeting," *Earth Summit Bulletin*, 1st June 1992, 1:2, and Canadian Delegation Report, Intergovernmental Negotiating Committee for a Framework Convention on Climate Change, Fifth Session, Second Part, 30th April–9th May 1992, New York.

258. "IMF and World Bank," 1:2.

259. Canadian Delegation Report, Intergovernmental Negotiating Committee, C.3.

260. *Globe and Mail*, 8th June 1992, A8:6.

261. *Ibid.*

262. Statement by Mostafizur Rahman, Foreign Minister, Bangladesh, UNCED, Rio, 10th June 1992.

263. Canadian Delegation Report, Intergovernmental Negotiating Committee.

264. *Economist*, 30th May 1992, 21.

265. *New York Times*, 22nd April 1993, A1.

266. *Christian Science Monitor*, 4th June 1992, 6.

267. *Globe and Mail*, 17th November 1992, A6:6.

268. *Christian Science Monitor*, 10th June 1992, 18:4.

269. Statement by Thorbjorn Berntsen, Minister of Environment, Norway, UNCED, Rio, 4th June 1992.

270. United Nations Document ENV/DEV/RIO/2, 3rd June 1992, Earth Summit Press Release, "Secretary-General in Statement to Environment/Development Meeting Stresses Need for Full Awareness of Planet's Fragility."

271. Statement by Flavio Cotti, Head of Delegation, Switzerland, UNCED, Rio, 10th June 1992.

272. Statement by Carl Bildt, Prime Minister, Sweden, UNCED, Rio, 13th June 1992.

273. Statement by Hans Alders, Minister for Environment, the Netherlands, UNCED, Rio, 5th June 1992.

274. Statement by Helga Steeg, Executive Director, International Atomic Energy Agency, UNCED, Rio, 4th June 1992.

275. U.N. Document, A/CONF.151/8, 1st June 1992. For text, see Robinson, *Agenda 21 and the UNCED Proceedings*, vol. 3, 1719.

276. Statement by Carlos Borrego, Minister of Environment, Portugal, UNCED, Rio, 3rd June 1992.

277. Policy Statement by John Major, Prime Minister, United Kingdom, to the House of Commons, 15th June 1992, No.599/92, British Information Services.

278. Statement by Helmut Kohl, Chancellor, Federal Republic of Germany, UNCED, Rio, 12th June 1992.

279. Statement by Per Stig Moller, Minister for the Environment, Denmark, UNCED, Rio, 9th June 1992.

280. Statement by Hans Alders, Minister for Environment, the Netherlands, UNCED, Rio, 5th June 1992.

281. "Convention on Climate Change Signed," *Environmental Policy and Law*, vol. 22, no. 4 (August 1992): 207.

282. *New York Times*, 14th June 1992, 10:3.

283. Statement by Michael Howard, Secretary of State for Environment, United Kingdom, UNCED, Rio, 9th June 1992.

284. Statement by Helmut Kohl, Chancellor, Federal Republic of Germany, UNCED, Rio, 12th June 1992.

285. Statement by Domingo L. Siazon, Jr., Director-General of UNIDO, UNCED, Rio, 4th June 1992.

286. Statement by Eiour Gounason, Minister for the Environment, Iceland, UNCED, Rio, 5th June 1992.

CHAPTER FIVE

International Law and the Preservation of Species: An Analysis of the Convention on Biological Diversity Signed at the Rio Earth Summit in 1992

At the very moment when this Earth—our only home—appears to be dying, thousands of men, women, and children around the world have rallied to save and preserve this precious planet which has sustained and nourished the human, and innumerable other, species for millennia. With the awakening to the horror of planetary degradation has come an ethic of human responsibility and human obligation to work quickly to reverse the damage already done and to prevent further deterioration of the environment. The cause of global conservation has, in the last few years, become mainstream thinking in many parts of the world. Today, there are few proponents of untrammelled development. Politicians, lawyers, and businessmen espouse the environmental ethic and proclaim their dedication to "green values" with the fervor of the recently converted. Environmentalism has acquired the status of a global religion. Those who do promote development at the expense of the environment frequently do so with assurances of their commitment to enhancing the quality of life or, in developing nations, cite poverty and lack of alternatives to rapid growth as justification. As with all religions, however, the precepts are more honored than observed—the level of action rarely matches the level of rhetoric. The ultimate challenge will be to take the leap from making verbal commitments to environmental goals to implementing these ideas.

The ethic of conservation has now become popular and widespread, especially among the young, who have demonstrated a commitment to the preservation of ecosystems and life forms which is encouraging for the future. Their

Reprinted with kind permission from an original version published by the *Dickinson Journal of International Law*.

lives will undoubtedly be affected for the better if we act quickly to protect what is left of our environment, for the worse if we fail to act. One of the most environmentally important tasks we can undertake now is to work to protect the other species which share this planet with us. Their preservation is not merely a matter of idealism; it is in our ultimate self-interest. Our survival demands that we do not inadvertently or through sheer human greed destroy these species, large and small, which sustain our very existence by forming part of the intricate food chain, replenishing the soil, clearing the water, feeding us directly, and even curing us of our illnesses, all the while providing us with a vision of the wonder, the beauty, and the amazing diversity of forms with which the principle of life can manifest itself. The realization that we human beings are ultimately only an aspect—albeit a significant one—in a vast process of Nature has evoked a global determination to save those other species whose very survival is in our all-too-callous hands. Long before the issue of preserving biological diversity became a subject for governmental negotiation, it had become an area of public concern.

This chapter will analyze the Convention on Biological Diversity (reprinted as Appendix 5 to this book) signed at the United Nations Conference on Environment and Development (UNCED) in June 1992. Rather than assessing the clauses seriatim, I have preferred to base my analysis on a consideration of various provisions grouped under topical subheadings. These separate sections will explore the role of developed and developing nations and discuss the diversity of opinion which emerged at the Rio Earth Summit. The emphasis will be on the most significant provisions of the Biodiversity Treaty, particularly those that are controversial. For the reader's convenience, the clauses of the Convention that are quoted will be in italics, followed by the number of the relevant article, in parentheses. Occasionally, the wording of a provision of the Convention may have to be repeated because of its relevance to different topics under consideration. Prior to the detailed analysis, one section of this chapter will present a brief summary of the Convention. It is important to stress that this chapter will *not* deal with the history of the complex negotiating process by which the Biodiversity Treaty was created. Rather, the emphasis will be on the Convention itself—on its significance, on its strengths and weaknesses, and on opinions expressed by various heads of government and state and by a number of environmentalists. The very controversial position of the Government of the United States under President George Bush will be discussed in detail, as this is of relevance to any analysis of the Treaty. The United States refused to sign this Convention in 1992, but it did accede after Bill Clinton became President. Hopefully, this chapter will provide readers with a comprehensive view of what is a very important development in international law, the beginning of a worldwide process aimed at averting species destruction on this planet.

Biodiversity Defined

According to the Convention on Biological Diversity, the term

> *"Biological diversity" means the variability among living organisms from all sources including,* inter alia, *terrestrial, marine and other aquatic ecosystems and the ecological complexes of which they are part; this includes diversity within species, between species and of ecosystems.* (Article 2)

More scientifically, Edward O. Wilson, in his book *The Diversity of Life*, proposes the following definition:

> The variety of organisms considered at all levels, from genetic variants belonging to the same species through arrays of species to arrays of genera, families, and still higher taxonomic levels; includes the variety of ecosystems, which comprise both the communities of organisms within particular habitats and the physical conditions under which they live.[1]

In layman's terms, biodiversity is "the total variety of life on earth,"[2] the web of life on this planet in its multiplicity of expressions—animal, plant, and even the minute organisms which inhabit the soil. This is the foundation for all life, and its interconnections are so complex and intricate that scientists have not yet penetrated all its mysteries, for the scope of the study is immense. "Even the known statistics of bio-diversity are humbling."[3] Some scientists have suggested that there are probably 1.4 million known species or organisms on Earth. Of these, "750,000 are insects, 41,000 are vertebrates and 250,000 are plants; the rest correspond to a complex of invertebrates, fungi and micro-organisms."[4] There are no definite statistics on the number of species when those not yet studied are considered. Estimates suggest a figure of 10 million species, including plants and animals;[5] the various ant species outnumber bird species.[6] As Cyril de Klemm of the Commission on Environmental Policy, Law and Administration (CEPLA) suggests, "Nobody even knows the approximate number of living species: 5, 10, 30 million, according to the most recent estimates, out of which less than two million have so far been described."[7] An even higher estimate of 80 million species has been proposed.[8]

One of the world's leading authorities on the subject of biodiversity, Edward O. Wilson, stresses that "[b]iodiversity is our most valuable but least appreciated resource."[9] Biodiversity can be affected by a variety of factors, "such as climate, number of organisms, topography, physical substratum, time and heredity."[10] The wealth of this resource is not evenly distributed throughout the planet.[11] "The natural ecosystems of forests, savannas, pastures and rangelands, deserts, tundras, rivers, lakes and sea contain most of the Earth's biodiversity. Farmers' fields and gardens are also of great importance as repositories, while gene banks, botanical gardens, zoos and other germplasm repositories make a small but significant contribution."[12] Although Canada, a large country, contains approximately 70,000 known species,[13] Madagascar, a much smaller state, is home to 150,000 species found in no other part of the world and contains about one quarter of the plant species of Africa.[14] Brazil, with a mere 6.3 per cent of the world's land area, enjoys 22 per cent of the planet's flowering plant species.[15] Tropical areas are most blessed with biodiversity, particularly those regions still covered with forests. Paul Harrison explains why the tropical rainforest is so rich in its diversity of life:

> Its leaf litter, ground herbs, shrubs, middle level and canopy offer a five-storey habitat. Each storey has multiple rooms: climbers, stranglers, clinging ferns, orchids, bromeliads and others. Each tree and plant has its personal range of insect lodgers, each insect its own minute parasites. Each storey and room has its distinct vertebrate predators, with different sets for day and night shifts.[16]

Estimates vary concerning the extent of forest cover still remaining on this planet, and the statistics on biodiversity are at times only the best guesses that scientists can produce. "Taken as a whole, dense tropical forests cover an area of 1,200,889,000 hectares. This total, however, represents only 7% of the Earth's surface and may contain . . . 90% of its biological diversity."[17] This accident of nature has significant political implications which will be discussed in some detail later.

The Convention on Biological Diversity distinguishes between biological diversity and biological resources, although the two terms are frequently used interchangeably in common parlance. According to the Convention, biological resources include

> *genetic resources, organisms or parts thereof, populations, or any other biotic component of ecosystems with actual or potential use or value for humanity.* (Article 2)

Biological diversity can be categorized by reference to "ecosystem diversity (the range of different interacting systems present in a region, nation or the world); species diversity (the range of species in a given area); and genetic biodiversity (the range of possible heritable characteristics [genes] found in a population or species)."[18] The lack of scientific knowledge about both the extensive range of species and the interactions among species makes it imperative that biodiversity research be considered a priority in all nations, particularly because of the threat of destruction faced by so many species today.

The Importance of Biodiversity

The significance of the world's biodiversity resource can never be overestimated. This vast and largely untapped treasure provides us with sustenance now and holds the possible key to improving the quality of life in the future. In providing new sources of food and medicine alone, the realm of biodiversity justifies its continuing existence and preservation. As John Ryan suggests, "The range of products hidden in forests, reefs, and other ecosystems is a powerful argument for their conservation."[19] At the present time, "humans use less than one-tenth of one percent of all naturally occurring species."[20] The United Nations estimates that "[o]nly a tiny fraction of species with potential economic importance [has] been utilized; 20 species supply 90 per cent of the world's food, and just three (wheat, maize, and rice) provide more than half."[21] Tropical forests are already yielding a wealth of "new foods, cosmetics, medicines, soaps, and other products."[22] The World Resources Institute estimates that "Indians dwelling in the Amazon Basin make use of some 1,300 medicinal plants, including antibiotics, narcotics, abortifacients, contraceptives, antidiarrheal agents, fungicides, anesthetics, muscle relaxants, and many others—most of which have not yet been investigated by researchers."[23] As Ben Jackson of the World Development Movement explains, "Half the doctors' prescriptions dispensed by chemists have their origins in wild organisms, worth around $14 billion in the United States alone."[24] The profits are considerable. "Over 7,000 medical compounds in western pharmacopoeia are derived from plants. Their retail value exceeds $40 billion a year."[25] A small soil organism from Spain which helps reduce cholesterol is worth $617 million annually.[26] The most publicized example of a vital medicinal plant is that of the rosy periwinkle, found in Madagascar, which has proved useful for the treatment of cancer.[27] In a speech made shortly before UNCED, the Canadian Prime Minister, Brian Mulroney, emphasized the significance of this important discovery in saving lives: "[Twenty] years ago, one in five children with leukaemia [was] successfully treated. Today the success rate is four out of five, thanks to a strain of rosy periwinkle found only in Madagascar."[28] The anti-cancer properties of taxol, extracted from the yew tree, have also been widely publicized.[29] According to an Associated Press report, the leaves of a newly discovered vine from the rain forest of Cameroon contain a chemical which is able to block reproduction of the AIDS virus in a test tube. Leaves of the vine have yielded an alkaloid called Michellamine B during laboratory tests at the National Cancer Institute in Frederick, Maryland.[30]

As delegates were shaping the environmental future of the planet at UNCED, in Canada an international team of scientists released a report indicating that fish oil could ease symptoms of rheumatoid arthritis.[31] Vanilla, now so commonly used in cooking, was developed from a biological resource discovered on the lands of indigenous people in Central America.[32] American scientists have found a painkiller two hundred times stronger than morphine in the skin of an Ecuadoran frog.[33]

In implementing the process now known as bioremediation, living organisms, bacteria, and fungi are squirted on toxic sites to metabolize the contaminants in an attempt to use green technology to clean up hazardous waste.[34] John Stackhouse reported in the *Globe and Mail* that in India the Ganga Action Plan, formulated in 1986, uses flesh-eating turtles to dispose of the human corpses which are ritualistically thrown into the Ganges, the nation's holy river. These "turtle patrols" have been given credit for cleaning up the river to the point where bathing in its waters, another ancient Hindu tradition, is once again safe.[35]

The value of the treasure trove of biodiversity is such that it yields benefits and seeming miracles almost daily. Ironically, even as the treasure is being destroyed, its real worth is being revealed. The multiple tasks carried out by different species make life on this planet sustainable. As Cyril de Klemm asserts:

> Another important aspect of the value of wild species is the essential role that many of them play within natural ecosystems as pollinators, seed dispersers, predators controlling the proliferation of other species, or decomposers of organic matter. The importance of the mutual services that species provide to one another and the serious disturbance that may result from interruption or impairment of these services is still poorly understood.[36]

De Klemm goes on to point out that "ecosystems themselves also provide mutual services to one another, the continued maintenance of which may be essential to the perpetuation of the ecological balance of entire regions."[37]

Despite Canada's developed nation status and the relative prosperity of its people, the Government admits the vital importance of biodiversity to the nation's economy: "Fisheries, forestry and agriculture are directly dependent and many other industrial sectors including the biotechnology industry derive their economic stability from a diversity of biological resources."[38] Indirectly, biodiversity benefits the " 'ecosystem services' of climate control, oxygen production, removal of carbon dioxide from the atmosphere, soil generation [and] freshwater supplies."[39]

For developing nations the dependence on the biodiversity resource is even greater. There is hardly a nation on this planet which does not derive monetary benefit from its biodiversity and its biological resources. As the United Nations Environment Programme points out, "Biological resources produce tangible benefits such as food, medicines, shelter and employment that can be readily translated into monetary terms."[40] Equally valuable for the economy of any nation is that "[b]iological diversity, as well as being vital for the functioning of ecosystems, does itself also provide valuable economic services, particularly tourism where visitors will pay money to experience the diversity of wildlife."[41]

As Paul Harrison suggests, "Only when we come to value that diversity as one of our human needs—and one that overrides many others—will we reverse the tide"[42] of destruction and devastation which threatens the survival of this treasure trove. The United Nations Environment Programme concludes that "[t]he dependence of humankind upon biodiversity and biological resources for its long-term well being is not yet fully appreciated by decision makers."[43]

The Global Problem Affecting Biodiversity

Colman McCarthy, writing in the *Washington Post*, comments that "[t]he Earth itself . . . is an ecological war zone in which human beings, when they aren't killing each other off in wars and homicides, are obsessed with doing in nearly everything else."[44] There is no doubt that the destruction of biodiversity is one of the most serious environmental problems facing us today. George Mitchell, Majority Leader of the United States Senate at the time of the Earth Summit, has termed this process an "ecological holocaust"[45] and comments:

> The rate of extinction today compared to the historic norm through geologic time is awesome and terrifying. More species of the earth's plants and ani-

mals may be lost in our lifetime than in the mass extinctions that swept the dinosaurs from the earth sixty five million years ago.[46]

John Ryan states that "[d]ifficult as it is to accept, mass extinction has already begun"[47] and adds that "biological impoverishment is occurring all over the globe."[48] At risk are all types of environments, lush tropical forests, coral reefs, islands, lakes, swamps, and even deserts. Don McAllister, Senior Biodiversity Advisor of the Canadian Centre for Biodiversity, explains that because the "hereditary part of diversity is coded in up to 3 billion molecules inside the chromosomes inside every cell" the loss of a species entails "losing up to 3 billion coded bits of information."[49]

The world's tropical forests, appropriately termed "cradles of life,"[50] are the natural habitat for much of this biodiversity. The former director of the United Nations Environment Program, Mostafa Tolba, estimates that though these forests cover only about 6 per cent of the Earth's land surface they contain over half the world's biodiversity.[51] These areas have probably already lost about a million species in the past two decades.[52] Future prospects are grim. "Some scientists say that about 60,000 of the world's 240,000 plant species—and perhaps even higher proportions of vertebrate and insect species—could become extinct in the next three decades unless tropical deforestation is slowed immediately."[53] Mostafa Tolba explained that "[s]pecies richness generally increases in magnitude as we move from the poles to the equator. In one 15 hectare area of Borneo rain forest, for example, approximately 700 species of trees have been identified."[54] James Speth of the World Resources Institute estimates that tropical deforestation increased by 50 per cent during the 1980s;[55] this percentage averages about 42 million acres annually,[56] equivalent to "an area . . . about the size of the state of Washington . . . lost each year; that's an acre and a half a second."[57]

Because tropical forests house so much biodiversity, international concern has emphasized this particular aspect of the ecological holocaust. "By an unfortunate quirk of geography, the way in which bio-diversity is distributed works against it. The tropics have the largest share of the earth's biological wealth. But they also have the fastest rates of human population growth, the greatest need to clear forest and drain wetlands, the heaviest pressures on coral reefs."[58] By one estimate, loss of habitat for wildlife has reached frightening proportions—four fifths or more in countries like Gambia, Liberia, Bangladesh, India, and Vietnam, with the continents of Africa and Asia registering a two-thirds loss.[59] "In the tropics developing countries as diverse as Chad, Bangladesh, India and Vietnam—wetlands destruction has exceeded 80 to 90 percent."[60]

The losses in developed nations are also significant. A mere 10 per cent of old-growth rain forest survives in the United States of America, largely in the Pacific Northwest.[61] An intense controversy has raged for the past few years in this area over the fate of the spotted owl, a conflict which has pitted loggers who believe jobs are at stake against environmentalists who want to preserve an endangered species. Commenting on this very vocal debate, Edward O. Wilson suggests that the proper question to ask is, "[W]hat else awaits discovery in the old-growth forests of the Pacific Northwest?"[62] The spotted owl, though better publicized than any other endangered avian form, is not the only bird facing a threat. Approximately 495 of the 650 species of birds which inhabit or fly through the United States every year are now considered endangered.[63] Globally, three quarters of the bird species are declining in numbers or facing extinction.[64]

The economic consequences of biological destruction are already being felt. The once prolific oyster population of Chesapeake Bay has "fallen by 99% since 1870."[65] These oysters perform the vital function of filtering the Bay's water, a task which once occurred every three days but now takes a year.[66] Overfishing has depleted fish stocks and created economic havoc in several parts of the world, including eastern Canada. Vice-President Al Gore points out that "since 1950 the total annual catch worldwide has increased by 500 percent and is now assumed to be higher than the replenishment rate in most areas."[67] He concludes that a "growing number of valuable food species are disappearing entirely."[68]

The United Nations Environment Programme (UNEP) has reported that the Earth will lose approximately one quarter of its estimated thirty million species in the next twenty to thirty years, a loss which "will compromise the ability of future generations to meet their needs."[69] Human activity is largely responsible for this situation. The burgeoning world population, over five and a half billion at the time of the Rio Summit,[70] is spreading into and taking over the habitat of millions of other species. The monetary orientation of human civilization demands constant development and incessant activity, both of which result in soil degradation, ocean resource depletion, and, with industrialization, air pollution and even the holes in the delicate ozone layer surrounding this planet.

Human beings are at once the problem and also, ironically, its ultimate victims. The unequal distribution of the world's monetary wealth has condemned the larger proportion of its population to a life of poverty which is in stark contrast to the relative comfort and even affluence enjoyed by a minority. The governments of developing nations are rushing to take steps to alleviate the misery and economic deprivation engulfing millions of their people. In the very process of seeking to give them a better life, however, they are destroying forever the prospects of future generations. Building factories to provide jobs is a positive action, but when these factories poison the air, water, and land with fumes and toxic waste the ultimate cost to society outweighs the economic benefits granted to a few employees. This is the basic problem confronting all governments today, most seriously in the poorer nations of the developing world but present in every country of the developed world as well. The economic destitution of vast populations is forcing governments to adopt rapid, short-term solutions which frequently sacrifice the environment in favor of development. This is the basic reason that 100 or more species are becoming extinct each day.[71] (Statistics vary because so little is known about the unrecorded species; one estimate suggests that the world loses at least 140 plant and animal species every day.)[72] Edward O. Wilson explains that "even with . . . cautious parameters, selected in a biased manner to draw a maximally optimistic conclusion, the number of species doomed each year is 27,000. Each day it is 74, and each hour 3."[73] George Mitchell eloquently explains the dilemma facing modern nations:

> The immediate demands of a swelling population to divert primary production of life into human food—an understandable drive in a world where starvation is all too common—collides directly with the need to maintain the biosphere as a rich and versatile caldron of life.[74]

There are numerous ways in which human activity poses a threat to the existence of biodiversity in any region. The International Union for the Conservation of Nature includes forest clearing, drainage of wetlands for farming, mining and quarrying, overgrazing, logging, urban and industrial development, tourist development and even the introduction of exotic species as reasons to explain the destruction of species.[75] To give only one example, each year Asia loses about half a million hectares of wetlands because of pollution and developmental projects, including irrigation and dams.[76] The United Nations Environment Programme suggests even more causes:

> [T]hreats may arise from natural hazards; from the indirect consequences of human processes, or externalities such as changes in agricultural commodity prices or the servicing of international debt; and from direct human activities such as shifting agriculture, logging, poaching or pollution.[77]

Clearly, "[w]hat we are dealing with . . . is not a simple process with just one or two causes. It is a massive onslaught firing with all barrels from many directions at once."[78] The conclusion is to recognize that "most threats are created by a potential beneficiary, normally the causal agent of the threat, and that actions for threat relief therefore involve an economic trade-off."[79] It was this issue which was to prove the greatest challenge to the negotiators of the Biodiversity Convention, not least because the economic trade-offs involved also imply a more equitable distribution of the world's wealth, with a larger share going to the poor nations (the South) from the rich countries (the North). I will explore this issue in detail later in this analysis.

In a speech to the Global Forum of Spiritual and Parliamentary Leaders, former Soviet President Mikhail Gorbachev emphasized that "[t]he stability of ecosystems, and hence, the quality of the environment, depend on the preservation and maintenance of biological diversity and equilibrium of the biosphere."[80] Gorbachev suggested in the same speech that "the hour of decision—the hour of historic choice—has come, and there is no reasonable alternative for man because he is not predisposed to suicide."[81]

Some Aspects of the International Response to the Crisis

It would be simplistic to assume that the degree of concern articulated over environmental destruction was immediately translated into international action to preserve and protect the Earth's biodiversity and biological resources. As with other environmental issues of pressing urgency, this one soon got tangled in a mesh of international politicking, nationalistic posturing, demands for money from those blessed with the preponderance of the resource, and vague, reluctant pledges of assistance from the rich nations which now espouse the cause of conservation. Clashes between the developed North and the developing South bedevilled efforts to deal effectively with the global problem. Given this background, the existence of a Convention is testament to the labor of the numerous negotiators. Implementing it is another matter altogether. Time alone will tell whether nations will abide by the commitments they have entered into by signing the Convention on Biological Diversity at UNCED in Rio.

A detailed history of the process by which the Biodiversity Treaty was formulated cannot, for space reasons, be given here. Accordingly, this section will provide only a brief assessment of some aspects of that history which are relevant to the detailed analysis of the Convention which will follow in later sections.

One of the most important issues to consider is that, for all its significance and importance to the global environment, biodiversity is subject to political structures established by the dominant species, structures which have divided this planet into a variety of nations, large, medium, and small. This division plagues solutions to every environmental problem facing us but it is an inescapable obstacle. Hence, the conservation of biodiversity has to be determined on the basis of mutual agreement among many nations which operate as free agents and are therefore at liberty to preserve or destroy this resource without accountability.

Fortunately, modern communications technology has expanded our horizons to such an extent that awareness of the problem is now global. At the level of public opinion, environmental activism has taken off to an extent undreamed of just two decades ago. One reason is the tremendous public relations success of the first global environmental conference, the United Nations Conference on the Human Environment, held at Stockholm in 1972. Although the level of rhetoric exhibited at that gathering far surpassed the extent of action, the message of concern and urgency did permeate large sectors of the population. The point was underscored in 1987, when the World Commission on Environment and Development published its report, *Our Common Future*.[82] From the many hearings conducted by the Commission, it soon became evident that the people of the world were demanding that their governments implement "sustainable development," development within a safe environmental framework, development which would provide a better life now but not at the expense of the future survival of humans and other species. *Our Common Future* also charts a path toward accepting the idea of the inevitable linkage between development and environment, reminding us that "the 'environment' is where we all live; and 'development' is what we all do in attempting to improve our lot within that abode."[83]

At the time of UNCED in Rio in June 1992, two decades after the Stockholm Conference, thousands of people, primed with knowledge of environmental problems in their own localities and around the world, watched on television as the parade of political leaders pledged their commitment to the ideals of environmentalism. There is no doubt that the awakening of people ranging from urban dwellers to indigenous tribes has been one of the most significant forces behind the Convention on Biological Diversity. Populations have moved their governments to commitment and even, at times, to action.

The problem is that all too often dedication to an environmental ethic is governed by the degree of self-interest involved. It is relatively easy to espouse a cause at someone else's expense. As the American spotted owl controversy indicates, saving species is all well and good provided one's own job is not in jeopardy. Governments are often pulled in different directions by various interest groups, each vigorously propounding or attacking the particular environmental issue at stake. Project this situation onto the much larger international stage and some idea of its complexity can be perceived. As the United Nations is not a world government and has no sovereign power, the formulation of any agreements which involve implementation at the national level can be a real challenge. Given the vast scope of this subject of biodiversity and the enormous attention it has received, one cannot, in a section of this length, do much more than provide a flavor of the historic significance of the earlier attempts to deal with the crisis, with emphasis on the few years preceding UNCED.

The United Nations Conference on the Human Environment, which convened in Stockholm in 1972, adopted a Declaration of Principles relevant to a number of aspects of the environmental agenda. Principle 4 of that Declaration states:

> Man has a special responsibility to safeguard and wisely manage the heritage of wildlife and its habitat which are now gravely imperilled by a combination of adverse factors. Nature conservation including wildlife must therefore receive importance in planning for economic development.[84]

The fact that burgeoning human population was imperilling other life forms was clearly recognized and the danger of extinction was appreciated. The point was again emphasized in the Charter for Nature, which the United Nations General Assembly adopted in 1982. Principle 2 of that important declaration specifies that

> [t]he genetic viability of the Earth shall not be compromised; the population levels of all life forms, wild and domesticated, must be at least sufficient for their survival, and to this end, necessary habitat shall be safeguarded.[85]

These principles form the foundation on which the Convention on Biological Diversity of 1992 is built. That it took two decades to translate the nonbinding obligations of these earlier declarations into the legally binding articles of an international Convention testifies both to the complexity of the subject of biodiversity and to the many political and economic issues which have inevitably become entangled with it. One can only agree with Cyril de Klemm, who, commenting on the duty of conservation undertaken by States, remarks that "[i]nternational conventions . . . contain binding conservation obligations entailing limitation of the sovereign rights of States over their natural resources. The fact that these limitations have been accepted voluntarily and without any counterpart, for the common good, is a concrete manifestation of the present consensus."[86]

Consensus on the significance of the problem generated a plethora of national and regional measures to raise global awareness, conserve biodiversity, and preserve endangered ecosystems. Biodiversity fits into so many categories of environmental concern that it is accommodated in a bewildering variety of conservation measures undertaken by governments and within specific regions. The flurry of regional and international activity has produced many instruments and action plans, like the Protocol for the Protection and Development of the Marine Environment of the Wider Caribbean Region (Protocol of 1989 to the Cartagena Convention).[87] One might also mention the World Conservation Strategy, the Global Network of Biosphere Reserves,[88] the Keystone Center's International Dialogues on Plant Genetic Resources, and the International Undertaking on Plant Genetic Resources sponsored by the United Nations Food and Agriculture Organization (FAO).[89]

Recognizing the loss of biological diversity as the ultimate consequence of all types of man-made pollution and human development, some nations have sought to tackle specific aspects of this crisis. However, not until UNCED in 1992 and the signing of the Convention on Biological Diversity has there been an attempt to create an international treaty which embraces almost the entire spectrum of the problem and focuses on the specific threat of extinction of species. Among the more prominent earlier international instruments which tried to deal with some facets of the problem is the Convention on International Trade in Endangered Species of Wild Flora and Fauna (CITES), adopted in 1973. This agreement tries to protect endangered species by restricting commerce in products made from them. Where a particular species faces extinction, the Convention prohibits all trade. In 1990, for example, a ban was placed on the ivory trade, largely to protect the world's rapidly dwindling elephant population.[90] A month before UNCED convened, the International Tropical Timber Council, meeting in Yaoundé, Cameroon, decided to encourage governments to coordinate activities concerning tropical timber listings in the CITES Appendices with the International Tropical Timber Secretariat.[91]

In March 1991, delegates representing member states of the United Nations Economic Commission for Latin America and the Caribbean adopted the Tlatelolco Platform on Environment and Development, which emphasizes the importance of conservation to that region "in order to protect its biological diversity, which constitutes a fundamental and sovereign part of the national patrimony of these countries endowed with such diversity."[92]

The European nations have been very active in the field of wildlife conservation through implementation of the Convention on the Conservation of European Wildlife and Natural Habitats (commonly referred to as the Berne Convention).[93] Extension of the Convention to eastern Europe and to some African nations was accepted in 1990 by European Environment Ministers representing member countries of the Council of Europe.[94]

Additional activity has been generated by conferences such as the one called by Al Gore in 1990 when he was a United States Senator. This international conference on the environment attracted delegates from forty-two countries and met in Washington, D.C., from 29th April to 2nd May 1990.[95] The parliamentarians attending this important meeting pledged themselves to pursue sustainable development. Furthermore, with respect to biodiversity, the delegates committed themselves to complete the Convention on Biological Diversity by 1992, preserve primary forests from further destruction by the year 2000, undertake national inventories of the biological resources, and prohibit trade in endangered species.[96] One of the most progressive ideas adopted by the conference was the proposal that "[n]ational law should require accountability by all institutions whose actions affect the survival of species and of habitats, including the effect of legislation itself upon the environment."[97] Although these decisions were not binding on the states which were represented at the conference, they do "provide a consensus basis for action for legislators in their own countries."[98]

In March 1992, just three months before the convening of UNCED, a Conference on Environment and Law in Asia drew delegates from more than twenty countries to Kathmandu, Nepal. They adopted the Kathmandu Declaration on Environment and Law, which includes support for strong international agreements to preserve biodiversity.[99]

The educative process was also stressed; one example is the project to develop expertise in biodiversity conservation and international law undertaken by the Environmental Law Centre. A seminar held in November 1991 in Harare, Zimbabwe, explored various aspects of biodiversity, with emphasis on the role of developing countries.[100] Involvement in the cause of biodiversity was not limited to scientists and international lawyers. The United Nations Centre on Transnational Corporations (UNCTC) developed criteria for sustainable development management by large industrial enterprises, including the suggestion that "[c]orporate practices must not threaten species diversity. . . . [E]xecutives should propose ways to conserve affected botanical and wildlife resources. . . . Diversity is essential for maintaining the planet's ecological balance."[101] As transnational corporations handle "one quarter of the world's productive assets; 70 per cent of the products in international trade; 80 per cent of the world's land cultivated for export crops;

and the major share of the world's technological innovations,"[102] their impact on the survival of biodiversity cannot be overemphasized. However, UNCTC also reported that while some transnational corporations engaged in some environmentally sound activities, much work remained to be done to persuade them to integrate environmental concerns into their corporate planning.[103]

Meanwhile, work was proceeding apace on the creation of an international, legally binding convention which would be signed at the Earth Summit in Rio in 1992. The Governing Council of the United Nations Environment Programme initiated the creation of a working group of experts to "investigate the desirability and possible form of an umbrella convention to rationalize current activities in" the field of biodiversity.[104] The first meeting of this group was convened in Geneva between 16th and 18th November 1988.[105] The challenge was to bring some order to the multiplicity of national and regional measures then in existence.

> The Working Group noted that there were a number of international agreements in this field that dealt with different aspects of a common purpose, i.e., conservation of the diversity of ecosystems, species, and genes. It was also noted that each convention had its particular purpose and that the Parties to each convention differed. The Working Group concluded that amendments to existing conventions for purposes of achieving "rationalization" or consolidation of resources would be difficult and time-consuming.[106]

The conclusions of this group would be of crucial significance to the Convention on Biological Diversity. Some of these proposals would eventually become formulations in the Treaty itself. In fact, the very existence of the international, legally binding Convention is probably due to the working group's agreement

> that even the totality of the existing conventions could not cover the full range of biological diversity. Action was therefore needed now either through a convention or through other measures. Most of the experts favoured the elaboration of a binding instrument, not excluding at the same time other measures for the conservation of biological diversity.[107]

Negotiations for the international agreement were intense and protracted, as is usual in such situations, particularly when the instrument under discussion will be legally binding. The final conference for the adoption of the agreed text of the Convention convened in Kenya on 22nd May 1992, days before the formal signing ceremony was to take place at Rio.[108] This last session was preceded by a number of meetings of technical experts and by several negotiating sessions held between November 1988 and May 1992.[109]

Dr. Mostafa Tolba, then Executive Director of the United Nations Environment Programme, continued to promote the process which eventually led to the adoption of an agreed text. He expressed his belief that "[t]he process of preparing a new legal instrument must be extensive and open to all Governments"[110] and suggested that "[w]ide-ranging consultations will help to develop the proposed framework convention, serve educational and promotional purposes, and stimulate active support from the countries that participate in the negotiations."[111] Additional support was provided to the process by the Preparatory Committee for UNCED.[112] The hope was that "the political momentum of UNCED might help generate rapid signature" of the Convention.[113] As Maurice Strong, Secretary-General of UNCED, pointed out, the negotiations for the Convention were complementary to the UNCED process and the organization of the Earth Summit in Rio. He also recognized that "biological diversity, which was central to a variety of concerns, was one of the most sensitive and difficult issues before UNCED."[114] However, despite all the prompting in favor of a Convention, the process was slow and tortuous. As Dr. Tolba reminded delegates to the February–March 1991 Meeting of Legal and Technical Experts on Biological Diversity, in the four years since the United Nations Environment Programme had called for the formulation of an international convention, over one million species had become extinct.[115]

Clearly, there is a tremendous amount of global concern over the biodiversity crisis, concern which has resulted in national, regional, bilateral, and multilateral measures to conserve this precious resource. Although, as outlined above, there has been a multiplicity of actions, the efforts have been piecemeal, not comprehensive. It was time to coordinate the universal enthusiasm into an international agreement which would focus the attention of governments, lawyers, environmentalists, and the general public on the need to save species from extinction. Responding to the pressures imposed on them, delegates adopted the final text of the Convention on 22nd May 1992, just days before UNCED's formal opening on Wednesday, 3rd June.[116]

Summary of the Convention on Biological Diversity

The principal objectives of the Biodiversity Treaty are to create a system for equitable sharing of the benefits of biological resources; to conserve biological diversity; and to utilize this resource in a manner which conforms to the concept of sustainable development. It sets forth both principles and specific actions to be undertaken by signatory nations and addresses the crucial issues of funding, technology transfer, and other forms of assistance to be allocated by the developed nations to the developing countries.

The rather lengthy Preamble affirms that *"the conservation of biological diversity is a common concern of humankind"*; reaffirms that *"States have sovereign rights over their own biological resources;"* but insists that *"States are responsible for conserving their biological diversity and for using their biological resources in a sustainable manner."*

The Convention links the reduction in biological diversity to human activity (Preamble), emphasizes the need for further research (Preamble), and adopts the precautionary principle that *"where there is a threat of significant reduction or loss of biological diversity, lack of full scientific certainty should not be used as a reason for postponing measures to avoid or minimize such a threat"* (Preamble). In line with other formulations adopted at UNCED, this Convention refers to the vital role of women in conservation (Preamble) and to the dependence of indigenous communities (Preamble). It also acknowledges the needs of developing countries, particularly those frequently classified as least developed (Preamble). It acknowledges that *"economic and social development and poverty eradication are the first and overriding priorities of developing countries"* (Preamble).

The Parties to the Convention undertake to perform a number of duties. These involve the development of national strategies for sustainable use of biodiversity (Article 6[a]); the integration of conservation into relevant sectoral programs (Article 6[b]); the identification of components of biodiversity (Article 7[a]) and of processes and activities which may have an adverse impact (Article 7[c]); the monitoring of components (Article 7[b]) and of the effects of possibly adverse activities (Article 7[c]); and the organization of data (Article 7[d]). Contracting Parties pledge to promote research and training *"for the identification, conservation and sustainable use of biological diversity"* (Article 12[a] and Articles 12[b] and 12[c]). Parties agree to promote international scientific cooperation (Article 18.1). Public awareness is also to be encouraged (Article 13[b]).

The Convention extends the responsibilities of signatory nations beyond their borders, as, for example, when certain activities are likely to have an adverse impact on a neighboring state. Transboundary obligations include notification and exchange of information (Article 14.1[c]), as well as action to prevent or minimize the danger to another state of any such activity (Article 14.1[d]). The attendant issues of liability and redress are left for examination by the Conference of the Parties (Article 14.2).

The Convention encourages conservation of biodiversity within its natural habitat, referred to as in-situ conservation. There are detailed provisions concerning the establishment of protected areas (Article 8[a]). Measures are required for the *"protection of ecosystems, natural habitats and the maintenance of viable populations of species in natural surroundings"* (Article 8[d]). The Convention emphasizes the need for national legislation and regulation (Article 8[k]). Ex-situ conservation is also considered important, and provision is made for establishing research facilities (Article 9[b]).

Contracting Parties pledge, as far as possible, to *"[i]ntegrate consideration of the conservation and sustainable use of biological resources into national decision-making"* (Article 10[a]) and to encourage cooperation between the public and private sectors to promote sustainable use of biodiversity (Article 10[e]).

In dealing with the rather controversial issue of access to genetic resources, the Convention firmly concedes to the nationalistic agenda of the developing nations and acknowledges that *"the authority to determine access to genetic resources rests with the national governments and is subject to national legislation"* (Article 15.1). However, Contracting Parties cannot restrict access in ways that run counter to the objectives of the Convention (Article 15.2). *"Access to genetic resources shall be subject to prior informed consent of the Contracting Party providing such resources"* (Article 15.5), and the Convention, in a very significant clause, specifies that *"[e]ach Contracting Party shall take legislative, administrative or policy measures . . . with the aim of sharing in a fair and equitable way the results of research and development and the benefits arising from the commercial and other utilization of genetic resources with the Contracting Party providing such resources"* (Article 15.7).

Technology transfer and access to biotechnology for developing nations are among the most controversial elements of the Biodiversity Treaty. The needs of the South are predominant in this section of the Convention. Contracting Parties undertake to *"facilitate access for and transfer to other Contracting Parties of technologies that are relevant to the conservation and sustainable use of biological diversity"* (Article 16.1). Such transfer is to be provided on *"fair and most favourable terms, including on concessional and preferential terms where mutually agreed"* (Article 16.2). With respect to the issue of patent rights—one of the most serious concerns of the United States of America—the Convention acknowledges that where the technology is subject to intellectual property rights *"such access and transfer shall be provided on terms which recognize and are consistent with the adequate and effective pro-*

tection of intellectual property rights" (Article 16.2). However, the Convention asks Contracting Parties to invoke legislative and administrative policy to provide technology, on mutually agreed terms, even where patent protection exists (Article 16.3). Most significant of all, Parties are to cooperate "*subject to national legislation and international law in order to ensure that such* [intellectual property] *rights are supportive of and do not run counter to*" the objectives of the Convention (Article 16.5). This issue will be discussed in detail below in the section dealing with the position of the United States.

Developing countries are to be encouraged to participate in biotechnological research (Article 19.1) and are to be given "*advance priority access on a fair and equitable basis . . . to the results and benefits arising from biotechnologies. . . . on mutually agreed terms*" (Article 19.2). The issue of safe handling of living modified organisms is left to a future protocol (Article 19.3).

Contracting Parties undertake to provide financial resources in accordance with their capabilities (Article 20.1), with developed nations providing "*new and additional financial resources*" to developing countries to enable the latter to fulfil the objectives of the Convention (Article 20.2). This obligation is made voluntary for the countries in transition (Article 20.2)—the former Soviet republics and the countries of Eastern Europe—because their economies are moving from the Communist to the capitalist system. With emphasis on "*adequacy, predictability and timely flow of funds*" (Article 20.2), the Convention also makes provision for bilateral, regional, and other multilateral financial aid (Article 20.3). Participation by developing countries is conditional on the availability of adequate funding (Article 20.4). The significance of this provision is such that it is quoted in full now and will be discussed in detail later.

> *The extent to which developing country Parties will effectively implement their commitments under this Convention will depend on the effective implementation by developed country Parties of their commitments under this Convention related to financial resources and transfer of technology and will take fully into account the fact that economic and social development and eradication of poverty are the first and overriding priorities of the developing country Parties.* (Article 20.4)

The Convention provides for a financial mechanism, functioning under the authority and guidance of and accountable to the Conference of the Parties (Article 21.1), operating within a "*democratic and transparent system of governance*" (Article 21.1). On an interim basis, following restructuring, the

Global Environment Facility of the United Nations Development Programme, the United Nations Environment Programme, and the International Bank for Reconstruction and Development were designated the financial mechanism for this Convention (Article 39).

The Conference of Parties would convene no later than one year after the Convention entered into force (Article 23.1), which could only occur ninety days after thirty states ratified it (Article 36.1). The Conference of Parties is assigned various duties, including the task of reviewing implementation of the Convention (Article 23.4) and adopting protocols (Article 23.4[c]) and amendments (Article 23.4[d]). The Conference of Parties can, more generally, *"[c]onsider and undertake any additional action that may be required for the achievement of the purposes of this Convention in the light of experience gained in its operation"* (Article 23.4[i]).

The Convention also establishes a Secretariat to perform various administrative functions (Article 24) and a Subsidiary Body on Scientific, Technical and Technological Advice (Article 25). The dispute settlement mechanism provides for conciliation (detailed in Annex II, Part 2), arbitration (detailed in Annex II, Part 1), negotiation, mediation, and submission of the problem involved to the International Court of Justice (Article 27). Amendments to either the main text or to any subsequent Protocols will be by consensus or, failing that, by a two-thirds majority vote of the Parties to the instrument in question present and voting (Article 29.3). "No reservations may be made to [the] Convention" (Article 37).

The Controversial Role of the U.S. Government

THE UNITED STATES AND THE RIO PROCESS

The role of the United States is pivotal in any analysis of the Convention on Biological Diversity, and this section will explore some of its facets. However, the question of the American position at Rio will also be discussed elsewhere in the chapter when particular clauses of the Treaty are being analyzed. The controversial attitude of the United States to the Biodiversity Convention is relevant to many aspects of this study and will be explored in considerable detail.

Much to the dismay of environmentalists both in the United States and throughout the world, the Administration of President George Bush turned his nation into the bête noire of the Earth Summit. Through much of the UNCED process—which included negotiations leading to the Rio Declaration, Agenda 21, the Framework Convention on Climate Change, and the

Convention on Biological Diversity—the United States, the world's only superpower, adopted a stance which insisted on the primacy of its own national interests at the expense of the universalist solutions which were proposed by environmentalists and by numerous other nations. This position earned the United States government an enormous amount of opprobrium both at home and abroad, making the Summit attendance of George Bush a public relations disaster and perhaps costing him votes in the 1992 presidential election. That George Bush began his presidency with an agenda to become known as the "environmental president"[117] makes this turnabout even more puzzling. After all, President Bush had said that "[e]nvironmental problems are global, and every nation must help in solving them."[118] Doubtless, scholars and academics will debate this inconsistency between his words and actions for years to come. Here, suffice it to say that the United States stood firmly against the development-oriented agenda the South proposed for the Rio Declaration and refused to agree to firm timetables to deal with the problem of global warming tackled by the Climate Change Convention. Indeed, according to a Reuters report in the *Globe and Mail*, President Bush refused to attend the Earth Summit in Brazil until other nations gave in and agreed to weaken the Climate Change Convention.[119] In an editorial, the *Vancouver (B.C.) Sun* was critical of the fact that President George Bush "predicated his presence in Rio on his ability to control the agenda."[120] The President told a meeting of business executives that "while he supported international cooperation to protect the environment, he would not sign any agreement that could curb long-term US economic growth."[121] He was also very cautious in pledging funding for environmental projects in developing nations and, most serious of all, decisively rejected and refused to sign the Convention on Biological Diversity.

These actions prompted Christopher Young of Southam News to comment critically that

> Bush proceeds to turn himself into a world-class villain by aiming to torpedo the Earth Summit, an attempt to address environmental dangers threatening the entire world, a motherhood issue if ever there was one. The politician who would refuse to sign up in this cause, must harbor a political death wish.[122]

Because the United States is such an important country politically, economically, and militarily, its actions at Rio were publicized throughout the world. There are few who doubt that the cause of environmentalism was dealt a serious blow by the approach taken by a nation which could have made itself environmental leader of the new world order. *Veja*, a Brazilian weekly, declared that "Bush comes to Rio as Earth Summit enemy."[123] Bill Walker of Greenpeace accused the United States of "blocking progress. They are trying to shape the summit agenda so that it reflects the United States' agenda, which has been to resist efforts to save the environment on a number of fronts."[124] Since the United States spends more annually than any other nation on environmental clean-up and improvement ($120 billion, by 1992 estimates)[125] and reportedly spent $800 billion for this purpose between 1982 and 1992,[126] it might quite easily have assumed a leadership role during the Earth Summit. That it rejected so significant an opportunity, and seriously damaged the Convention on Biological Diversity by not signing it, have to be explored by reference to the country's internal political and economic considerations.

First, if one assumes (as did an editorial in the *Christian Science Monitor*) that "the occupant of the White House sets the tone for environmental protection"[127] both in the United States and to some extent globally, then one can only conclude that President Bush and his advisors calculated that environmentalism was a risky cause to propound in an election year. The Republican reticence about environmental issues could be perceived as a reaction to a recessionary economy and as a calculation that the American voter would insist on jobs rather than ecological reform. Accordingly, it was inevitable, though unfortunate, that U.S. officials, according to a report in *Time*, kept insisting that "the American life-style is not up for negotiation."[128] This unfortunate remark was to be thrown back at American delegates, and it echoed and re-echoed in the world's media throughout the Earth Summit.

David Holstrom of the *Christian Science Monitor* reported that a research team at the University of San Paulo had concluded that the globalization of the American lifestyle, with its large homes, numerous appliances, and two or three cars per family, would "be impossible in terms of the energy required to sustain it."[129] From the perspective of environmental conservation, the statistics on this lifestyle are frightening. The United States population, a mere 5 per cent of the planet's total, manages to consume 25 per cent of global energy and to emit 22 per cent of global carbon dioxide annually.[130] "One-fifth of American households now own three or more vehicles and 90 percent of new cars have air conditioning."[131] In 1987 Americans discarded 3.4 million tons of appliances; statistics for other refuse—including clothing, newspapers, and yard wastes—amounted to 1,429 pounds for every adult in the United States.[132] The American lifestyle is envied in much of the world largely because it encourages upward mobility regardless of origin, the acquisition of wealth, and its unrestricted consumption. Unfortunately, this way of life is contrary to the concept of sustainable development, is very costly to the American and global environment, and would, if emulated around the world, probably result in the environmental destruction of this planet and all its species, including human beings.

To some leaders of poor countries in the developing world, who are attempting to satisfy the consumer cravings of their vast populations and yet protect the global environment, the American emphasis on the sanctity and non-negotiability of this lifestyle appeared to be the worst form of hypocrisy. According to the United Nations, "the richest 20 percent of the world's population uses 83 percent of its output."[133] The Prime Minister of Malaysia, Dr. Mahathir bin Mohamad, told delegates at UNCED that "if we begin by saying that our life-style is sacred and not for negotiation then it would be meaningless to talk of development and the environment."[134] He went on to explain that as one fourth of the world's population—the rich countries—consume 85 per cent of its wealth and produce 90 per cent of its waste, a reduction of 25 percent in this consumption would reduce global pollution by 22.5 per cent.[135] The American stand at Rio was perceived as an aggressive assertion of self-interest in a forum where nations were expected to subsume their national agendas to a larger ethic of environmental benefit for the whole world. As the editor of the *Vancouver Sun* suggested, "the fate of all cannot be mortgaged for the comfort of some."[136]

Cynics suggested that the priority in George Bush's mind was the election, not his status as potential world environmental leader. "Bush is wary of putting any limits on industry that could jeopardize the fledgling economic recovery and put his own re-election at risk."[137] Tommy Koh of Singapore, one of the dynamic mainsprings of the Rio process, commented, "This will teach the United Nations not to hold a conference in an American election year."[138] Martin Fletcher of the *Times* (London) suggested that Bush had "consistently put the demands of business and industry before environmental concerns."[139] It could also be assumed that environmentalism was perceived in the White House as unnecessary baggage which would not impress the American people. The *Christian Science Monitor* commented:

> [E]nvironmental preservation and protection is unlikely to be a determining issue in this year's presidential election. It is difficult to focus on spotted owls and fouled streams when Dad has been laid off at the tractor factory and Mom's job as a supermarket cashier falls short in meeting the bills.[140]

At the very moment when delegates were gathering in Rio to attend UNCED, the Government of the United States reported an unemployment

rate of 7.5 per cent, the highest in eight years.[141] Bob Dederick, Chief Economist at Chicago's Northern Trust Company, suggested shortly after UNCED
that the unemployment issue was an albatross around the President's neck.[142]
Joel Naroff, Chief Economist for First Fidelity Bancorporation in Philadel-
phia, explained that the annual rate of increase in jobs (0.3 per cent) was the
lowest since the Second World War.[143] Two months before Rio, in a survey
taken in April 1992, only 3 per cent of American voters believed that the
environment was the most important issue.[144] However, a *USA Today*/CNN/
Gallup poll released in January 1992 showed that 60 per cent of Americans
considered the environment a very important campaign issue, although it
ranked eleventh out of sixteen issues.[145] The domestic situation in the United
States could not but have an international impact, and these depressing economic statistics, arising in an election year, would influence U.S. policy
towards the Rio formulations. On a more positive note, a Gallup poll released
in June 1992 (the month of the Earth Summit) demonstrated that 65 per cent
of those surveyed in the United States would "trade higher prices for a safer
environment," and 59 per cent indicated that "they would accept slower
economic growth to protect the environment."[146]

There was a tendency as well to suggest (as in the *Economist*) that the
mainspring of United States resistance to the ideals espoused at Rio was
Vice-President Dan Quayle, specifically his Council on Competitiveness.[147]
This Council was established by President Bush to ensure that "federal regulations do not impede economic progress."[148] The Council was accused in an
editorial in the *Christian Science Monitor* of weakening "the impact of the
Clean Air Act."[149] Vice-President Quayle is thought to have been anxious to
protect the biotechnology industry in the United States from any threat to
intellectual property rights posed by an international convention on biodiversity.[150] Critics of the Council alleged that this body "pushed changes that
weakened the Clean Air Act, tried to rewrite government protections for wetlands, and forced the Environmental Protection Agency to drop a proposal
that would have reduced municipal incineration by 25 percent."[151] The
Christian Science Monitor stated in an editorial that "US leadership has been
driven primarily by concerns that agreements to be signed at Rio don't infringe too heavily on American economic prerogatives."[152] Vice-President
Quayle argued that restrictive environmental legislation was "harming initiative and the performance of the American economy."[153]

According to Thomas Jorling, New York State Commissioner of Environmental Conservation, the Council on Competitiveness had "stymied" the efforts of the United States Environmental Protection Agency (EPA).[154] The

Christian Science Monitor also suggested that "[t]he Republican record of the past 12 years has generally been one of constant battle with environmental regulation."[155] However, Administration officials suggested that the environmental influence of the Vice-President and his Council on Competitiveness was overrated.[156]

President Bush perceived his environmental record optimistically. "I came to this office," he said, "committed to extend America's record of environmental leadership and I've worked to do so in a way that is compatible with economic growth."[157] Among his environmental achievements, the President pointed to the market-oriented clean air legislation; the moratorium on offshore oil and gas development in both the eastern and western United States; improved enforcement of regulations; acceleration of CFC phaseout to 1995; the ban on driftnet fishing; support for a ban on the ivory trade; and an extensive tree-planting program.[158] Michael Deland, Chairman of the White House Council on Environmental Quality, also pointed to the allocation of $1 billion for acquisition of new parkland and areas for recreation and a tripling of the rate of cleanup of Superfund toxic-waste sites.[159] Additionally, supporters of the President stressed the Administration's achievement in raising the enforcement budget of the EPA by 72 per cent, doubling federal spending for the acquisition of wetlands, doubling spending on research in energy and conservation subjects, and halting construction of the Two Forks Dam in Colorado.[160]

Internationally, the United States played a very significant role during the entire UNCED process, a fact often overlooked and ignored because negative publicity appeared to accentuate only the controversial positions adopted by the Bush Administration. As Russell S. Frye commented in a presentation to the International Bar Association in September 1992,

> Many of the key initiatives at UNCED were led by the United States. The US brokered compromise on energy issues with Saudi Arabia and other OPEC nations. . . . the US supported the African countries in seeking a treaty on desertification. American representatives pushed for stronger provisions for monitoring, restrictions on overfishing, and the like to protect the nation's oceans. The US was a leader on forestry issues, much to the consternation of Malaysia and other countries opposed to any restriction whatsoever on their harvesting of tropical rainforests.[161]

The United States also committed itself to a Biodiversity Research Initiative to encourage "the development of biodiversity inventories and surveys to create the information base necessary for the protection of species."[162] Brad

Knickerbocker of the *Christian Science Monitor* quoted one American delegate to UNCED as commenting, "What we do is never considered enough, and we've come to live with that."[163]

The crucial question is whether the American presidential election influenced the universal acceptance and success of the Convention on Biological Diversity. There can be no doubt that ratification by signatories, implementation, and action at the national, regional, and international levels would have proceeded far more rapidly with respect to this and other Rio formulations had the American President lent the weight of his government's support to the process instead of becoming its most prominent obstacle. The answer to the question will have to be a tentative affirmative. So great is the influence of the United States in world affairs that it ought not to be surprising that its rejection of this Convention would have a tremendous impact.

Arguably, the United States position at Rio was a product of many factors: the ideology of Vice-President Quayle and some members of the White House staff, to say nothing of President Bush himself; the recessionary economy and the need to assure Americans that jobs were a top priority; the desire to protect American industry and business interests, even at the expense of global environmental considerations; the apparent lack of interest in environmental matters during the presidential campaign—a signal to the candidates that the economy mattered more than ecology; and, finally, an unwillingness or inability on the part of President Bush to seize the initiative in the war against planetary devastation with the same firmness he had displayed in his war against Iraqi aggression in Kuwait. He lost an opportunity which could have assured him a reputation as one of America's great presidents and possibly as an outstanding world leader. That he may have sacrificed permanent glory for an election victory which eluded him in any event is ironic and, for the United States, the Rio process, and the entire world, lamentable if not tragic.

THE UNITED STATES OPPOSITION TO THE CONVENTION
ON BIOLOGICAL DIVERSITY

Disparaged as a "squiffy little treaty,"[164] the Convention on Biodiversity aroused considerable opposition in the Bush Administration. According to the *Economist*, the U.S. Department of the Interior played a significant role in the negotiations and "disliked the whole idea of the biodiversity convention. It would probably have preferred the talks to collapse long before Rio, leaving no treaty for President Bush to sign."[165] That the talks did not collapse, and were indeed successful, is testimony to the determination of the rest of the world to pursue the environmental agenda regardless of the opposition of the United States. The *New York Times* suggested that E. U. Curtis Bohlen, Assistant Secretary of State for Oceans and International Environmental and Scientific Affairs, who led the U.S. delegation in the biodiversity negotiations, was in favor of acceptance, as was William K. Reilly, Administrator of the EPA.[166] A memo of 14th April 1992 by David M. McIntosh, Executive Director of the President's Council on Competitiveness, is thought to have been decisive in influencing the Bush Administration to reject the Convention. According to McIntosh, "[E]conomic harm from signing the treaty would substantially outweigh environmental benefits."[167] McIntosh denied that he expected his memorandum to "stop a treaty," insisting that "[w]e thought people should be made aware of it and let them adjust it in negotiations."[168]

There were a variety of objections on the American side to the provisions and priorities of the Convention. The entire concept of biodiversity was probably perceived as a "hot potato" in the domestic arena, largely because "[b]iodiversity issues can spark major political battles in the US, as the spotted owl controversy shows."[169]

Testifying before the Standing Committee on Environment of the Canadian House of Commons in November 1992, Walter Reid, Vice-President of the World Resources Institute, of Washington, D.C., explained that

> the reason why the U.S. didn't sign appears to be, in a word, politics. The U.S. misjudged the potential effects of the convention and based an ill-considered decision on what appeared to be politically expedient rather than environmentally and economically desirable.[170]

Although political expediency undoubtedly played a role, it cannot be perceived as the only reason for the United States rejection of the Convention. Although the controversial aspects of the Convention will be discussed later, it is important here to explain why the American delegation felt so strongly about the weaknesses of the Biodiversity Convention. In seeking to regulate corporate activities to some extent, the Treaty may have been more than Bush could ideologically tolerate,[171] given his preference for the untrammelled operation of the free market. The Administration had committed itself to less, not more, regulation of technology, as evidenced by the White House release in February 1992 of a "scope document" which "established that biotechnology products do not pose any inherent risk and therefore [would] not be regulated differently than products of other technologies."[172] In May 1992, shortly before the Earth Summit, the U.S. Food and Drug Administration announced that "genetically engineered food products [would] not receive separate or special regulatory attention."[173]

The concept of enforced sharing of benefits derived from the genetic resources of the South was probably too threatening for these proponents of the former system, under which the South derived no benefit from its genetic resources while Northern companies which carried out the research and manufactured the product derived from the Southern resource made enormous profits. The Biodiversity Treaty seeks essentially to rectify this situation, which is perceived as unjust in the developing world. In its Preamble the Convention encourages the *"desirability of sharing equitably benefits arising from the use of traditional knowledge, innovations and practices relevant to the conservation of biological diversity and the sustainable use of its components."* The Convention also states that *"access to and sharing of both genetic resources and technologies are essential"* (Preamble). Because the American biotechnological industry rose to global prominence in an era when the source country of the original genetic material was not even recognized, the changed situation created by the Convention was probably perceived as threatening to the status quo. The economic stakes for the United States are very high. With respect only to the pharmaceutical industry, the World Resources Institute estimates that drugs derived from plants have a retail value of approximately $43 billion a year and "the industry is growing at more than 10 percent per year."[174] At UNCED, President Bush couched his opposition in the language of principle:

> We come to Rio prepared to continue America's unparalleled efforts to preserve species and habitat. And let me be clear. Our efforts to protect biodiversity itself will exceed—will exceed—the requirements of the treaty. But that proposed agreement threatens to retard biotechnology and undermine the protection of ideas. . . . [I]t is never easy, it is never easy to stand alone on principle, but sometimes leadership requires that you do. And now is such a time.[175]

William Stevens reported in the *New York Times* that according to the White House the process of international regulation, once under way, could have a serious impact on the United States. The Administration argued that the Convention "would lead to international regulation of the genetic engineering industry, an area in which the United States would like to maintain competitive leadership."[176] The Bush White House was subjected to intense and well-organized pressure from such groups as the Association of Biotechnology Companies, the Pharmaceutical Manufacturers' Association, and the American Intellectual Property Law Association. Lisa Raines, Vice-President of Industrial Relations for the Industrial Biotechnology Association, advised the United States negotiators to reject provisions in the Convention concerning intellectual property rights. She complained about the "highly objectionable provisions permitting developing countries to ignore or restrict intellectual property rights in the field of biotechnology."[177] It was also feared that accepting the provisions of the Convention could "set a bad precedent for negotiations on the General Agreement on Tariffs and Trade (GATT) and other trade agreements."[178] William Reilly, Head of the United States Delegation to the Earth Summit, agreed, explaining that "[w]e have negotiated in the Uruguay Round of GATT to try to protect Intellectual Property Rights. We're not about to trade away here in an environmental treaty what we worked so hard to protect there."[179]

One of the most important achievements of this Convention lies in its clear recognition that biodiversity is a *national* resource and not part of the common heritage of mankind. Inevitably, those biotechnological companies which have profited by the common heritage concept cannot but resent the new situation. In its Preamble, the Convention declares that *"States have sovereign rights over their own biological resources."* The endorsement of national rights goes further: *"States have, in accordance with the Charter of the United Nations and the principles of international law, the sovereign right to exploit their own resources pursuant to their own environmental policies . . ."* (Article 3). Because the Contracting Parties are nation states, the enormous activity generated by the Convention will, inevitably, take place primarily at the national level when parties *"[d]evelop national strategies, plans or programmes for the conservation and sustainable use of biological diversity"* (Article 6[a]) and *"[d]evelop or maintain necessary legislation"* to protect threatened species (Article 8[k]). Hence the Convention brings biodiversity within the sphere of national control and directs individual governments to take all kinds of measures, from research and training (Article 12) to undertaking environmental impact assessments (Article 14.1[a]). Walter Reid, Vice-President of the World Resources Institute, explains this aspect of the significance of the Convention:

> From the standpoint of global biodiversity conservation, the most important thing is that it confirms under international law that biodiversity is a sovereign national resource and that governments have the authority to determine the conditions under which access to that resource is granted. The distinction between this treatment of biodiversity and its previous treatment as the common heritage of mankind could not be sharper or its implications for conservation more profound.[180]

The Bush Administration was also reluctant to support the financial stipulations of the Convention, because it appeared that the Treaty made financial contributions mandatory.[181] President Bush apparently objected to the funding clauses[182] and refused to sign any document which would give "developing countries too much say in funding decisions."[183] The essential objections were therefore largely grounded in economics, the U.S. Government stating that the Convention "could enable developing countries to extract limitless funds from wealthy nations for preserving endangered species, and could hinder the continued access of America's biotechnology and pharmaceutical industries to those species found in Third World countries."[184] In an oft-repeated remark, quoted in *U.S. News and World Report*, President Bush said that "we are going to stay out front, but we are not going to act like we have an open checkbook."[185] In a press conference, he also insisted, "We don't have an open pocket book and we cannot enter into anything if we don't keep the commitment. The financial arrangements are too open-ended for us."[186] The President repeated these objections about the financial provisions in his speech at UNCED, stating that the Convention's "financing scheme will not work."[187]

Financial considerations were therefore primary, although the Bush Administration voiced strong objections to what it considered to be the all-out assault on the concept of intellectual property rights in the Convention. The President suggested, in fact, that this was the biggest problem with the Convention: "the question of intellectual property rights—the proposal that companies in the developed world making use of the resources of poorer nations should have to pay them royalties on the products they developed."[188]

To cynics, it appeared that the United States wanted to continue to have free and unfettered access to the South's treasure trove of biodiversity without any consideration of sharing the benefits. The environmentalist Vandana Shiva voiced the frustration of developing nations: "Most of the bio-diversity exists in the South. Two-thirds of it exists in the South. It is a Third World resource. By calling it a global resource, by calling it a common heritage of mankind, the North is basically preparing the ground to assure raw material supply for the emerging biotechnology industry which needs this diversity as input."[189] Shiva went on to explain that "[b]asically, behind the bio-diversity conflicts is a conflict over who will control this future raw material and whether that bio-diversity will be able to sustain life in the Third World or [whether] it will only sustain profits for Northern corporations."[190] The Centre for Science and Environment in India perceived a double standard in the negotiating process for the Biodiversity Convention and argued that "the high sounding plea of the common heritage of humankind is a rhetorical device to disguise the continued exploitation of the poorer countries and their farmers."[191] On a kinder, gentler note, Kenya's Minister for Environment, Philip Leakey, urged the United States to sign the Convention and cautioned that "failure to sign . . . will be an opportunity missed, and will have far reaching effects."[192] It was possibly the realization of these long-term consequences that prompted Dr. Henry Shands, who directs the American genetic resources program and was a member of the United States negotiating team, to repeat the commitment of his nation "to the concept of open exchange of genetic resources with all nations of the world."[193] Geoffrey Hawtin, Director of the International Board for Plant Genetic Resources (IBPGR), in Rome, minimized the impact of the Bush Administration's rejection of the Convention, explaining that "[t]he United States has been one of the most liberal countries in terms of making material available. I would not expect that position to change."[194]

An ancillary problem is that the sheer poverty of developing nations compels them to utilize their biological resources in the most profitable way open to them, by destroying the timber and expanding development into the forests. An editorial in the *Times* (London) stated that "[u]nless poor countries are compensated for conserving these resources they will, they say, have to put short-run before long-run development."[195] In the developing world, the issue of intellectual property rights did not appear to be a valid justification for the American rejection of the Convention, particularly as President Bush appeared to believe that biotechnology companies ought to utilize the resource without compensation to its owners, an argument which violates the fundamental premise of the legal rights of property ownership. So it seemed to environmentalists in the developing world that the United States was suggesting that it was important to protect its own property rights but acceptable to violate those of developing nations. Although this may have been something of an exaggeration, it became part and parcel of the hostile mood which developed against the United States before and during UNCED.

THE LEAKED MEMO

Much to the embarrassment of the United States, a bizarre event involving the Convention on Biological Diversity unfolded while delegates were gathering in Rio to attend UNCED. According to reports in the *New York Times*, William Reilly, Chief Administrator of the Environmental Protection Agency, wrote on 3rd June 1992 to Clayton Yeutter, President Bush's domestic-policy chief, informing him that "Brazil [had] offered to 'fix' the Biodiversity Convention so that the United States could sign it."[196] Reilly voiced his "serious doubt whether the Brazilians can get others to accept a fix,"[197] but he was willing to allow them to try. The Brazilian part of this quid pro quo was to be a definite commitment by the United States to sign the Convention. Reilly explained to Yeutter that "the U.S. refusal to sign the Biodiversity Convention is the major subject of press and delegate concern here"[198] and went on to suggest that "[t]he changes proposed, while not making everyone in the U.S. Government totally happy, would address the critical issues that have been identified. They are worth a last examination."[199] The proposed changes concerned minor linguistic adjustments and the deletion of provisions "that create anxieties about intellectual property rights, technology transfer, concessional terms [and] the regulation of biotechnology."[200]

This confidential memo was leaked to the *New York Times*, which published extracts. Sean Cronin, Washington correspondent of the *Irish Times*, alleged that the leak was carried out by an aide to Vice-President Quayle.[201] Assertions of high-level involvement were vehemently denied by Jeffrey Nesbit, spokesman for the Vice-President.[202] However, Keith Schneider of the *New York Times* maintained that "[a]n Administration official, concerned that Mr. Reilly might persuade the White House to sign the treaty, gave a copy of the memorandum to the *New York Times*."[203] President Bush threatened to dismiss the official who had leaked the memo, provided he could be found.[204] However, he did not change his mind about the Convention, and shortly after Reilly proposed the changes, Samuel Skinner, White House Chief of Staff, rejected the idea.[205] The *New York Times* reported:

> Administration officials said the changes did not come close to resolving
> the more fundamental concerns they had about the treaty. "It caused a little
> commotion at the White House," another Administration official said. "The
> response was what you'd expect. A flat no." [206]

The entire episode turned out to be a public relations fiasco and left Reilly in a position which was "barely tenable."[207] He was forced to defend "his reputation as someone who had the confidence of his Government"[208] because, as one Brazilian diplomat asked, "What is the White House trying to do to their man here?"[209] The editor of *Environmental Policy and Law* commented that Reilly's "situation was confusing, to say the least. He has long been well-known in the environment movement as a progressive, and many people could not understand his continued patience in accepting such a situation."[210] During this episode David McIntosh, who had written the earlier memorandum which determined the American Government's rejection of the Convention, was reported to be "at the center of a group of White House staff members that met to determine how to react to Mr. Reilly's proposal, and then quickly decided to reject it."[211] Although President Bush was careful to praise the EPA Chief Administrator and accord him "full support,"[212] Reilly's embarrassment was compounded by the fact that he had suggested to reporters that a compromise might be forthcoming, only to be informed by the media that the White House had rejected his proposal.[213] At UNCED, William Reilly, who was the chief negotiator for the United States, could only issue a statement proclaiming that it was "most unfortunate that someone within our government chose to leak information about these most important efforts that demanded diplomatic discretion."[214] Al Gore, who was in Rio at the head of a delegation of United States Senators, commented that "[o]nce again the principal official in charge of U.S. environmental policy has been overruled—and this time the whole world is watching."[215]

The debacle of the leaked memorandum exposed to all the world that the Government of the only superpower was grappling with its own internecine struggles over the issues of development versus environment. Ironically, this episode only highlighted the fact that there are vast chasms to cross not just in the United States but everywhere in attempting to harmonize these seemingly polarized positions. In a very real sense, the rift within the U.S. Administration mirrored the dilemma facing the delegates at UNCED, who were, after all, seeking to find an international bridge between development and the environment. Although sustainable development has been articulated as the ultimate nexus between these two ideas, global implementation will be something of a challenge.

THE CLINTON PRESIDENCY AND THE BIOLOGICAL DIVERSITY CONVENTION

There were great environmental expectations when Bill Clinton entered the White House, largely because Al Gore, a renowned environmentalist, became Vice-President of the United States. Brad Knickerbocker of the *Christian Science Monitor* reported that "[w]hen the returns from the presidential election came in, environmentalists around the country cheered."[216] During the election campaign the Democratic Party had committed itself to preserving the Earth's biodiversity.[217]

Internationally, environmentalists were encouraged by Al Gore's vocal criticism of President Bush's performance at Rio. He remarked, for example, that "countries were looking for strong leadership from the U.S. on environmental policy,"[218] leadership he believed that the United States had failed to provide.[219] Environmentalists pinned their hopes on the new Vice-President to lead his nation in conservation efforts. The Sierra Club, a prominent environmental group, had endorsed the Clinton-Gore ticket,[220] but it did so cautiously, emphasizing Clinton's proposals and intentions rather than his environmental record as Governor of Arkansas.[221] As James Rusk, environment reporter of the *Globe and Mail*, commented shortly after the election,

> What activists expect is that, as vice-president and Mr. Clinton's chief adviser on environmental issues, Mr. Gore will ensure that environmental concerns do not get lost as the new administration sets economic and foreign policies.[222]

Rusk also expressed his belief that the United States would sign the Biodiversity Convention.[223] Abraham Lowenthal, Director of the Center for International Studies at the University of Southern California, suggested that Clinton's Administration would probably "focus on global environmental issues . . . with the goal to narrow the gap between the US position and that of other major nations reflected at June's Earth Summit in Rio de Janeiro."[224]

Clinton made a number of significant environmental promises during the election campaign, including a commitment to sign the Convention on Biological Diversity.[225] However, he also admitted to favoring economic considerations over environmental ones when compelled by circumstances to make a choice, although he "now rejects the notion that the environment must be sacrificed for jobs."[226] According to Walter Reid, Vice-President of the World Resources Institute, Clinton had stated during the Earth Summit "that had he been president, he would have negotiated a better convention and would have signed it in Rio."[227]

President Clinton announced on 21st April 1993 that the United States would sign the Convention on Biological Diversity, explaining that "[w]e cannot walk away from challenges like those presented by the biodiversity treaty. We must step up to them."[228] The U.S. Administration indicated that a statement of interpretation would accompany the American signature. According to a *New York Times* report, "White House officials said the interpretive statement . . . would assure American companies that they would not have to share technology with developing countries that provide resources for products manufactured by those companies."[229] President Clinton added that "[w]e think we have done the work necessary to protect the intellectual property of American industries and to protect the environment."[230] The President expressed his belief that the difficulties had been worked out and that the effort would be supported by both the business and environmental communities.[231] Despite this positive action, President Clinton has yet to assume the role of international environmental leader. Hopefully, he will not discard this opportunity as cavalierly as did President Bush.

As regards Vice President Al Gore, in his book *Earth in the Balance* he sketched out a visionary framework for an action plan to save the environment which included "the rapid creation and development of environmentally appropriate technologies"[232] and their quick transfer to developing and other nations.[233] Such an idea would obviously be viewed very favorably in Africa, Asia, and Latin America. However, in another section Gore also proposed "[b]etter protection for patents and copyrights [and] improved licensing agreements."[234] The latter proposal would place him more within the policy parameters of the U.S. negotiating team during the UNCED process. These two ideas are not necessarily contradictory and could well become part and parcel of United States policy concerning international biodiversity. In an important article written for the *Journal of NIH Research* in October 1992, then Senator Gore explained that "[a]ccess to native resources and protection of intellectual property are complementary concerns"[235] and defined his priorities with respect to biodiversity:

> Sustainable use of the Earth's biodiversity requires three actions; (1) conservation of the diversity, (2) protection of intellectual property rights earned by those who invest in research on new uses of native species, which then become more valuable and more likely to be conserved because of their enhanced value, and (3) maintenance of equitable and open access to native resources.[236]

Gore went on to blame the Bush Administration for refusing "to engage in meaningful negotiations on the issue of intellectual property protection."[237] This refusal, in his view, resulted in the inclusion of language opposed to the interests of the biotechnology industry in the Convention. A very significant statement followed: "The Clinton-Gore administration will work to present a coherent plan for protecting biodiversity and intellectual property rights in a way that enhances conservation and facilitates global research efforts by U.S. businesses and universities."[238]

Being keenly aware of the interests of the American biotechnology industry, the Clinton Administration appeared anxious to preserve an area which has both economic and environmental benefits for the United States and ultimately for the whole world. Clearly, Clinton did not want to appear isolated on global environmental issues as Bush was. The international public relations debacle occasioned by the Bush Administration's refusal to sign the Convention may have propelled Clinton to accept it as a way of announcing to the world that the United States is back in the fold, as it were, and pursuing the correct path environmentally.

THE PROVISIONS OF THE CONVENTION AND THE INTERESTS
OF THE UNITED STATES

Opposition to the Biodiversity Treaty was strong among President Bush's advisors and was quite vocal in the U.S. biotechnological industry. Accordingly, it is worthwhile to examine some provisions of the Convention with these attitudes in mind. It is important to stress that the opposition to the Biodiversity Convention was not exclusive to the Bush White House or to Vice-President Quayle's entourage. The provisions of the Convention drew forth the sustained and determined efforts of a number of individuals in the United States who were extremely apprehensive about American involvement in this Treaty.

Any analysis of this nature has perforce to focus on the United States. Given its economic, political, and military influence around the world and the growing influence of its pre-eminent biotechnological industry, the country had to become part of this Convention if the ideal of conserving biodiversity was to succeed. Without the United States, the Treaty could have been hobbled, if not crippled.

Employment opportunities arising from the Convention could become very significant. Arguably, the action plans of the Convention, the series of measures to identify, monitor and protect biological diversity—might provide an important opportunity for American scientists and researchers to act as advisors to developing nations which are now committed to the many tasks directed by the Convention. Further benefits could accrue to international lawyers who would be called upon to negotiate the agreements between corporations and foreign governments concerning the sharing of biodiversity.

United States isolation at Rio had some immediate negative consequences. According to Brian Boom, Vice-President for the Botanical Sciences at the New York Botanical Garden, the Venezuelan Government "suspended the signing of new agreements for scientific collaboration with U.S. institutions as a response to the U.S. position on the UNCED Biodiversity Convention."[239] It was soon evident that the dictates of "the global market . . . [could] force US biotechnology firms to fall in line with the rest of the world."[240] Kenton Miller of the World Resources Institute commented that "[w]e are either in the club or out of the club."[241] As Russell Frye explained to the International Bar Association, developing nations might prefer to sell their biodiversity resource to companies which agree to pay them royalties, and the United States may face trade barriers imposed against American biotechnological companies by developing nations.[242]

Article 15 of the Convention considers the issue of access to genetic resources. It recognizes the *"sovereign rights of States over their natural resources"* (Article 15.1) and stipulates that national governments have the *"authority to determine access to genetic resources"* (Article 15.1). However uncomfortable the multinational corporations based in the United States might feel about this tribute to national sovereignty, the article is in effect only stating the obvious by specifying that national legislation rules access to genetic resources.

The following provision of the Convention militates against any nation holding an isolationist, aloof position vis-à-vis this instrument on biodiversity. This clause could also have expedited President Clinton's decision to sign the Convention:

> *Each Contracting Party shall endeavour to create conditions to facilitate access to genetic resources for environmentally sound uses by other Contracting Parties and not to impose restrictions that run counter to the objectives of this Convention.* (Article 15.2)

The mild language ought not to disguise the underlying premise that the Convention implies greater cooperation between Contracting Parties. The reverse—less cooperation with non-party states—could also be implied, though it is not stated. *"Access, where granted, shall be on mutually agreed terms"* (Article 15.4). This provision makes it clear that the developing nations are determined to have a say in the utilization of a resource which they have, until recently, been literally giving away with little or no benefit to their people. The next clause underscores this aspect of the new world environmental order:

> *Access to genetic resources shall be subject to prior informed consent of the Contracting Party providing such resources, unless otherwise determined by that Party.* (Article 15.5)

There is already evidence that mutually beneficial agreements can be formulated by developing nations and large corporations. "In 1991 Merck and Company, the world's largest pharmaceutical firm, agreed to pay Costa Rica's National Institute of Biodiversity $1 million" for the collection and identification of organisms for possible medical use.[243] The agreement allows for a two-year transfer of samples from plants, animals, and insects.[244] Once a medical function is discovered and the substance marketed, the Costa Rican Government will earn a share of the royalties, funds which will subsidize conservation programs. The financial benefits for both the corporation and the originating country could be extensive. In 1990, to cite only one example, Merck's sales of the drug Mevacor (used for lowering cholesterol levels) amounted to $735 million;[245] Mevacor was originally derived from a fungus. With 12,000 plant and 300,000 insect species within its borders,[246] Costa Rica stands to derive enormous benefits from this type of agreement, as indeed does Merck. It should also be noted that Thomas Eisner of Cornell University, who brokered this agreement, is an ardent supporter of the Convention and accuses its opponents of "antediluvian thinking."[247] Arthur Campeau, Personal Representative of the Prime Minister of Canada to UNCED, commented on the Merck agreement with Costa Rica before the House of Commons Standing Committee on Environment:

> Developing countries are already choosing to cooperate only with those researchers who will offer that type of agreement. As a result, other biotechnology and pharmaceutical companies are responding by offering similar agreements. This presents an opportunity for Canadian biotechnology companies.[248]

Mostafa Tolba, former Executive Director of the United Nations Environment Programme, pointed out that, far from being a penalty on industry, funding by the private sector to support conservation is in its ultimate self-interest "to ensure business sustainability."[249]

The Convention also stipulates that

> *[e]ach Contracting Party shall endeavour to develop and carry out scientific research based on genetic resources provided by other Contracting Parties with the full participation of, and where possible in, such Contracting Parties.* (Article 15.6)

This provision seeks to redress the imbalance in research facilities between the North, which can afford to provide funding for its scientists, and the South, which cannot afford a research grants system on the same scale, given the basic requirements for food, shelter, and other bare necessities of a large proportion of its huge population. The result of this imbalance is a continuing brain drain from the South to the North as the best scholars and scientists are lured away by the possibilities available to fully realize their potential.

The existence of better facilities and more funding for research in developing nations would be an important asset for those countries and would probably result in slowing the exodus of their most talented people. In an editorial, the *Times* (London) commented that "[i]t is right that poor countries should start to share in such profits and in genetic research."[250]

The South apparently assumed that the Northern nations which would benefit from their biodiversity could simply legislate equitable sharing of the results of research and development and take administrative or policy measures to direct that the benefits of commercial use be shared with the country of origin, on mutually agreed terms. The relevant provision states:

> *Each Contracting Party shall take legislative, administrative or policy measures, as appropriate . . . with the aim of sharing in a fair and equitable way the results of research and development and the benefits arising from the commercial and other utilization of genetic resources with the Contracting Party providing such resources. Such sharing shall be upon mutually agreed terms.* (Article 15.7)

Although policy measures to promote sharing are possible, there would probably be considerable resistance to the enforcement of blanket proposals directing the transfer of technology or results of research. Kirk Raab, President of the biotechnology company Genentech, expressed his apprehensions to President Bush, suggesting that "[t]he vague language relating to 'technology transfer' and equitable sharing appear to be code words for compulsory licensing and other forms of property expropriation."[251] Issues of compensation would also have to be considered, as would the extent to which the research and technology could be replicated or its results marketed in developing nations. It is also significant that the more mandatory word "shall" is used, clearly directing governments to take the steps enumerated.

With respect to the issue of intellectual property rights and technology transfer, the Convention specifies in Article 16 that, first, Contracting Parties undertake *"to provide and/or facilitate access for and transfer to other Contracting Parties of technologies that are relevant to the conservation and sustainable use of biological diversity or make use of genetic resources and do not cause significant damage to the environment"* (Article 16.1). Second, developing countries shall be provided access to technology *"under fair and most favourable terms, including on concessional and preferential terms where mutually agreed"* (Article 16.2). Third, *"[i]n the case of technology subject to patents and other intellectual property rights, such access and transfer shall be provided on terms which recognize and are consistent with the adequate and effective protection of intellectual property rights"* (Article 16.2). Thus far, the provisions would not appear to be too threatening to American interests. It is now conceded by almost every Northern government that the transfer of environmentally safe technology to the developing nations will be of mutual benefit, because the rapidly growing Southern industrial base can convert to less-polluting methods of production which can only improve the quality of life for all the inhabitants of this planet. With respect to this provision, the Committee on International Environmental Law of the Association of the Bar of the City of New York noted that intellectual property rights are protected. The same group pointed out that other sections regarding technology transfer specify that it will occur on "mutually agreed terms," as was required by U.S. negotiators. Hence, the Committee concluded that the objections of the biotechnology industry were "based on a misinterpretation of the text."[252]

However, the same clause goes on to insist on adherence to its own consistency with three provisions which follow. The first of these controversial stipulations states:

> *Each Contracting Party shall take legislative, administrative or policy measures, as appropriate, with the aim that Contracting Parties, in particular those that are developing countries, which provide genetic resources are provided access to and transfer of technology which makes use of those resources, on mutually agreed terms, including technology protected by patents and other intellectual property rights.* (Article 16.3)

It is unclear whether this implies that developed countries should legislate away intellectual property rights to meet their obligations under the Convention. The language is, to say the least, ambiguous and rather clumsy in clarifying the responsibilities of the Parties. This stipulation is also linked to the subsequent provisions and to the articles on financial resources and mechanisms in what appears to be a classic example of United Nations "internationalese," the language of apparent consensus which ignores clarity. The *Times* (London) commented that this provision "explicitly drives a coach and horses through existing international patent law" and complained that "[n]o more effective disincentive to gene-splicing and other forms of bio-research could have been devised."[253] However, an alternative perspective was provided by the Ambassador of Ghana, who told delegates at UNCED that "developing countries demand that the Convention on biodiversity include legally binding commitments to ensure the link between the access to the genetic materials of developing countries and the transfer of bio-technology and research capabilities from developed countries."[254] President Ali Hassan Mwinyi of Tanzania observed that "[m]any of us in developing countries find it difficult to accept the notion that biodiversity should be considered as the common heritage of mankind, while the flow of biological products from the industrialised countries is patented, expensive and considered as the private property of the firms that produce them. This asymmetry," he said, "expresses the inequality of opportunity and is unjust."[255]

Even more controversial is the next stipulation of the Convention, which states:

> *Each Contracting Party shall take legislative, administrative or policy measures, as appropriate, with the aim that the private sector facilitates access to, joint development and transfer of technology . . . for the benefit of both governmental institutions and the private sector of developing countries.*
> (Article 16.4)

Implementing this provision could be quite a challenge. Many developing nations are accustomed to the idea of governmental regulation of the economy to a degree which would be totally unacceptable in the United States, and this provision reflects a monumental lack of comprehension on the part of delegates from developing nations about the limited ability of governments in the North to order the private sector in the manner specified. Geoffrey Hawtin, Director of the International Board for Plant Genetic Resources (IBPGR), in Rome, wondered how countries could influence the private sector: "it's hard to imagine U.S. legislation that could encourage the private sector"[256] to implement this clause of the Convention. Unfortunately, such provisions demonstrate how large is the chasm between North and South. Although this stipulation meets the Southern agenda, its implementation would cause problems in any market-oriented society which favors private sector production, distribution, and exchange. As Michael Roth, Corporate Patent Counsel of Pioneer Hi-Bred International Inc., explained, "we rely heavily on intellectual property rights to protect our products." He went on to remark that "[t]here is too much language in the convention about making technology available to developing nations, in effect, on an unrestricted basis to make us comfortable that our rights would be protected."[257] The text of this provision of the Convention is so ambiguous that the reader is left wondering whether it really is as threatening as it appears or whether it seeks merely to encourage technology transfer and utilize governmental persuasion to effect that aim. After all, it is hard to dispute the fact that there "is a need for gene-rich developing countries to work in tandem with technology-rich developed countries. It forms the basis for a real 'win-win' deal between

North and South, based on mutual benefits from cooperation."[258] It is important to note that although this stipulation in the Convention is linked to the
initial recognition of the protection of intellectual property rights specified in
Article 16.2, it is also connected with the possible limitation of such rights
implicit in Article 16.3. The convoluted language and the references to seemingly divergent aims reflected in different clauses make analysis of the technology transfer provisions a real challenge.

Mostafa Tolba, former Executive Director of the United Nations Environment Programme, provided a rationale for including a provision affecting the
private sector:

> New technologies are mostly patented. International transfer of biotechnol-
> ogy has been limited largely to the transfer of technology from the public
> sector in developed countries to the public sector in developing countries.
> Given the substantial private sector investments largely in funding Re-
> search and Development in modern biotechnology, the private sector un-
> doubtedly possesses new technologies that could be usefully applied or
> adapted to developing country needs.[259]

Even more challenging is the next part of this very controversial article on
technology transfer:

> *The Contracting Parties, recognizing that patents and other intellectual prop-
> erty rights may have an influence on the implementation of this Convention,
> shall cooperate in this regard subject to national legislation and interna-
> tional law in order to ensure that such rights are supportive of and do not
> run counter to its objectives.* (Article 16.5)

This stipulation, when considered with the previous directions to states to
take legislative and administrative policy measures to facilitate implementation of the Convention, appears to suggest that intellectual property rights
ought eventually to sustain the fundamental principles of the Convention and
not restrict or hinder them. If this is the intent, the phrase *"subject to national
legislation"* presumably refers to future legislation which would accommodate the provisions of this Convention. The implication is that eventually the
Biodiversity Treaty, being international law, will take priority over national
legislation unless the latter is brought into line with its provisions. The
clause, as it stands, would pose quite a challenge to anyone seeking to implement it because it appears to be almost self-contradictory. As Russell Frye
comments, "The uncertainty regarding how the Biodiversity Convention
might be implemented may be at the root of the critics' objections."[260]

The Convention's silence on the crucial issue of ownership of genetic resources collected before its enactment could pose some problems in the future. The issue was considered by Geoffrey Hawtin, Director of the International Board for Plant Genetic Resources, in Rome, who suggested that, on the basis of one interpretation, such earlier collections would simply be outside the Convention, or, alternatively, "if you sign the Convention, the only materials you can provide are those that originated in your country or arrived there under the terms of the Convention."[261] A further complication might occur when some "breeding materials could combine germplasm from seeds collected in many nations."[262]

Another problem which will have to be considered is whether the governments of democratic, capitalist states can restrict intellectual property rights to the extent required by this Convention without arousing considerable hostility from their legislatures and from the affected community of scientists, academics, and corporations engaged in research. The hurdles are by no means insurmountable. As Al Gore suggests, "[T]he model . . . comes from the U.S. government itself, which has technology-transfer programs that provide the federal government royalty-free use of inventions developed by companies from government-funded and owned research and data."[263] Walter Reid of the World Resources Institute also sees a positive future because the Convention will "stimulate the adoption of stronger intellectual property rights within developing countries." He explains:

> One of the reasons for the U.S. decision not to sign the convention was that it would weaken intellectual property rights. However, by creating conditions for increased use of biodiversity in developing countries, many countries will now choose to establish and strengthen intellectual property rights for biological resources, because now they have something to gain. This will now take place not because they're being forced through trade agreements to strengthen intellectual property rights but because now they can see it's in their own self-interest.[264]

Arthur Campeau, Canada's Ambassador for Sustainable Development, said of the issue of intellectual property rights that

> the convention specifically works to protect, not diminish, the intellectual property rights of manufacturers. I raise this point because it was the major reason given by the United States for its decision not to sign this convention. In deference, I believe that they are wrong and that their own industries and the new administration will see this issue differently.[265]

Ever the optimist, Maurice Strong, Secretary-General of UNCED, indicated that he had not lost hope because President Bush had not signed the

Convention, and added, "I hope that they will take a second look at it."[266] Certainly, if a number of developing countries proceed to tighten their legislation protecting intellectual property rights it would go far toward alleviating many of the anxieties in the United States about this Convention. Richard Wilder, a patent attorney and the spokesman for the Association of Biotechnology Companies, called it unfortunate that the delegations from countries with a very negative view of intellectual property, particularly those from India and Malaysia, had the most influence over the wording of the Convention on Biological Diversity.[267] An alternative viewpoint comes from M. S. Swaminathan, World Food Prize Laureate, who explained that at Rio "[p]atent protection rather than promoting a better quality of life for the poor became the major obsession."[268] If the developing world can in future be perceived as cooperating in the process of protecting patents and copyrights to a much larger degree than is now the case, then opponents of this Convention in the United States will have much less to complain about. Technology transfer, which acknowledges intellectual property rights under a system of international law which is mutually respected by the donor and the recipient, would be the best way to facilitate the concept of sharing, so important an aspect of the Biodiversity Convention. As Thomas Lovejoy of the Smithsonian Institution suggested, the idea is "to start thinking about the problem as a joint venture in which both sides have property rights."[269]

The concerns of the United States negotiators involved "first, intellectual property rights; second, biotechnology safety regulation; and third, the financial mechanism."[270] Vice-President Quayle was accused in the *Economist* of objecting "passionately to the treaty's attempts to regulate the handling of genetically modified organisms."[271] However, the Convention only stresses a cautionary approach to ensure that such organisms are safe before they are released into the environment.[272] With respect to safety regulations, the Convention postpones the issue:

> *The Parties shall consider the need for and modalities of a protocol setting out appropriate procedures, including, in particular, advance informed agreement, in the field of the safe transfer, handling and use of any living modified organism resulting from biotechnology that may have adverse effect on the conservation and sustainable use of biological diversity.* (Article 19.3)

Meanwhile, it also provides for notification about the safety risks involved in the use of any agent:

> *Each Contracting Party shall, directly or by requiring any natural or legal person under its jurisdiction providing the organisms referred to in paragraph*

3 above, provide any available information about the use and safety regulations required by that Contracting Party in handling such organisms, as well as any available information on the potential adverse impact of the specific organisms concerned to the Contracting Party into which those organisms are to be introduced. (Article 19.4)

Contracting Parties are to *"[p]revent the introduction of, control or eradicate those alien species which threaten ecosystems, habitats or species"* (Article 8[h]) and regulate the *"risks associated with the use and release of living modified organisms resulting from biotechnology which are likely to have adverse environmental impacts"* (Article 8[g]).

Because the financial provisions of this Convention caused serious consternation in the governments of other developed countries as well as arousing considerable hostility in the Bush Administration, I will consider those stipulations in the next section. It will also deal with the opinions of some of the other nations of the North.

The Developed Nations and the Convention on Biological Diversity

Although the United States bore the brunt of global hostility for its rejection of the Biodiversity Treaty, it was not alone in expressing its apprehensions and concerns over aspects of the agreement. A number of developed countries, particularly the United Kingdom, had serious reservations about this Convention, especially its financial provisions. It was the Canadian delegation, headed by Arthur Campeau, which ultimately guaranteed that the Convention would be accepted by the North. Canada not only signed the document but promised to (and did) ratify it within the year.[273] As Canadian Prime Minister Brian Mulroney said in an address at Harvard University: "Canada's signal leadership on the biodiversity accord gave the agreement significant impetus. And why did we support it? Because it was in our national interest to do so, because we consider that biodiversity is an important aspect of sustainable development and because, quite simply, it covers vital dimensions of life on earth."[274]

The North was clearly divided both before and during Rio. The environmental problems were viewed equally seriously by delegates from all developed countries. They could not agree, however, on the same solutions. Eventually, most Northern nations preferred to participate in the Convention despite their misgivings rather than be isolated from the important process of environmental clean-up. The United States was the only significant holdout with respect to the Biodiversity Convention.[275] It is important to emphasize that many nations of the North had initial problems with the Convention. As Anne McIlroy of Southam News reported on 1st June 1992, the Biodiversity Convention was in serious trouble because of the concerns of the United States, Great Britain, France, and Japan,[276] among the developed nations. With the exception of the United States, these nations eventually agreed that it was preferable to sign the Treaty and become participants in the process of conserving biodiversity. Basically, in the Rio process one could not speak of the "North" and imply the same degree of cohesiveness and identity of interest as propelled the developing nations, the "South." The developed countries did not share an agenda or international priorities in terms of the ideals of UNCED, nor did they speak as a unit (length constraints preclude a detailed study of each Northern nation's views). They shared misgivings, for example, regarding the issue of open-ended funding required by the poor nations, but it is obvious that each developed country pursued its own interests during the Rio process. The United States, used to being the leader among developed nations, was clearly upset by the break-away positions some of its allies adopted, especially with respect to the Biodiversity Treaty. This section will analyze the financial provisions of the Convention, since these aroused considerable concern among a number of Northern countries. It will then very briefly demonstrate some facets of the divergence in opinion which prompted the Europeans, Canadians, and Japanese to follow a path different from that of the United States over the biodiversity issue.

As the financial provisions of the Convention were the main stumbling block to universal acceptance, it would be worthwhile to examine them in order to understand both the views of the North and the extent of the political chasm between the North and the South over this all-important issue of funding. With respect to financial provisions, the South established the parameters within which it would be prepared to conserve its valuable biodiversity resource. The Convention engenders a vast amount of activity involving identifying, monitoring, conserving, and protecting biodiversity. The developing nations know that they can ill afford to fund such extensive measures, many of which are simply beyond their technological capacity, particularly in the case of the least developed countries. Accordingly, in demanding funding from the North so that the whole world would continue to enjoy the benefits of biodiversity, the South was only expressing its own inability to finance the actions which alone can keep developers and populations out of protected areas. Even though they retain much of the world's biodiversity, the developing nations are already fighting a defensive war to preserve this resource. Their burgeoning populations cry for space, food, shelter, clothing, and consumer products—all of which demand the clearing of forests (the greatest threat to biodiversity), the conversion of land to agriculture, and the building of factories. Hence the war (and it is that now) to save biodiversity will involve restricting development in the South, or at least ensuring that development proceeds in a manner which is sustainable within the environment. Southern nations argue that, as the North will benefit greatly from the conservation of biodiversity, it ought to pay to preserve the resource so that the South can for once benefit as well.

The Convention on Biological Diversity acknowledges the primacy of the funding issue in its Preamble, stating that

> *the provision of new and additional financial resources and appropriate access to relevant technologies can be expected to make a substantial difference in the world's ability to address the loss of biological diversity,*

and reiterating that

> *special provision is required to meet the needs of developing countries, including the provision of new and additional financial resources and appropriate access to relevant technologies.* (Preamble)

The issue of *"appropriate funding"* (Article 1) is again referred to in the Objectives of the Convention. Contracting Parties are also asked to *"[c]ooperate in providing financial and other support for in-situ conservation . . . particularly to developing countries"* (Article 8 [m]). A similar provision refers to funding for ex-situ conservation and for the *"establishment and maintenance of ex-situ conservation facilities in developing countries"* (Article 9[e]). The obligations of States are qualified with phrases like *"as far as possible"* and *"as appropriate."* Hence, adhering to the Convention is not as mandatory as its critics allege. Walter Reid of the World Resources Institute provides one explanation for this:

> [T]he language in the convention is rather weak. I think this is largely due to the U.S. negotiating position. Almost all the obligations accepted by nations are prefaced with the wording, "as far as possible" and "as appropriate," which gives a tremendous amount of wiggle room for governments to avoid doing some of what is intended by the convention.[277]

It could be argued that the financial stipulations, which have caused so much anxiety in developed country governments, are probably too weak to be really effective. Contracting Parties will fund their national activities on the basis of their capability, a provision which allows countries to plead economic reasons for nonperformance.

Each Contracting Party undertakes to provide, in accordance with its capabilities, financial support and incentives in respect of those national activities which are intended to achieve the objectives of this Convention, in accordance with its national plans, priorities and programmes. (Article 20.1)

In an extremely clumsy paragraph, the Convention specifies the main financial obligations of the North to the South:

The developed country Parties shall provide new and additional financial resources to enable developing country Parties to meet the agreed full incremental costs to them of implementing measures which fulfil the obligations of this Convention and to benefit from its provisions and which costs are agreed between a developing country Party and the institutional structure referred to in Article 21, in accordance with policy, strategy, programme priorities and eligibility criteria and an indicative list of incremental costs established by the Conference of the Parties. . . . The implementation of these commitments shall take into account the need for adequacy, predictability and timely flow of funds and the importance of burden-sharing among the Contributing Parties. . . . (Article 20.2)

Article 21 (referred to above) includes the provisions for the financial mechanism and will be analyzed later in this section. The Convention leaves the door open for bilateral, regional, and other multilateral schemes to fund projects for developing countries (Article 20.3). It also emphasizes the *"specific needs and special situation of least developed countries"* (Article 20.5), *"small island States"* (Article 20.6), and those developing country regions *"that are most environmentally vulnerable, such as those with arid and semi-arid zones, coastal and mountainous areas"* (Article 20.7).

In a very significant provision, the Convention makes Southern implementation contingent on Northern funding:

The extent to which developing country Parties will effectively implement their commitments under this Convention will depend on the effective implementation by developed country Parties of their commitments under this Convention related to financial resources and transfer of technology and will take fully into account the fact that economic and social development and eradication of poverty are the first and overriding priorities of the developing country Parties. (Article 20.4)

This provision is weighted heavily in favor of developing countries, clearly at the expense of the North. It also reflects the new reality of the post–Cold War era, which has seen rich nations pitted against poor countries in an intense confrontation which springs up every time an issue of global significance is debated. Indeed, the North-South confrontation now dominates most international fora and has become a major factor to consider in environmental issues. The poor countries, with the majority of the world's population, are simply not willing to remain poor any longer. Moreover, they do not see why they ought to curb their development to enable the rich nations—the world's worst polluters—to continue to degrade the environment without paying monetarily for the right to do so. Therefore, "[t]he poor countries say they will only tailor their development to meet environmental goals if the rich pay the costs of doing so."[278] The entire issue of protecting biodiversity became bogged down over the political ramifications of the North-South funding controversy. As Robert C. Cowen explained in the *Christian Science Monitor*, "[t]he poor nations ask the rich for substantial financial help with the research and conservation needed to preserve biodiversity. Rich nations balk at paying the lion's share of the bill."[279] The provision in the Biodiversity Convention on contingent implementation reflects the attempt to link ecological improvement with eradication of poverty as twin approaches in the cause of creating a new world order.

Although the North-South conflict still prevails in international economic and environmental discussions, there is a growing public realization in the developed world that measures have to be taken to alleviate the misery and poverty of the majority of the world's population. This is no longer merely an idealistic aim. It is now increasingly perceived as the ultimate form of self-interest for the developed world. It was this recognition, at the popular level, which compelled governments like that of the United Kingdom to sign the Convention "against their better judgment."[280] The challenge for the future will be to implement its financial provisions, particularly when governments in the developed nations have to allocate funding and decide whether priorities should dictate domestic economic improvement or Third World aid with, hopefully, global environmental clean-up as a bonus. Brad Knickerbocker also suggests that "there is a vast difference between the 'needs' of those in developing countries and the 'wants' that people in more advanced industrialized nations have gotten used to."[281]

The most controversial financial clauses were those which delineated the mechanism for the provision of funding from the North to the South. United States officials objected that these provisions could "enable developing countries to extract limitless funds from wealthy nations for preserving endangered species."[282] The Treasury Department of the Government of the United Kingdom was equally apprehensive about provisions which would "allow the world's poorer nations to present industrialised countries with an open-ended bill for preserving their plant and animal life."[283] It would be worthwhile to examine the controversial clauses to assess the validity of the fears of the developed countries.

> *There shall be a mechanism for the provision of financial resources to developing country Parties for purposes of this Convention on a grant or concessional basis the essential elements of which are described in this Article. The mechanism shall function under the authority and guidance of, and be accountable to, the Conference of the Parties for purposes of this Convention. The operations of the mechanism shall be carried out by such institutional structure as may be decided upon by the Conference of the Parties at its first meeting. For purposes of this Convention, the Conference of the Parties shall determine the policy, strategy, programme priorities and eligibility criteria relating to the access to and utilization of such resources. The contributions shall be such as to take into account the need for predictability, adequacy and timely flow of funds . . . in accordance with the amount of resources needed to be decided periodically by the Conference of the Parties and the importance of burden-sharing among the contributing Parties. . . . The mechanism shall operate within a democratic and transparent system of governance.* (Article 21.1)

Essentially, this provision brings the financial mechanism under the authority of the Conference of the Parties, where, presumably, developing countries would dominate by sheer numbers. The Conference would decide its own priorities concerning utilization of the funding, so developing countries would obviously have considerable weight in determining how the money would be spent and where the emphasis of biodiversity conservation would lie. The developed world is obliged to consider that its funding has to meet the standards of predictability, adequacy, and timeliness. Most controversial of all, the Conference of the Parties determines the amount of resources required. The financial mechanism would be operated democratically, which again would imply a considerable amount of input from the developing world. A further provision gives the Conference of the Parties authority to *"determine the policy, strategy and programme priorities, as well as detailed criteria and guidelines for eligibility for access to and utilization of the financial resources including monitoring and evaluation on a regular basis of such utilization"* (Article 21.2). The financial mechanism is subject to periodic review (Article 21.3). On an interim basis, the Convention entrusted financial implementation to a restructured Global Environment Facility of the United Nations Development Programme, the United Nations Environment Programme, and the International Bank for Reconstruction and Development (Article 39).

President Bush complained that "the text gave the developing countries too much say in deciding the treaty's aid provisions. There is concern that the money intended to preserve rare species might be misused by Third World regimes."[284] The *Times* (London) reported that British Prime Minister John Major "would like to ensure that money for environmental purposes went to the underdeveloped countries and not to their politicians."[285] United States opposition was clearly having an influence with the British Government. This led environmentalists and the Opposition in the British Parliament to insist that the United Kingdom sign. Simon Hughes, spokesman for the Liberal Democrats, tabled a motion of no confidence in the Conservative Government.[286] David Blunkett, shadow environment minister for the Labour opposition, said bluntly that "if the British government follows the US in refusing to sign . . . it will have committed an act of betrayal against future generations, for which it must stand condemned."[287] Bryan Gould, environment spokesman for the Labour Party, "accused the United Kingdom of colluding with the Americans in 'frustrating the world's environmental agenda.' The move was part of a complex game plan to wreck the conference and ensure that nothing came out of the summit other than hot air, 'the last thing the planet now needs,' " he is reported to have said.[288]

Gordon Shepherd of the World Wide Fund for Nature also insisted that "Mr. Major must sign the bio-diversity treaty, regardless of what President Bush does. The world can live without the US signing this treaty, but it cannot live with the extinction of species which is continuing all the time."[289] Robin Pellew, Director of the World Conservation Monitoring Centre, Julie Hill of the Green Alliance, and Fiona Reynolds of the Council for the Protection of Rural England voiced serious concern about the hesitation of the British Government and suggested that "[f]or the UK to surrender its participation would be to abrogate its responsibilities in the international environmental arena."[290] Pellew and his colleagues expressed an apprehension that nonparticipation would marginalize the United Kingdom in the implementation process.[291] Ultimately, the Prime Minister, who was really "over a political barrel,"[292] was forced to swallow his objections as he found it "politically impossible to avoid signing the treaty."[293]

According to a report by Jill Sherman in the *Times* (London), the British Government rejected the accusation that it was attempting to protect American interests.[294] Britain explained that it was mainly concerned that the financial mechanism, controlled by the Conference of the Parties, would entail major funding demands in a forum where developed countries would always be outvoted. Prime Minister Major confirmed that his nation could not support an "open-ended commitment" of financial aid for developing nations.[295]

His Secretary of State for Environment, Michael Howard, warned that the United Kingdom "was not prepared to sign a blank cheque for the biodiversity deal."[296] At UNCED, Howard clarified his government's position: "When a government signs a convention it is not simply committing itself to broad principles. It is committing itself to a text which entails binding obligations to act. So we have had to scrutinize that text with care."[297]

The governments of many developed countries believed that the vague language of the financial mechanism provisions could entail an enormous obligation for the approximately nineteen aid donor nations, members of the Organization for Economic Cooperation and Development (OECD).[298] With respect to the controversial provisions of Article 21, nineteen members of the OECD issued a declaration after negotiations for the Convention ended in Nairobi in May 1992 "giving their view that this provision should not be interpreted as enabling the Conference of the Parties to the Convention to determine the amount of individual contributions to be provided by the donor countries."[299] According to Robin Oakley in the *Times* (London),

> Britain remained nervous about the overall financial implications of the bio-diversity treaty. It was still seeking the attachment of a financial protocol limiting the power of the Third World effectively to raise an environmental levy on the industrialised nations at a level of its choice.[300]

A spokeswoman for the United Kingdom Environment Department explained that Britain "could go on pouring money into a bottomless pit."[301]

Another apprehension related to the absence of a veto for developed nations on financial decisions of the Conference. "Nor will the piper call the tune when the money thus extracted is spent," commented the *Times*, going on to say that "[d]ecisions on conservation strategies and priorities, on who gets this money and how it is used would also be taken by 'democratic' means."[302] Despite its critical approach, the *Times* argued that "[e]ven weakened conventions can lead to stronger ones" and stated that

> [t]he way would be open under the bio-diversity convention to reward poor countries for preserving the natural gene pool, including the possible payment of royalties on commercially useful development of genetic resources. The principle . . . must be that [it is] not a back door for yet more aid but a payment for better world conservation.[303]

Michael Howard, the United Kingdom's Secretary of State for Environment, attempted to put an optimistic face on the British position by suggesting that "[w]e are now satisfied that means can be found within the

Convention to ensure that no country is obliged to contribute an open-ended blank cheque to implement it."[304] The way out of the dilemma was found in another provision of the Convention, Article 23, which states that the *"Conference of the Parties shall by consensus agree upon and adopt rules of procedure for itself and for any subsidiary body it may establish."* Despite being small in number, the developed nations could utilize this consensus rule to ensure an effective voice in the proceedings of the Conference of the Parties. Walter Reid of the World Resources Institute explains how this consensus provision could protect the interests of the North: "By establishing, in Article 23, rules of procedure based on consensus, the convention protects any country from financial commitments that are not in its own interest. So the U.S. concern about the financial mechanism also is not founded."[305] In an official news release the Government of the United Kingdom noted with reference to the provision on rules of procedure that "it will be possible to argue that decisions on financial issues arising under the Convention should themselves be subject to consensus so as to avoid the danger of financial obligations being imposed on us beyond what we are prepared to agree."[306]

Faced with a storm of hostility from environmentalists and elements of the media, the Americans and their British allies expressed their criticism of those developed nation governments which were supportive of the Convention. The Bush Administration "accused Germany and Japan in particular of capitulation to 'political correctness' and the 'guilty developed-world' argument that wealthy nations somehow 'owe the rest of the world.' "[307] According to Michael McCarthy and Martin Fletcher in the *Times* (London), White House officials were reported to have "attacked America's 'holier-than-thou' allies for paying lip service to environmentalism, signing treaties they had no intention of complying with and making America the world's scapegoat."[308] There was resentment in the British Government as well about some European Community nations being "willing to sign up blithely to agreements which will never be implemented and which they have no intention of honouring rather than putting work into securing deals which have some chance of actually being put into effect."[309] An official at Whitehall was quoted by Robin Oakley of the *Times* (London), as complaining about the cavalier attitude of some other countries.[310]

The *Irish Times* interpreted this emerging tension among the developed nations as a conflict "over influence in the Third World after the end of the Cold War" and commented in an editorial that "[a] notable feature of the conference has been the developing friction between the major western powers attending it, especially the U.S., the European Community and Japan. White House spokesmen have been vitriolic in their criticisms of Japan and Germany for their 'politically correct' approach to some of these issues."[311] Richard Godown, President of the Industrial Biotechnology Association, suggested that the signatures by other developed countries were "just a public relations ploy."[312] It was clear that in signing the Biodiversity Convention nations like Canada, Germany, and Japan had "broken with the United States."[313] This angered American Republicans, one of whom commented bitterly that

> Japan's out there killing whales and running driftnets, for God's sake, while we've got the world's toughest environmental laws and we're twisting ourselves into knots over how many jobs to abolish to save a subspecies of owl. . . . And these guys presume to lecture us about environmental responsibility.[314]

Some of the United States criticism was unjustified, because serious efforts were made by the Europeans to find some way of accommodating the American objections to the Convention. According to Paul Lewis of the *New York Times*, Member States of the European Community offered "the Bush Administration a face-saving way of signing the convention on biodiversity by agreeing to make a statement setting out their own interpretation of contentious clauses."[315] William Stevens of the *New York Times* reported that the initiative for this effort came from Britain, France, and Japan, the plan including signature and a statement of interpretation.[316] However, Lewis reported that the White House showed no interest in the idea.[317] The European Community Spokesman, Laurens Jan Brinkhorst, said in Rio that the Community regretted the isolation of the United States and added, "We don't want a slugging match of everybody against the United States."[318]

The decision by other developed nations to break ranks with the United States was only partly motivated by the domestic pressure they felt from their own environmental activists. Ultimately, these nations signed because they realized that it was in their self-interest to do so. In the early days of UNCED, it was assumed that Japan might not sign the Convention because of the restrictions it placed on its pharmaceutical industry.[319] However, any doubts about the financial provisions or the issue of intellectual property rights were of less significance than the fact that nations like Japan and Germany are world leaders in environmental technology[320] and could gain financially when the developing world proceeded towards implementation of the Convention. Clearly, there would be enormous profits for companies developing clean "technologies that could be adopted throughout the world."[321] It was most unfortunate that the Government of the United States did not perceive the economic benefits of participation in the Treaty and looked instead only at its apparent flaws.

In vivid contrast to the American role at UNCED, the Japanese were able to parlay their attendance at Rio into a public relations success. Former Prime Minister Noboru Takeshita, Head of the Japanese delegation, suggested that "[d]emonstrating leadership to solve the global environmental problem must be the pillar of Japan's role in the international community."[322] The Japanese delegation (over 100 officials) to UNCED was more than twice the size of the American delegation (45 officials),[323] indicative of the different emphasis placed on environmental issues by these two great economic powers. With an obvious eye on the economic benefits for Japanese companies, Shozaburo Nakamura, Minister in Charge of Global Environment, announced that Japan would host the United Nations Environment Programme's Environmental Technology Center. The "Center's main objective," he informed delegates at UNCED, "is to promote the transfer of environmentally sound technologies and know-how to developing countries."[324] The market for environmental goods and services is estimated to be worth at least $200 billion, and forecasts by the Organization for Economic Cooperation and Development project an increase in value to $300 billion by the year 2000.[325] Japan also proudly announced the passage of domestic legislation, an Endangered Species Law.[326]

Despite their overt enthusiasm for environmental reform on a global scale, the delegates of most developed countries and the numerous environmentalists who were watching the formulation and acceptance of the Convention from the sidelines shared some misgivings about the Treaty. The Japanese, for instance, "wanted clearer terms on how money would be spent under the treaty."[327] An environmental lobbyist for the World Wildlife Fund, Alistair Graham, bemoaned the fact that the Biodiversity Treaty lacks real teeth and pointed to the innumerable qualifying phrases in the text,[328] especially prominent in the activity-oriented clauses. The Convention makes no provision for sanctions against any Contracting Party which violates its terms. Given the politicized nature of the negotiating process, any punitive measures were unlikely. Implementation will therefore be largely voluntary, given the large number of qualifiers added to the obligations of the Convention. Bruce Babbitt, U.S. Secretary of the Interior in the Clinton Administration, lamented the absence of sanctions, noting that "[t]he very concept of a biodiversity treaty necessarily implies some limitation of a nation's claim to an absolute right to slaughter whales, burn forests, or drive animal species to extinction."[329] While this view has some validity, its feasibility in the international law scenario is somewhat questionable. Initial formulations of international law tend to be rather vague and permissive, with the expectation that clear objectives and targets are negotiated in subsequent protocols. The immediate challenge was to ratify the Convention and begin the process of implementation—as Arthur Campeau, Canada's Ambassador for Sustainable Development, said, "to get on with the job."[330] Wolfgang Burhenne, Legal Advisor to the International Union for the Conservation of Nature and Natural Resources (IUCN), articulated the thoughts of many delegates and environmentalists when he observed:

> It was difficult to conclude the Convention; it will, perhaps, be more difficult to implement it meaningfully. One thing is certain; if the developed world does not honour what it has (reluctantly) agreed to, the Convention will remain a paper tiger. We should do all we can to make a success of this unique chance to view biological diversity globally.[331]

Generally, the developed countries appeared to share the view that, despite its obvious flaws, the Convention on Biological Diversity committed the world to an environmental conservation activity which was far more important than either political or financial considerations. Rob Storey, New Zealand's Minister for the Environment, underlined the practical advantage of the Convention when he told his colleagues at UNCED that "[o]ur economic development is based almost entirely on species and genetic material introduced from other countries. For this reason, we understand the need for the closest possible cooperation between all countries."[332] Per Stig Moller, Denmark's Minister for the Environment, emphasized the positive achievements of the Convention:

> For the first time we have been able to address the issues related to access to genetic resources and related to bio-safety on a global scale and in legal language. We also think the convention reflects a good and fair balance between the interests of the developing world and the developed world. It has good prospects of starting a process that will gradually lead to stronger commitments.[333]

Chancellor Helmut Kohl of Germany stressed the role of the Conventions on Biological Diversity and Climate Change in contributing to "a more effective global protection of the environment."[334] Spain, seeing yet another opportunity arising from the Treaty, applied to host the Secretariat of the Convention.[335] Ruth Feldgrill-Zankel, Austria's Minister of Environment, admitted that the Convention was very general in some provisions but nevertheless thought it represented "a first important step in the right direction."[336] Expressing his support for the Convention, President Mitterrand of France still made known his preference for "bolder commitments."[337] Though regretting the absence in the Biodiversity Treaty of a list of ecological zones requiring protection,[338] the French Environment Minister, Segolene Royal, expressed support for the Convention provided it was seen "only as the first stage of an effort to protect the diversity of the world's plant and animal life."[339] This perception of the Treaty as a first step was shared by Michael Smith, Ireland's Minister for the Environment.[340] Naoya Sugio and Alex Steffen of the *Japan Times* commented that "[f]ive years ago, the idea of global leaders signing [a] convention . . . to . . . slow the extinction of species seemed unrealistic. Now such actions are considered insufficient by most leaders."[341] President Mitterrand also emphasized the responsibility of the countries of the North to "restore their own environment" and to "refrain from doing anything detrimental to the environment of the countries of the South."[342] Finland's Minister for Foreign Affairs, Paavo Vayrynen, pointed to the need for "internationally binding agreements which redirect development in individual nations."[343] Thorbjorn Berntsen, Norway's Minister of Environment, expressed his belief that the resort to national action plans in the Convention would provide "the necessary flexibility to be able to adapt measures, strategies and policies to national and local circumstances."[344] Calling the Biodiversity Treaty an "important milestone," Berntsen went on to add that "[t]his convention is not just an agreement on protection, but also covers the *use* of biological resources. Equally important, the Convention creates a framework for fair and equitable sharing of scientific results and other benefits arising from genetic resources. This gives an economic value to biological resources, which will constitute a powerful incentive to protect these resources."[345] Andreas Gavrielides, Minister of Agriculture and Natural Resources for Cyprus, stressed the vital international concern involved in the issue,[346] while Sweden's Minister of the Environment and Natural Resources, Olof Johansson, concentrated on the importance of a more "equitable distribution of resources between nations."[347]

Speaking as President of the European Community, the Prime Minister of Portugal, Anibal Cavaco Silva, admitted that the Convention fell short of initial expectations but suggested, optimistically, that

it incorporates a series of measures which constitute a solid basis for future progress in this area. In addition to its fundamental role in the conservation of biological diversity, it is very positive that the Convention in question has recognized the principles of safeguarding legitimate national interests and of the common, shared responsibility which should prevail in relations between states in matters regarding the very survival of the planet.[348]

In signing the Convention, the European Community attached a statement of interpretation expressing regret about the inadequacy of environmental objectives. The statement also called for "further tightening-up later; the need for strict respect of the financing provisions; and the need to respect intellectual property rights."[349] Despite some misgivings, developed nations found a sufficient number of advantages to participating in the process of conservation established by this Convention.

The Developing Nations and the Convention on Biological Diversity

There can be no doubt that the Biodiversity Treaty marked a significant break-through for the poor nations of the world in terms of international recognition of their rights environmentally, economically, and developmentally. The Convention on Biological Diversity was, in terms of the agenda of the developing countries, the strongest document to emerge from the entire Rio process for creating a new world environmentally. The South has now obtained acknowledgment from the North that its rich biodiversity is a vital resource for the world and that developing nations have property rights in this resource, rights which extend beyond their national boundaries to encompass any external utilization of biodiversity. Most important of all, the rich nations have pledged financial assistance to enable the poor countries which own the resource to develop their own conservation and protection programs and to share in the benefits of any development of biodiversity. It is clear that

> [a]s a sovereign national resource, biodiversity now becomes an asset for developing and developed countries alike. The convention thus creates an economic incentive to conserve biodiversity that can be added to the ethical imperative that most nations share.[350]

The most important benefit of the negotiating process which led to the formulation of this Convention was in publicizing the cause of conservation of this vital resource. At UNCED, delegates from an amazing variety of developing nations emphasized the significance of their biodiversity as a vital resource. The Governor-General of Papua New Guinea, Sir Wiwa Korowi, told delegates that 80 per cent of his population depend on the utilization of these resources and hence his country has included a statement recognizing the importance of biodiversity in its National Constitution.[351] Vincent Perera, Sri Lanka's Minister of Environment, highlighted the significance of the southwestern area of his nation because it "has one of the highest concentrations of biological diversity seen anywhere in the world," though the "gene-rich southwest part of the country has been subject to heavy timber exploitation and deforestation in the past."[352] He indicated that Sri Lanka has taken action to protect its remaining forests.[353] Joeli Kalou, Fiji's Minister of State for Environment, similarly explained that his nation was actively protecting its unique species.[354] Princess Sonam Chhoden Wangchuck, representing Bhutan at UNCED, informed delegates that her government "has decided to maintain 60 percent of our country's land area under forest cover. Twenty percent or more of the total land area will consist of parks and sanctuaries for the preservation of bio-diversity."[355] One indirect benefit of the Rio Summit has been the nearly universal acknowledgment that this precious resource requires assertive measures to protect and conserve it for future generations. As Princess Sonam Chhoden Wangchuck explained, "Through these measures we hope to pass on to our future generations not only a rich cultural heritage but also a diverse natural heritage of which they can be proud."[356]

Environmentalists from Asia, Africa, and Latin America were able, because of the global interest in UNCED, to educate their populations about both the significance of the resource and its potential value. As Philip Leakey, Kenya's Minister for Environment and Natural Resources, commented at UNCED, biodiversity "is an area where for once developing countries are able to make significant and important international contributions as equal partners in the family of nations."[357] More cynically, *Newsweek* commented that "[w]hile few delegates know a fungus from a mold, they do know the most important thing about biodiversity: the rich North needs, the poor South has it."[358] Avani Vaish, an Indian delegate, remarked: "The most important thing is that the value of genetic resources will be really appreciated. Resources were a free commodity, like air and water, but [by the terms of the Treaty] they're under international jurisdiction."[359] In 1954, when Gordon Svoboda of Eli Lilly and Co. extracted cancer-fighting alkaloids from Madagascar's rosy periwinkle, the drug company's sales were in the millions of dollars. Madagascar got nothing.[360]

For developing nations it seems only fair that "countries of origin should have access to biotechnology developed through the use of genetic material they provide."[361] The awareness of the value of the resource has also brought recognition of the serious amount of damage done to biodiversity. Resio Moses, Secretary of Micronesia's Department of External Affairs, told the delegates at UNCED that "[w]e are now concerned about the future of many species of marine and plant life that we had thought to be virtually inexhaustible."[362] Guy Willy Razanamasy, Prime Minister of resource-rich Madagascar, lamented the fact that his country's biodiversity is "extraordinarily rich, yet gravely threatened" and mentioned the destruction caused by deforestation, shifting agriculture, erosion, and marine damage, the last attributable to Madagascar's position on "a major route for petrol tankers which proceed to wash out their tanks along" its shores.[363]

The awareness of the value of biodiversity and of the threat it now faces inevitably led to greater interest in biotechnology and the promise that science holds for the future. The entire preparatory process for UNCED was itself important in publicizing the issues for the North and the South. R. Olembo, Deputy Assistant Executive Director of the United Nations Environment Programme, explained to the Sub–Working Group on Biotechnology that

> [b]iotechnology represents a power to change, a potential to lift living standards, to sustain and accelerate development. Its application to crops, livestock husbandry, forestry, fisheries, hazardous waste treatment and other areas holds tremendous potential.[364]

As Olembo pointed out, the poorer countries lack the financial resources to develop their own technologies, but their ownership of the genetic resource gives them "tremendous leverage to strike a new global deal."[365] The nations of the South were not slow to seize the initiative, and they retained it throughout the UNCED process. This is why so many of the instruments associated with the Earth Summit bear the distinct stamp of the Southern agenda. The Centre for Science and Environment in India had cautioned delegates that "[d]eveloping countries must not sign the biodiversity convention unless it reduces the existing asymmetries in access to knowledge and technology."[366]

The opposition of President Bush to the Convention dramatically increased the amount of publicity accorded to it both in the North and in the South. As Walter Reid points out, "[O]ne ironic benefit of President Bush's decision not to sign was that for two weeks everybody in the United States was subjected to newspaper articles and news stories about biodiversity. The effect of that decision was a tremendous public relations coup for the issue of biodiversity."[367] In the developing world, it appeared as though the United States was interested only in reaping enormous profits from a Southern resource without sharing the benefits with the country of origin. The refusal of the United States to participate in the Biodiversity Treaty and its determined opposition to firm timetables for the sister Framework Convention on Climate Change did serious damage to its reputation in the nations of the South.

From the perspective of the developing world, the Convention provides a positive opportunity for future conservation and development within a framework which offers benefits for both the North and the South. The Preamble affirms that *"the conservation of biological diversity is a common concern of humankind."* The Convention also acknowledges that *"special provision is required to meet the needs of developing countries, including the provision of new and additional financial resources and appropriate access to relevant technologies"* (Preamble); notes the *"special conditions of the least developed countries and small island States"* (Preamble); recognizes that *"economic and social development and poverty eradication are the first and overriding priorities of developing countries"* (Preamble); and asks Parties to *"[c]ooperate in providing financial and other support for in-situ conservation . . . particularly to developing countries"* (Article 8[m]). The provisions referring to ex-situ conservation of biodiversity express a preference for such conservation to take place in the country of origin (Article 9) and provide for the *"establishment and maintenance of ex-situ conservation facilities in developing countries"* (Article 9[e]).

The stipulations concerning research and training (Article 12) emphasize *"the special needs of developing countries"* with promotion of research *"which contributes to the conservation and sustainable use of biological diversity, particularly in developing countries"* (Article 12[b]). Developing countries which provide the genetic resource are to be encouraged to participate in biotechnological research, which ought to be undertaken, where feasible, within the country of origin (Article 19.1). Developing countries will also benefit from priority access to the results of biotechnological research from their own genetic resources, on the basis of mutually agreed terms (Article 19.2).

As we have seen, more-controversial clauses specify that access to technology for developing countries will be under *"fair and most favourable terms, including on concessional and preferential terms where mutually agreed"* (Article 16.2). Parties are also to take legislative measures so that their private sector *"facilitates access to, joint development and transfer of technology . . . for the benefit of both governmental institutions and the private sector of developing countries"* (Article 16.4). E-Hyock Kwon, Minister of Environment for the Republic of Korea, emphasized the significance of sharing technology fairly when he addressed delegates at UNCED. He declared that "in . . . light of the great cause of global environmental preservation, it is contradictory that many countries do not have access to the environmental technology which is essential to meet the regulations and obligations of international environmental conventions."[368] He also pointed out that although intellectual property rights may facilitate the development of technology, such rights, in hindering technology transfer, could "result in the failure of global environmental protection."[369] Consequently, he made two proposals: first, that an international mechanism be established to facilitate technology transfer and to provide compensation for intellectual property rights; second, that an appropriate mechanism be introduced to purchase essential environmentally oriented technology, which could then be supplied to some poor nations on a noncommercial basis, on the understanding that "[t]he policies, information and technologies for sustainable development should be available and accessible to all countries."[370] Li Peng, Premier of the People's Republic of China, explained that technology transfer under concessional terms "is only wise for the developed countries to do, for it serves their own interests as well as those of developing countries."[371] Ahmad Mattar, Singapore's Minister for the Environment, emphasized that the "timely application of appropriate technology will help developing countries to avoid environmental pitfalls experienced by the developed countries."[372]

The concept of shared benefits was very important to the developing nations, which did not want to be excluded from the vast profits made from the utilization of biodiversity by the North. In April 1992, shortly before UNCED, Malaysia hosted a ministerial conference of developing countries. This meeting resulted in the Kuala Lumpur Declaration on Environment and Development, which is a forthright assertion of the position of the South:

> [T]he Convention on Biological Diversity must include legally binding commitments to ensure the link between the access to the genetic material of developing countries and the transfer of biotechnology and research capabilities from developed countries, as well as sharing of commercial profits and products derived from genetic material.[373]

One encouraging note for the developing nations was sounded by William Reilly, Head of the United States Delegation to UNCED. He told delegates that his nation "strongly supports technology cooperation with developing countries to help them find sustainable paths to economic development."[374]

The Convention specifies that developing countries shall also benefit from the *"exchange of information, from all publicly available sources, relevant to the conservation and sustainable use of biological diversity"* (Article 17.1) and from measures taken to promote inter-Party technical and scientific co-operation to implement the Convention (Article 18.2).

There is a growing apprehension on the part of some developing nations that this biological resource may fall prey to the voracious appetite of the biotechnology industry. The President of Tanzania, Ali Hassan Mwinyi, expressed his concern that "certain herbs and other forest resources in high demand for biotechnology have often been extracted to depletion in developing countries, thus threatening the continued existence of plant varieties and undermining biodiversity."[375] He was also worried that "[g]enetic engineering could . . . be ecologically damaging to animal life from excessive use of particular species, such as monkeys, rats and birds for research and experimentation."[376]

Financial provisions are of considerable importance to the environmental agenda of the South. The Convention states that *"developed country Parties shall provide new and additional financial resources to enable developing country Parties to meet the agreed full incremental costs to them of implementing measures which fulfil the obligations of this Convention"* (Article 20.2). Indeed, developing country implementation is contingent on funding from the developed world:

> *The extent to which developing country Parties will effectively implement their commitments under this Convention will depend on the effective implementation by developed country Parties of their commitments under this Convention related to financial resources and transfer of technology and will take fully into account the fact that economic and social development and eradication of poverty are the first and overriding priorities of the developing country Parties.* (Article 20.4)

There could be no stronger statement of the developing country agenda than this provision which declares that funding must come from the North if biodiversity is to be protected. The stipulation emphasizing the importance of economic and social development also caters heavily to the Southern sense of priorities. The particular needs of the least developed nations, small island states, and environmentally vulnerable areas are also recognized by the Convention (Articles 20.5–20.7). As we have seen, the financial mechanism *"shall function under the authority and guidance of, and be accountable to, the Conference of the Parties for purposes of this Convention"* (Article 21.1), with emphasis on *"predictability, adequacy and timely flow of funds . . . with the amount of resources needed to be decided periodically by the Conference of the Parties,"* the mechanism to operate *"within a democratic and transparent system of governance"* (Article 21.1).

The developing countries organized a strong lobby, led by India and Malaysia, to ensure the dominant status of the Conference of the Parties regarding financial decisions, because this would provide for one-country, one-vote decisions, resulting in more influence for the South.[377] In a critical editorial, the *Times* (London) commented that Third World governments "want money with no strings" and denounced the fact that "[a] sensible principle that poor countries should be rewarded for protecting species has been turned into a binding obligation on the West to provide a grandiose, multi-course free lunch."[378] An alternative view would hold that without economic incentives developing countries will simply be unable to undertake conservation of their biodiversity.

Fortunately for the implementation of the Convention, many developed nations recognize the responsibility they will have to bear in assisting the developing world to preserve biodiversity. There is now a growing awareness among policy makers in the rich countries that poverty poses the greatest threat to the global environment, that "[p]overty degrades not only those who suffer it, but also those who tolerate it."[379] This is coupled with the realization that unless economic improvement—via development—occurs in those areas, the environment of the entire planet is doomed simply because the people of the South do not have many alternatives or options. President Ali Hassan Mwinyi of Tanzania pointed out that in the developing countries approximately 150 million people are suffering from malnutrition, 1.5 billion people do not have access to health care facilities, 1.75 billion people do not have clean water, 2.8 billion people lack sanitation facilities, and 4 million children die, while 750 million children suffer acute diarrhea—every year.[380] As he explained, "The reality of poverty-driven degradation of the environment is felt throughout the third world, where the incidence of environmental degradation and poverty is always a case of untold misery."[381] Gro Harlem Brundtland, Prime Minister of Norway, asserted that "[w]e should not be surprised that developing nations are approaching the Rio Summit with open economic demands. For them, it is essentially a conference about development and justice."[382]

Developed countries made generous pledges at Rio but, given the enormous range of environmental tasks envisaged by the Convention on Biological Diversity, the Framework Convention on Climate Change, and the massive Agenda 21, there obviously will never be as much financial assistance as the world will require. The Global Environmental Facility has estimated that biodiversity protection could cost approximately $35,000 per square kilometer.[383] The World Resources Institute predicted that during the

1990s environmental and natural resource conservation would cost between $20 billion and $50 billion annually.[384] The challenge of the future implementation process will inevitably be to allocate funding where it can achieve the greatest results and, more importantly, to aim the initial grants at conservation, clean-up, or protection problems which simply cannot wait. Unfortunately, the "[c]ountries with the richest diversity of life on Earth are also the ones least able to afford conservation measures."[385]

During the Earth Summit Japan pledged 1 trillion yen (7 to 7.7 billion U.S. dollars) in aid for global environmental protection, a "doubling of the nation's environment-related aid."[386] Although the Japanese funds will be utilized for a number of programs, protection of the rain forest is definitely included in their agenda.[387] According to William K. Stevens, writing in the *New York Times* in June 1992, "Japan indicated that it would substantially increase its foreign aid, which has been running at about $10 billion a year, to help poor countries pay for environmental programs. Japanese officials said Tokyo would offer $7 billion for the environment, but it was unclear how much of this would be new money beyond what Japan had already intended to spend."[388] Former Prime Minister Noboru Takeshita explained at UNCED that "[t]he destruction of the global environment has been caused in large part by the activities of developed countries," and he suggested that this therefore requires "a commitment by developed countries" in dealing with the "problem of poverty in developing countries, a problem closely linked to the issue of environmental protection."[389] It was no surprise that Japan came away from Rio with its environmental reputation in excellent shape despite a record tainted by deforestation, whale hunting, and being "the world's biggest importer of wildlife."[390] Hideo Obara, Director-General of the Nature Conservation Society of Japan, observed that "[m]any countries expect some financial support from the Japanese government, so they have decided not to criticize it."[391] Naomi Kamei, Coordinator at Friends of the Earth in Japan, suggested (somewhat cynically) that the promised environmental aid would mainly assist "Japanese companies and Japanese politicians."[392]

Developing nations displayed considerably less enthusiasm for the aid package proposed by the United States. President Bush promised that

> [w]e come to Rio with an extensive program of technology cooperation. We stand ready, government and private sector, to help spread green technology and launch a new generation of clean growth. We come to Rio recognizing that the developing countries must play a role in protecting the global environment, but will need assistance in pursuing these cleaner growths. So we stand ready to increase U.S. international environmental aid by 66 percent

above the 1990 levels, on top of the more than $2.5 billion dollars that we provide through the world's development banks for Agenda 21 projects.[393]

The President also promised $150 million per year for reforestation projects in developing countries, a sum dismissed by environmentalists as "a drop in the ocean."[394] The American Forests for the Future Initiative doubled total assistance for international forest conservation, from $1.35 billion to $2.7 billion.[395] "US officials indicated that countries supporting the US forest initiative would be more likely to receive forestry aid than those who opposed it."[396]

John Major, Prime Minister of the United Kingdom, asserted that "money is the root of all progress"[397] and promised to contribute to save the rain forests and wildlife. The bulk of Britain's extra £100 million was allocated for work involving climate change, and another £25 million for other projects.[398] Michael Howard, British Secretary of State for Environment, declared that the new funding would be channelled through the Global Environment Facility (GEF).[399] Of greater significance in the long term is the British project to encourage international studies of natural resources, set research goals, and establish inventories of the most significant species and habitats. This project, called the Darwin Initiative, would take advantage of Britain's recognized expertise, as demonstrated at centers like the Royal Botanic Gardens at Kew, the Natural History Museum at London, and the World Conservation Monitoring Centre at Cambridge.[400] The Darwin Initiative would help countries both to monitor and to utilize their biodiversity resources.[401] Most important among its objectives is the training it would provide for professionals from developing countries.[402] Funding for the project is relatively modest at £6 million over three years,[403] but as Environment Minister David Maclean pointed out: "The developed world is in recession. If the developing world thinks the major nations of the world in the middle of a very, very bad recession have the resources to be able to proceed quickly, then I am afraid that is just not possible."[404] However, to demonstrate its commitment to the implementation of the Convention, the United Kingdom offered to host the first Conference of the Parties.[405]

With respect to environmental projects in developing nations, a number of Northern countries made generous promises at UNCED to contribute more 2funding to ecological improvement in the poor nations. Germany proposed additional funding and debt remission for poor nations, and Chancellor Helmut Kohl asserted his country's readiness to assume its share of the burden.[406] President Mitterrand offered to double the French contribution to the

World Environment Fund, among other pledges.[407] Austria, which in 1990 spent 1.9 per cent of its gross national product on environmental programs, joined the ranks of the top third of contributors with increased pledges for the developing world.[408] Explaining that his country already spends more than 1 per cent of its GNP on development assistance, Thorbjorn Berntsen, Norway's Minister of Environment, announced the allocation of additional financing for global environmental problems.[409] The Swedish Government was also generous, allocating funding specifically for financing UNCED commitments.[410] Even Ireland, "one of the least developed economies in a developed part of the world,"[411] promised to increase its development assistance.[412] Although these pledges were encouraging for the South, most were phrased rather vaguely, and it is not easy to pin down the actual amounts of *new* funding with precision. Government leaders were obviously at Rio to lend a glow of summitry to the occasion; accordingly expansive, they possibly competed to some extent to outdo each other in appearing generous.

The United Nations has indicated a target of 0.7 per cent of the GNP of developed countries to be allocated as aid for developing nations.[413] Unfortunately, few industrialized nations have met this objective. The following statistics, published in an editorial in the *Globe and Mail* in March 1993, record the percentage of gross national product allotted to foreign aid: Canada, 0.40 per cent; France, 0.52 per cent; Germany, 0.42 per cent; Italy, 0.32 per cent; Japan, 0.31 per cent; U.K., 0.27 per cent; and U.S.A., 0.19 per cent.[414] A more positive contribution was made by smaller European countries like Denmark, (0.93 per cent), the Netherlands (0.93 per cent), and Sweden (0.90 per cent).[415] Developing nations attempted throughout the Rio process to persuade and pressure the North about the need to reach the U.N. aid target. However, although most developed nations (with the exception of Switzerland and the United States)[416] indicated a willingness to accept the target, most of them (with the notable exception of France)[417] continued to resist any attempt to set a deadline for achieving that goal.[418] Prime Minister Carl Bildt of Sweden suggested that meeting the target "should be a moral duty for all of the countries of the rich world."[419]

Canada's Environment Minister, Jean Charest, confirmed during the Earth Summit that his country was prepared to increase its foreign aid to 0.7 percent of the gross domestic product.[420] He did not specify a deadline for meeting this commitment,[421] probably because it would increase Canadian aid by almost $2 billion.[422] Prime Minister Brian Mulroney, after reminding delegates at UNCED that Canada had contributed $1.3 billion for sustainable development projects in the South over the previous five years, promised to provide $25 million to the Global Environment Facility and $50 million for humanitarian assistance and drought relief in Southern Africa, and to eliminate $145 million in Latin American debt to his nation.[423] The reality is that "Canada's growing debts have put the country on the threshold of a major crisis,"[424] which forced the Canadian Cabinet to decide in December 1992 to cut approximately $600 million in aid by 1995;[425] the cuts in aid were expected to total $4.4 billion in five years.[426] The process of cutting aid to developing nations is not recent. Since the 1980s, Canada has reduced its aid to Africa by more than half. The lavish promises made at Rio notwithstanding, "Third World programs will get a smaller proportion of an already shrinking pie."[428]

The United Nations has estimated that it would cost approximately $600 billion a year to implement sustainable development throughout the world, with $125 billion required as an annual contribution from the industrialized countries.[429] In the months leading up to UNCED, developing countries attempted frantically to persuade the North to pledge the approximately $70 to $75 billion extra per year estimated to be the amount required to fund the main environmental programs.[430] Despite intense negotiating, they were not successful, nor could they secure deadlines (the proposed target year was 2000)[431] for the developed countries to reach the United Nations target of an aid allocation of 0.7 per cent of gross national product. The *Times* (London) commented that "[a] Martian would find it hard to believe that an obscure debate about completely arbitrary numbers is dominating the biggest meeting ever held of world political leaders, a meeting allegedly called to deal with dire environmental emergencies threatening the very existence of Planet Earth. Yet so it is."[432]

Domestic economic problems in developed countries dominate and frequently overwhelm their foreign aid agendas, and these nations were soon reconsidering their promises to developing countries. Madeleine Drohan of the *Globe and Mail* explained that there were a number of reasons for this:

> With total unemployment in the United States, Canada, Japan, Germany, France, Italy and the United Kingdom topping 23 million, trade tensions on the rise, Russia in crisis and the major European economies plus Japan's in a serious slump,[433]

there were few alternatives available for the countries of the North. Just a year after Rio, global environmental problems already seemed to be a lesser priority in the minds of government leaders than the domestic economic crises facing the developed countries. The dissolution of the Soviet Union converted its component member states into major foreign aid recipients rather than donors. The Americans, Canadians, and western Europeans made sustaining democracy in Russia a major priority in foreign aid. This could not but have a negative impact on the amounts available to be donated as foreign aid to the developing nations.

One significant consequence of UNCED was the growing realization that the world simply has to re-think its economic system and re-define the value of any commodity. Every nation has to consider the hitherto hidden costs in terms of pollution and environmental damage in any production or development process. The citizens of all countries have collectively borne the price of industrialization in the North. A forthright advocate for the perspective of the developing nations, Prime Minister Dr. Mahathir bin Mohamad of Malaysia, told delegates at UNCED:

> When the rich chopped down their forests, built their poison-belching factories and scoured the world for cheap resources, the poor said nothing. Indeed they paid for the development of the rich. Now the rich claim a right to regulate the development of the poor countries. And yet any suggestion that the rich compensate the poor adequately is regarded as outrageous. As colonies we were exploited. Now as independent nations we are to be equally exploited.[434]

The degradation of air, land, and water is evident for all to see. The Industrial Revolution of the West continues to exact a human toll every day of every year as people pay with their health and frequently with their lives for the enjoyment of consumer goods. The planet simply cannot afford the same pace of environmentally ignorant development in the South.

Economically, once a new system of enumerating the real cost of a product becomes the norm, governments might finally realize that it would be preferable to adopt environmentally safe development now than to pay much higher costs later. Only then perhaps will the funding issues be addressed seriously. The process of internalizing environmental costs was explained by Jose Esquinas-Alcazar, Secretary of the United Nations Food and Agriculture Organization's Commission on Plant Genetic Resources, who said, "When you buy an apple, you should pay not only for the cost of production, but also for the cost of conservation of natural resources that would allow future generations to continue having apples."[435] Arthur Campeau, who headed the Canadian delegation in negotiations for the Biodiversity Convention, commented, "In our society, what we value economically, we conserve. What we don't value economically, unfortunately, we waste."[436] He continued:

[T]he biggest hurdle to the conservation of biological diversity is outmoded concepts of economic value. What are now considered economic externalities must be taken into account. We must evolve more sophisticated methods of calculating our national accounts, our GNP, which take into account the maintenance and depreciation of our biological resources. To redress that, we need comprehensive methods for assessing the worth of biodiversity. We need tools to prove the costs of inaction grossly exceed the relatively small price of action.[437]

The crucial importance of financing the provisions of the Rio Summit's various Conventions was stressed by Maurice Strong, Secretary-General of UNCED, who emphasized the need for

developing new sources of funding, because the steps we have taken still do not promise to meet the larger needs. We should consider, for example, new taxes, user charges, emissions permits, citizen funding, all based on the polluter-pays principle. I believe the amounts of money available simply from funds wasted in existing subsidies to non-environmentally-sound activities could alone provide all the money necessary as an indispensable investment in environmental security.[438]

Within months of the Earth Summit, the enthusiasm generated by the UNCED process to protect and conserve biodiversity had begun to wear thin. Towards the end of 1992, Maurice Strong found that the reaction to Rio in developed countries was not very encouraging,[439] and Mostafa Tolba of the United Nations Environment Programme commented that "[t]he pace of governmental action has faltered."[440] The fate of all the millions of species with which we share this planet is now more threatened than ever. Although the will to save them may still be there, finding the financial resources to achieve that objective will be a greater challenge than ever. For the developing nations, the time has come to realize that biodiversity must be conserved, whether the developed world provides adequate funding or not. The imperative behind this cause has not declined, however low a priority environmental action may now be in the funding agendas of Northern governments. To say that the future of the planet is at stake is no exaggeration.

Obligations under the Convention

The Biodiversity Treaty is very activity oriented, at both the national and the international levels. As we have seen, it will be an extremely expensive undertaking to implement its provisions, and this is why there are so many qualifying phrases to the obligations to make them less onerous for all Contracting Parties. The extensive resort to qualifying phrases like "as far as possible" and "as appropriate" could, however, hinder environmentalists seeking rapid implementation of the provisions of the Convention.

One benefit of the Convention is that it has carefully mapped out measures which need to be taken and has charted the necessary course of action for any Member State which is interested enough to undertake these activities. Obviously, each nation will implement the Convention according to its capacity and, with respect to developing nations, on the basis of the amount of aid received from the North. However, it is important briefly to explain the obligations, because a knowledge of these provides some insight into the multifaceted approach which will have to be taken by governments if the cause of biodiversity conservation is to succeed. There could, incidentally, be numerous opportunities for technical experts, scientists, researchers, and lawyers, whose skills will be required around the world to activate the commitments of Member Nations. As Kenton Miller, Program Director for Forests and Biological Diversity for the World Resources Institute, suggests, "Countries that get involved in biotechnology and other spin-offs of biological resources can enjoy new jobs."[441] Russell Mittermeier, President of Conservation International in Washington, D.C., asserts that "[b]iodiversity is the new Silicon Valley," referring by comparison to the technological revolution in computers which occurred in that region of California.[442]

The obligations of Contracting Parties are enumerated below to explain the many duties required by the Convention on Biological Diversity. Articles which have already been analyzed in detail will not be repeated in this section. The duties include

1. The development of *"national strategies, plans or programmes for the conservation and sustainable use of biological diversity"* (Article 6[a])
2. The integration of *"sustainable use of biological diversity into relevant sectoral or cross-sectoral plans, programmes and policies"* (Article 6[b]
3. The identification of *"components of biological diversity"* with reference to a list of categories in Annex I to the Convention (Article 7[a])
4. The monitoring of components of biodiversity, *"paying particular attention to those requiring urgent conservation measures and those which offer the greatest potential for sustainable use"* (Article 7[b])
5. The identification of *"processes and categories of activities which have or are likely to have significant adverse impacts on the conservation and sustainable use of biological diversity"* (Article 7[c]) and the monitoring of the effects (Article 7[c])

6. The maintenance and organization of data derived from the identification and monitoring activities (Article 7[d]).

   With respect to in-situ conservation, the Convention asks that Contracting Parties

7. *"Establish a system of protected areas or areas where special measures need to be taken to conserve biological diversity"* (Article 8[a])

8. *"Develop, where necessary, guidelines for the selection, establishment and management of protected areas or areas where special measures need to be taken to conserve biological diversity"* (Article 8[b])

9. *"Regulate or manage biological resources important for the conservation of biological diversity whether within or outside protected areas, with a view to ensuring their conservation and sustainable use"* (Article 8[c])

10. *"Promote the protection of ecosystems, natural habitats and the maintenance of viable populations of species in natural surroundings"* (Article 8[d])

11. *"Promote environmentally sound and sustainable development in areas adjacent to protected areas with a view to furthering protection of these areas"* (Article 8[e])

12. *"Rehabilitate and restore degraded ecosystems and promote the recovery of threatened species"* (Article 8[f])

13. Establish ways to regulate risks *"associated with the . . . release of living modified organisms resulting from biotechnology . . . taking also into account the risks to human health"* (Article 8[g])

14. *"Prevent the introduction of, control or eradicate those alien species which threaten ecosystems, habitats or species"* (Article 8[h])

15. Provide necessary conditions for *"compatibility between present uses and the conservation of biological diversity and the sustainable use of its components"* (Article 8[i])

16. Preserve, maintain, and promote the wider application of the knowledge and practices of indigenous and local communities concerning sustainable use of biological diversity and encourage the equitable sharing of the benefits of this information (Article 8[j])

17. *"Develop or maintain necessary legislation and/or other regulatory provisions for the protection of threatened species and populations"* (Article 8[k])

18. Regulate processes *"where a significant adverse effect on biological diversity has been determined"* (Article 8[l])

19. *"Cooperate in providing financial and other support for in-situ conservation . . . particularly to developing countries"* (Article 8[m]).

     As a complement to the in-situ measures outlined above, the Convention asks Contracting Parties to undertake ex-situ conservation activities as well. Each Contracting Party shall

20. *"Adopt measures for the ex-situ conservation of components of biological diversity, preferably in the country of origin of such components"* (Article 9[a])

21. *"Establish and maintain facilities for ex-situ conservation . . . preferably in the country of origin of genetic resources"* (Article 9[b])

22. *"Adopt measures for the recovery and rehabilitation of threatened species and for their reintroduction into their natural habitats under appropriate conditions"* (Article 9[c])

23. Regulate the collection of biological resources for ex-situ conservation *"so as not to threaten ecosystems and in-situ populations of species"* (Article 9[d])

24. *"Cooperate in providing financial and other support for ex-situ conservation . . . and in the establishment and maintenance of ex-situ conservation facilities in developing countries"* (Article 9[e]).

     There are also measures to be taken to ensure sustainable use of components of biological diversity. Each Party shall

25. "Integrate consideration of the conservation and sustainable use *of biological resources into national decision-making"* (Article 10[a])

26. *"Adopt measures relating to the use of biological resources to avoid or minimize adverse impacts on biological diversity"* (Article 10[b])

27. *"Protect and encourage customary use of biological resources in accordance with traditional cultural practices that are compatible with conservation or sustainable use requirements"* (Article 10[c])

28. *"Support local populations to develop and implement remedial action in degraded areas where biological diversity has been reduced"* (Article 10[d])

29. *"Encourage cooperation between its governmental authorities and its private sector in developing methods for sustainable use of biological resources"* (Article 10[e])

30. Adopt *"economically and socially sound measures that act as incentives for the conservation and sustainable use of components of biological diversity"* (Article 11).

     The research and training provisions emphasize the special needs of developing countries and direct that the Parties shall

31. *"Establish and maintain programmes for scientific and technical education and training in measures for the identification, conservation and sustainable use of biological diversity and its components"* (Article 12[a])

32. *"Promote and encourage research which contributes to the conservation and sustainable use of biological diversity"* (Article 12[b])

33. *"Cooperate in the use of scientific advances"* to develop methods for sustainable use of biological resources (Article 12[c]).

    With respect to public education and awareness, Contracting Parties shall

34. *"Promote and encourage understanding of the importance of, and the measures required for, the conservation of biological diversity, as well as its propagation through media, and the inclusion of these topics in educational programmes"* (Article 13[a])

35. *"Cooperate . . . with other States and international organizations in developing educational and public awareness programmes"* for its conservation and sustainable use (Article 13[b]).

    There are provisions for impact assessment, directing each Party to

36. *"Introduce appropriate procedures requiring environmental impact assessment of its proposed projects that are likely to have significant adverse effects on biological diversity."* The aim is to minimize ill effects and to allow for public participation in the process (Article 14.1[a])

37. *"Introduce appropriate arrangements to ensure that the environmental consequences of its programmes and policies that are likely to have significant adverse impacts on biological diversity are duly taken into account"* (Article 14.1[b]).

    Transboundary responsibilities are also specified in Article 14, again with qualifying phrases attached. Each Party shall

38. *"Promote, on the basis of reciprocity, notification, exchange of information and consultation on activities under [its] jurisdiction or control which are likely"* to have a significant adverse impact on the biological diversity of other States. The Convention encourages the formulation of bilateral, regional, or multilateral agreements to further this aim (Article 14.1[c])

39. In a situation of imminent danger, *"notify immediately the potentially affected States . . . as well as initiate action to prevent or minimize such danger or damage"* (Article 14.1[d])

40. *"Promote national arrangements for emergency responses to activities*

. . . which present a grave and imminent danger to biological diversity" (Article 14.1[e]).

There are detailed provisions referring to exchange of information and technical and scientific cooperation. Contracting Parties shall

41. *"Facilitate the exchange of information, from all publicly available sources"* (Article 17)

42. *"Promote international technical and scientific cooperation in the field of conservation and sustainable use of biological diversity"* (Article 18.1)

43. Develop and implement national policies for the promotion of technical and scientific cooperation, particularly with developing countries (Article 18.2)

44. *"Promote cooperation in the training of personnel and exchange of experts"* with the objective of encouraging the development and use of technologies, *"including indigenous and traditional technologies"* (Article 18.4)

45. *"Promote the establishment of joint research programmes and joint ventures for the development of technologies"* (Article 18.5).

It is evident from this detailed account of the tasks to be undertaken that developing nations will need considerable financial and technical assistance to accomplish the objectives of the Convention. It will be up to them to continue to press on developed countries the primacy of the cause of biodiversity conservation so that the issue remains, as it were, "on the front burner."

Although implementation could result in a frenetic round of activity, the Convention is not really as firm about mandating the obligations of Contracting Parties as it ought to have been. Aside from the frequent resort to qualifying phrases already mentioned, the Contracting Parties will largely be self-monitoring:

> *Each Contracting Party shall, at intervals to be determined by the Conference of the Parties, present to the Conference of the Parties, reports on measures which it has taken for the implementation of the provisions of this Convention and their effectiveness in meeting the objectives of this Convention.* (Article 26)

Walter Reid of the World Resources Institute commented that "the negotiators discarded one powerful tool that could have been used for monitoring and instigating action. This was a mechanism to list globally threatened and endangered species and habitats."[443] Should a Contracting Party renege, even on the ambiguous and qualified obligations of the Convention, there are no measures specified. The Convention does state, however, that *"[t]he Conference of the Parties shall keep under review the implementation of this Convention"* (Article 23.4) and adds that the Conference shall inter alia *"[c]onsider and undertake any additional action that may be required for the achievement of the purposes of this Convention in the light of experience gained in its operation"* (Article 23.4[i]).

It is also important to note that while there is an extensive, almost exhaustive, activity list,

> specific conservation actions are lacking. While it does require countries to undertake conservation planning, to develop legislation to protect threatened species and populations, to monitor the status of biodiversity, and to establish a system of protected areas, it doesn't establish criteria for measuring progress toward biodiversity conservation, and it requires no specific action to slow the loss of species and habitats.[444]

Time alone will tell how effective this Convention will be in the future and how dramatic a difference it will make to the survival of other species on this planet. The analysis in this chapter has explored both its strengths and its flaws with as much impartiality as possible. All the faults of the Convention notwithstanding, the fact of its existence is a tremendous achievement in itself. For so many different nations with varied viewpoints and divergent perspectives to come together to agree to save biodiversity testifies to the seriousness with which the crisis of species destruction is perceived around the world. If some delegates attempted to wrest gains for their own national or regional agendas, that is frequently the case in international negotiations of this nature. One of the world's most articulate supporters of the cause of saving biodiversity is Arthur Campeau, formerly Canada's Ambassador for Sustainable Development and leader of his country's delegation during the negotiations for the Convention. Speaking from that experience, he asserted confidently that

> the Convention and the principles that underlie it are sound. They will bring benefits, not just to Canadians but to all peoples, and in fact to all species. Quite simply, it represents the turning of a new page in our understanding of the world and of the impact our species is having on the others, on ecosystems we must co-inhabit, of the economic implications that flow from the conservation and non-conservation of biodiversity and from its sustainable and non-sustainable use. It is also about the intrinsic value of life itself.[445]

Conclusion

Maurice Strong, Secretary-General of UNCED, was cautiously optimistic as he concluded the Conference on Environment and Development. He told the delegates:

> If we have reason for satisfaction, we certainly do not have reason for complacency. The real measure of our success will be in what happens when we leave here, in our own countries, in our own organizations, in our own lives. Will this Summit merely be a high point in our expressions of good intentions and enthusiasm and excitement, or will it really be the start of the process of fundamental change which we absolutely need?[446]

Although the Earth Summit at Rio was the largest environmental gathering ever held, and although it generated great enthusiasm around the world for the cause of ecological betterment, it is still too early to determine its real effectiveness. That verdict awaits the test of time and the extent of dedication with which governments and environmentalists fulfil the verbal commitments they made at Rio. Two decades after the Stockholm Conference, people were lamenting the lack of achievement, despite national and regional accomplishments. If, two decades from now, there are active plans to conserve and protect forests in every continent, detailed inventories of species, fair and equitable agreements on sharing genetic resources, and some assurances that poor nations can enjoy the benefits of the latest environmental technologies, that will be a testament to the successful implementation of the Convention on Biological Diversity. But, two decades from now, if the pressure of burgeoning populations in the developing world has pushed wildlife out of its present shrunken habitat; if the South remains poor and the North continues to enjoy the overwhelming share of the world's wealth; if the forests are gone and the air everywhere is foul to breathe; if ecological devastation has affected food supplies and, indeed, disrupted the food chain which sustains us; if the world continues as it is now with pockets of privilege surrounded by vast areas of deprivation and misery—then they will look back at this Convention and declare it a great promise which remained unfulfilled. Arguably, the world expected far too much from the extravaganza called the Earth Summit. As the author Walter Russell Mead commented: "The old, pre-Rio diplomacy was important, but simple. . . . The new diplomacy isn't simple. There are more than 175 countries, and on issues like trade, the environment and nuclear proliferation, many have to be consulted. Worse still, the new diplomacy deals with far more complicated issues."[447]

If UNCED raised awareness about the crucial environmental problems of our time, that alone should make it a big success. If it generated debate in every nation of this world about the need to clean up the planet, then it will have served a great purpose. If it stirred our collective conscience about the plight of those forms of life which are vulnerable, that rates as an achievement. If it provided world leaders with a stage on which they made pledges, then it provides the public with a standard to which they can hold their governments, and that too is an indication of some success.

As regards the instruments of international law (whether soft law or binding law) formulated for this Earth Summit, the fact that so many governments, willingly or otherwise, signed the Conventions testifies to the amount of public interest that has been generated. This also is a measure of success. After all the bickering between rich and poor, North and South; after all the quibbling about bracketed texts and delicately balanced verbal compromises; after lawyers and politicians and diplomats and world leaders have ended their flood of rhetoric and self-serving statements; the real issue still remains: whether we believe that the amazing, wondrous diversity of life on this incredibly beautiful planet is worth preserving, or whether, through sheer indifference, or because of other priorities or economic considerations, we will just let it all die forever? The governments have produced the law. Now it is up to the public in every part of this world to ensure that biodiversity remains a live issue, no pun intended!

Notes

1. Edward O. Wilson, *The Diversity of Life* (Cambridge, Mass.: Harvard University Press, 1992), 393.

2. John C. Ryan, "Conserving Biological Diversity," in *State of the World, 1992* (New York: W. W. Norton, 1992), 9.

3. Paul Harrison, *The Third Revolution* (London: I. B. Tauris, 1992), 59.

4. Luiz Fernando Soares de Assis, *A Regional View of Negotiations on Biodiversity,* United Nations Economic Commission for Latin America and the Caribbean—ECLAC, Document LC/L.610, 12th February 1991, 6.

5. *Time*, 1st June 1992, 26.

6. Harrison, *The Third Revolution*, 59.

7. Cyril de Klemm, "The Conservation of Biological Diversity: State Obligations and Citizens' Duties," *Environmental Policy and Law*, vol. 19, no. 2 (April 1989): 50.

8. Ryan, "Conserving Biological Diversity," 9.

9. Wilson, *The Diversity of Life*, 281.

10. Soares de Assis, *Regional View of Negotiations on Biodiversity*, 1.

11. Harrison, *The Third Revolution*, 59.

12. United Nations Document, *Agenda 21*, Chapter 15, Conservation of Biological

Diversity, Introduction. Also see Nicholas Robinson et al., *Agenda 21 and the UNCED Proceedings* (New York: Oceana Publications, 1992), vol. 1, 429–54.

13. Office of the Prime Minister of Canada, "Canada Ratifies Two Historic Environmental Conventions," 4th December 1992.

14. *Time*, 1st June 1992.

15. Ryan, "Conserving Biological Diversity," 11.

16. Harrison, *The Third Revolution*, 60.

17. Soares de Assis, *Regional View of Negotiations on Biodiversity*, 7.

18. Office of the Prime Minister of Canada, "Canada Ratifies Two Historic Environmental Conventions," 4th December 1992.

19. Ryan, "Conserving Biological Diversity," 20.

20. Statement by Mostafa Tolba, Executive Director, United Nations Environment Programme, second session, Ad Hoc Working Group of Experts on Biological Diversity, Geneva, 19th–23rd February 1990 (hereinafter cited as Tolba statement, Geneva, 19th–23rd February 1990).

21. *Global Outlook 2000* (New York: United Nations, 1990), 95.

22. Ryan, "Conserving Biological Diversity," 20.

23. Kenton Miller, Walter Reid, and Charles Barber, "Deforestation and Species Loss: Responding to the Crisis," in J. T. Matthews, ed., *Preserving the Global Environment* (New York: W. W. Norton, 1991), 97.

24. Ben Jackson, *Poverty and the Planet* (London: Penguin, 1990), 29.

25. Statement by Mostafa Tolba, Executive Director, United Nations Environment Programme, third session, Ad Hoc Working Group of Experts on Biological Diversity, Geneva, 9th July 1990 (hereinafter cited as Tolba statement, Geneva, 9th July 1990).

26. Don McAllister, Senior Biodiversity Advisor, Canadian Centre for Biodiversity, Canadian Museum of Nature, Testimony before House of Commons Standing Committee on Environment, 23rd November 1992, Third Session, 34th Parliament, 1991–92, Issue no. 47, 47:21.

27. *Economist*, 30th May 1992, 17.

28. Prime Minister Brian Mulroney, Address at the Canadian Museum of Civilization, Hull, Quebec, 1st June 1992 (Office of the Prime Minister).

29. de Klemm, "Conservation of Biological Diversity," 50.

30. *Evening Telegram* (St. John's, Nfld.), 7th April 1993, 38:5.

31. *Vancouver (B.C.) Sun*, 5th June 1992, A10:2–4.

32. *Time*, 1st June 1992, 26.

33. *Vancouver Sun*, 6th June 1992, A6:5.

34. *Globe and Mail* (Toronto), 18th July 1992, D8:2–5.

35. *Globe and Mail*, 25th September 1992, A9:1–5.

36. de Klemm, "Conservation of Biological Diversity," 50.

37. *Ibid.*, 51.

38. Office of the Prime Minister of Canada, "Canada Ratifies Two Historic Environmental Conventions," 4th December 1992.

39. *Ibid.*

40. United Nations Environment Programme, *Guidelines for Country Studies on Biological Diversity*, Version 2.07, 22nd March 1993, Nairobi, 5.

41. *Ibid.*

42. Harrison, *Third Revolution*, 72.
43. United Nations Environment Programme, *Guidelines for Country Studies*, 6.
44. *Washington Post*, 16th January 1993, A23 (Source: LEXIS).
45. George Mitchell, *World on Fire* (New York: Scribner's, 1991), 116.
46. *Ibid.*, 117.
47. Ryan, "Conserving Biological Diversity," 9.
48. *Ibid.*
49. Don McAllister, Senior Biodiversity Advisor, Canadian Centre for Biodiversity, Canadian Museum of Nature, Testimony before House of Commons Standing Committee on Environment, 23rd November 1992, Third Session, 34th Parliament, 1991–92, Issue no. 47, 47:7.
50. Mitchell, *World on Fire*, 118.
51. Tolba statement, Geneva, 19th–23rd February 1990.
52. *Time*, 1st June 1992, 29.
53. *Christian Science Monitor*, 2nd June 1992, 10–11.
54. Statement by Mostafa K. Tolba, Executive Director, United Nations Environment Programme, fourth session, Intergovernmental Negotiating Committee for a Convention on Biological Diversity, Nairobi, 23rd September 1991, Speech 199/30 (hereinafter cited as Tolba statement, Nairobi, 23rd September 1991).
55. *Christian Science Monitor*, 2nd June 1992, 4.
56. *Ibid.*, 10–11.
57. *Ibid.*, 4.
58. Harrison, *The Third Revolution*, 60.
59. *Ibid.*, 68.
60. Tolba statement, Nairobi, 23rd September 1991, Speech 199/30.
61. Ryan, "Conserving Biological Diversity," 11.
62. Wilson, *The Diversity of Life*, 259.
63. Mitchell, *World on Fire*, 117.
64. Ryan, "Conserving Biological Diversity," 13.
65. *Ibid.*, 12–13.
66. *Ibid.*
67. Al Gore, *Earth in the Balance* (Boston: Houghton Mifflin, 1992), 143.
68. *Ibid.*
69. *Globe and Mail*, 20th May 1992, A2:5.
70. *Times* (London), 30th April 1992.
71. *Christian Science Monitor*, 2nd June 1992, 10.
72. Sandra Postel, "Denial in the Decisive Decade," in *State of the World, 1992* (New York: W. W. Norton, 1992), 1.
73. Wilson, *The Diversity of Life*, cited in *Christian Science Monitor*, 22nd October 1992, 11.
74. Mitchell, *World on Fire*, 121.
75. International Union for the Conservation of Nature, *The IUCN Plant Red Data Book* (IUCN: Gland, Switzerland, 1978), 13–17, cited in Harrison, *The Third Revolution*, 72.
76. Tolba statement, Geneva, 19th–23rd February 1990.
77. United Nations Environment Programme, *Guidelines for Country Studies*, 14.

78. Harrison, *The Third Revolution*, 72.

79. United Nations Environment Programme, *Guidelines for Country Studies*, 14.

80. Mikhail Gorbachev, Speech, Moscow, 19th January 1990, cited in Mitchell, *World on Fire*, 123.

81. *Ibid.*

82. World Commission on Environment and Development, *Our Common Future* (Oxford: Oxford University Press, 1989).

83. *Ibid.*, xi.

84. United Nations Document, Declaration of the United Nations Conference on the Human Environment, A/CONF.48/14, 16th June 1972.

85. United Nations Document, Charter for Nature, 1982, cited in de Klemm, "Conservation of Biological Diversity," 52–53.

86. de Klemm, "Conservation of Biological Diversity," 53.

87. For text, see "Cartagena Convention," *Environmental Policy and Law*, vol. 19, no. 6 (December 1989): 224.

88. Tolba statement, Geneva, 19th–23rd February 1990.

89. *Diversity*, vol. 8, no. 2 (1992): 6.

90. Hilary F. French, "Strengthening Global Governance," in *State of the World, 1992* (New York: W. W. Norton, 1992), 156.

91. "ITTO: Cooperation with CITES," *Environmental Policy and Law*, vol. 22, nos. 5/6 (December 1992): 363.

92. For text, see "Tlatelolco Platform on Environment and Development," *Environmental Policy and Law*, vol. 21, nos. 3/4 (July 1991): 182.

93. See generally, Cyrill de Klemm, "Berne Convention," *Environmental Policy and Law*, vol. 20, nos. 1/2 (March 1990): 25.

94. "Council of Europe: Strategy and Action Plan Adopted," *Environmental Policy and Law*, vol. 20, no. 6 (1990): 215–16.

95. "International Parliamentarians Discuss Environment," *Environmental Policy and Law*, vol. 20, no. 3 (June 1990): 87.

96. *Ibid.*, 88.

97. Text of Decisions of the Interparliamentary Conference on the Global Environment, Washington, D.C., 29th April–2nd May 1990, *Environmental Policy and Law*, vol. 20, no. 3 (June 1990): 113.

98. "International Parliamentarians Discuss Environment," *Environmental Policy and Law*, vol. 20, no. 3 (June 1990): 88.

99. "Asia: Environment and Law," *Environmental Policy and Law*, vol. 22, no. 2 (April 1992): 106.

100. "IUCN/ELC: Biological Diversity Conservation and International Law," *Environmental Policy and Law*, vol. 22, no. 1 (February 1992): 26.

101. "UNCTC: Criteria for Sustainable Development Management," *Environmental Policy and Law*, vol. 20, nos. 4–5 (September–October 1990): 187; news report on 139.

102. "Transnational Corporations and Sustainable Development," *Environmental Policy and Law*, vol. 20, nos. 4–5 (September–October 1990): 139.

103. "UNCTC: Corporate Environmental Management Practices," *Environmental Policy and Law*, vol. 22, no. 1 (February 1992): 15.

104. United Nations Environment Programme Document, UNEP/Bio.Div.1/3, 9th November 1989, 1.

105. "Working Group on Biological Diversity," *Environmental Policy and Law*, vol. 19, No. 1 (March 1989): 5.

106. United Nations Environment Programme Document, UNEP/Bio.Div.1/3, 9th November 1989, Report of the Ad Hoc Working Group on the Work of Its First Session, 3.

107. *Ibid.*, 5.

108. United Nations Environment Programme, "Nairobi Final Act," Conference for the Adoption of the Agreed Text of the Convention on Biological Diversity, Nairobi, Kenya, 22nd May 1992.

109. *Ibid.*, 3.

110. United Nations Environment Programme Document, UNEP/Bio.Div.2/2, 21st December 1989, Note by the Executive Director, 4.

111. *Ibid.*

112. "Preparatory Committee for the UN Conference on Environment and Development," *Environmental Policy and Law*, vol. 20, nos. 4–5 (September–October 1990): 132.

113. "UNEP: Preparations for 1992," *Environmental Policy and Law*, vol. 20, nos. 4–5 (September–October 1990): 126.

114. United Nations Environment Programme Document, UNEP/Bio.Div/N5-INC.3/4, 4th December 1991, III, Second Plenary Meeting, 5.

115. "INC Biological Diversity Convention: No Progress," *Environmental Policy and Law*, vol. 21, no. 2 (May 1991): 48.

116. "UNCED Highlights," *Earth Summit Bulletin*, 1st June 1992, 1:1.

117. *Christian Science Monitor*, 28th April 1992, 1:1.

118. *Ibid.*, 19th May 1992, 20:2.

119. *Globe and Mail*, 30th March 1992, A8:4.

120. *Vancouver Sun*, 2nd June 1992, A12:1.

121. *Christian Science Monitor*, 23rd April 1992, 8:1.

122. *Edmonton Journal*, 12th June 1992, A17:5.

123. *New York Times*, 2nd June 1992, A10:1.

124. *Globe and Mail*, 30th March 1992, A8:5.

125. *Christian Science Monitor*, 3rd June 1992, 2:2.

126. *Times* (London), 6th June 1992, 1:3.

127. *Christian Science Monitor*, 22nd September 1992, 20:1.

128. *Time*, 1st June 1992, 32:1.

129. *Christian Science Monitor*, 30th April 1992, 7:3.

130. *Time*, 1st June 1992, 22:1.

131. *Christian Science Monitor*, 30th April 1992, 7:3.

132. *Ibid.*, 7:3–4.

133. *Globe and Mail*, 20th May 1992, A2:3.

134. Statement by Dr. Mahathir bin Mohamad, Prime Minister, Malaysia, UNCED, Rio, 13th June 1992.

135. *Ibid.*

136. *Vancouver Sun*, 2nd June 1992, A12:1.

137. *Christian Science Monitor*, 23rd April 1992, 2.

138. *Ibid.*, 16th June 1992, 20:1.

139. *Times* (London), 6th June 1992, 1:3.

140. *Christian Science Monitor*, 22nd September 1992, 20:1.

141. *Globe and Mail*, 6th June 1992, B2:2.

142. *Christian Science Monitor*, 10th August 1992, 7:3.

143. *Ibid.*

144. *Ibid.*, 28th April 1992, 1:1.

145. *Ibid.*, 28th January 1992, 6:3.

146. *Vancouver Sun*, 8th June 1992, A11:2–3.

147. *Economist*, 13th June 1992, 93:3.

148. *Christian Science Monitor*, 1st October 1992, 10:3.

149. *Ibid.*, 22nd September 1992, 20:1–2.

150. *Economist*, 13th June 1992, 93–94.

151. *Christian Science Monitor*, 1st October 1992, 10: 3–4.

152. *Ibid.*, 19th May 1992, 20:1.

153. *New York Times*, 6th June 1992, 6:1.

154. *Christian Science Monitor*, 3rd March 1992, 9:3.

155. *Ibid.*, 22nd September 1992, 20:1.

156. *Ibid.*, 28th January 1992, 6:3.

157. *New York Times*, 2nd June 1992, A10:3.

158. *Christian Science Monitor*, 28th April 1992, 2:3.

159. *Ibid.*, 28th January 1992, 6:2.

160. *Ibid.*, 1st October 1992, 11:2.

161. Russell S. Frye, Presentation to the 24th Biennial Conference of the International Bar Association, Cannes, France, 23rd September 1992, published as "Uncle Sam at UNCED," *Environmental Policy and Law*, vol. 22, nos. 5/6 (December 1992): 342.

162. Paul Raeburn, "The Convention on Biological Diversity: Landmark Earth Summit Pact Opens Uncertain New Era for Use and Exchange of Genetic Resources," *Diversity*, vol. 8, no. 2 (1992): 3.

163. *Christian Science Monitor*, 10th June 1992, 3:4.

164. *Economist*, 13th June 1992, 93:3.

165. *Ibid.*

166. *New York Times*, 5th June 1992, A6:2.

167. *Ibid.*

168. *Ibid.*, 6th June 1992, 6:1.

169. *Christian Science Monitor*, 19th May 1992, 20:2.

170. Testimony of Walter Reid, Canadian House of Commons Standing Committee on Environment, 23rd November 1992, House of Commons, Issue No. 47, 34th Parliament, 3rd Session, 1991–92, 47:61 (hereinafter cited as Reid testimony, 23rd November 1992).

171. *Vancouver Sun*, 10th June 1992, A12:5.

172. Steve Usdin, "Biotech Industry Played Key Role in U.S. Refusal to Sign BioConvention," *Diversity*, vol. 8, no. 2 (1992): 7.

173. *Ibid.*

174. Al Gore, "Essentials for Economic Progress: Protect Biodiversity and Intellectual Property Rights," *Journal of NIH Research*, October 1992.

175. Statement by George Bush, President, United States of America, UNCED, Rio, 12th June 1992.

176. *New York Times*, 6th June 1992, 6:2.

177. Frye, "Uncle Sam at UNCED," 344.

178. Usdin, "Biotech Industry Played Key Role," 7.

179. "What They Are Saying: First Reactions to the Biodiversity Convention," *Diversity*, vol. 8, no. 2 (1992): 8.

180. Reid testimony, 23rd November 1992, 47:60.

181. *New York Times*, 6th June 1992, 6:4.

182. *Edmonton Journal*, 15th June 1992, A5:6.

183. *Times* (London), 3rd June 1992, 1:2.

184. *Ibid.*, 2nd June 1992, 1:5–6.

185. *U.S. News and World Report*, 22nd June 1992, 20.

186. *Times* (London), 8th June 1992, 1:1.

187. Statement by George Bush, President, United States of America, UNCED, Rio, 12th June 1992.

188. *Times* (London), 8th June 1992, 1:2.

189. Vandana Shiva, interview by Stephen Bradshaw, BBC News and Current Affairs, telecast on the Canadian Broadcasting Corporation's News Network, June 1992.

190. *Ibid.*

191. Centre for Science and Environment, "The CSE Statement on Global Environmental Democracy to Be Submitted to the Forthcoming UN Conference on Environment and Development," New Delhi, India.

192. Statement by Philip Leakey, Minister for Environment and Natural Resources, Kenya, UNCED, Rio, 12th June 1992.

193. Raeburn, "The Convention on Biological Diversity."

194. *Ibid.*

195. *Times* (London), 3rd June 1992, 15:1.

196. *New York Times*, 5th June 1992, A6:3.

197. *Ibid.*, A6:3–4.

198. *Ibid.*, A6:4–5.

199. *Ibid.*, A6:5.

200. *Ibid.*, A6:1.

201. *Irish Times*, 6th June 1992, 9:4.

202. *New York Times*, 6th June 1992, 6:1.

203. *Ibid.*, A6:1.

204. *Ibid.*, 8th June 1992, A5:1.

205. *Ibid.*

206. *Ibid.*, 5th June 1992, A6:1–2.

207. *Times* (London), 6th June 1992, 1:2.

208. *New York Times*, 6th June 1992, 6:5.

209. *Ibid.*

210. Editorial, *Environmental Policy and Law*, vol. 22, no. 4 (August 1992).

211. *New York Times*, 5th June 1992, A6:2.

212. *Irish Times*, 6th June 1992, 9:4.

213. *Times* (London), 6th June 1992, 1:3.

214. *Ibid.*
215. *New York Times*, 6th June 1992, 6:5.
216. *Christian Science Monitor*, 12th November 1992, 14:1.
217. *Ibid.*, 22nd September 1992, 20:1.
218. *Globe and Mail*, 13th June 1992, A4:3–5.
219. *Ibid.*, A4:4.
220. *Ibid.*, 6th November 1992, A8:3.
221. *Christian Science Monitor*, 1st October 1992, 11:1.
222. *Globe and Mail*, 6th November 1992, A8:1.
223. *Ibid.*, A8:2.
224. *Christian Science Monitor*, 29th October 1992, 18:1.
225. *Ibid.*, 12th November 1992, 14:4.
226. *Ibid.*, 28th April 1992, 2:2–3.
227. Reid testimony, 23rd November 1992, 47:62.
228. *Globe and Mail*, 22nd April 1993, A10:5–6.
229. *New York Times*, 22nd April 1993, A1:6.
230. *Ibid.*, A4:3.
231. *Ibid.*
232. Gore, *Earth in the Balance*, 306.
233. *Ibid.*
234. *Ibid.*, 320.
235. Al Gore, "Essentials for Economic Progress."
236. *Ibid.*
237. *Ibid.*
238. *Ibid.*
239. Dr. Brian Boom, Vice-President, New York Botanical Garden, letter, cited in Gore, "Essentials for Economic Progress."
240. Frye, "Uncle Sam at UNCED," 344.
241. Richard Stone, "The Biodiversity Treaty: Pandora's Box or Fair Deal?" *Science*, vol. 256, 19th June 1992.
242. Frye, "Uncle Sam at UNCED," 344.
243. Wilson, *The Diversity of Life*, 320.
244. Frye, "Uncle Sam at UNCED," 344.
245. Wilson, *The Diversity of Life*, 320–21.
246. *Ibid.*, 321.
247. Usdin, "Biotech Industry Played Key Role," 7.
248. Testimony of Arthur Campeau, Q.C., Personal Representative of the Prime Minister of Canada to UNCED, before Canadian House of Commons Standing Committee on Environment, 23rd November 1992, House of Commons, Issue no. 47, Third Session, 34th Parliament, 1991–92, 47:43 (hereinafter cited as Campeau testimony, 23rd November 1992).
249. Tolba statement, Geneva, 9th July 1990.
250. *Times* (London), 6th June 1992, 15:1.
251. *New York Times*, 14th June 1992, 6:4.
252. Usdin, "Biotech Industry Played Key Role," 7.
253. *Times* (London), 3rd June 1992, 15:2.

254. Statement by Kofi N. Awoonor, Ambassador of Ghana, UNCED, Rio, 5th June 1992.

255. Statement by Ali Hassan Mwinyi, President, United Republic of Tanzania, UNCED, Rio, 13th June 1992.

256. Paul Raeburn, "IBPGR Director Hawtin Encouraged by BioConvention but Uncertainties Remain," *Diversity*, vol. 8, no. 2 (1992): 4.

257. Usdin, "Biotech Industry Played Key Role," 7.

258. Tolba statement, Geneva, 9th July 1990.

259. Tolba statement, Geneva, 19th–23rd February 1990.

260. Frye, "Uncle Sam at UNCED," 344.

261. Raeburn, "IBPGR Director Hawtin Encouraged," 4.

262. *Ibid.*

263. Gore, "Essentials for Economic Progress."

264. Reid testimony, 23rd November 1992, 47:61.

265. Campeau testimony, 23rd November 1992, 47:43.

266. *Times* (London), 3rd June 1992, 12:7.

267. Usdin, "Biotech Industry Played Key Role," 7.

268. "What They Are Saying: First Reactions to the Biodiversity Convention," *Diversity*, vol. 8, no. 2 (1992): 8.

269. *Time*, 1st June 1992, 26:3.

270. Reid testimony, 23rd November 1992, 47:62.

271. *Economist*, 13th June 1992, 94.

272. *Ibid.*

273. Statement by Brian Mulroney, Prime Minister, Canada, UNCED, Rio, 12th June 1992.

274. Address by Brian Mulroney, Prime Minister of Canada, to the Kennedy School of Government, Harvard University, 10th December 1992 (Office of the Prime Minister).

275. *New York Times*, 6th June 1992, 6:6.

276. *Vancouver Sun*, 1st June 1992, 1:1.

277. Reid testimony, 23rd November 1992, 47:60.

278. *New York Times*, 3rd June 1992, A12:1.

279. *Christian Science Monitor*, 25th March 1992, 13:3.

280. *Times* (London), 11th June 1992, 17:2.

281. *Christian Science Monitor*, 2nd June 1992, 11:3.

282. *Times* (London), 2nd June 1992, 1:5–6.

283. *Ibid.*, 3rd June 1992, 1:2.

284. *Ibid.*, 1st June 1992, 1:1–2.

285. *Ibid.*, 8th June 1992, 12:3.

286. *Ibid.*, 3rd June 1992, 1:3.

287. *Ibid.*, 1st June 1992, 1:3.

288. *Ibid.*, 4th June 1992, 8:7.

289. *Ibid.*, 1st June 1992, 1:3.

290. *Ibid.*, 4th June 1992, 15:4.

291. *Ibid.*

292. *Ibid.*, 3rd June 1992, 15:2.

293. *Ibid.*, 8th June 1992, 12:3.

294. *Ibid.*, 4th June 1992, 8:6–7.

295. *Ibid.*, 3rd June 1992, 1:3.

296. *Ibid.*, 4th June 1992, 8:6.

297. Statement by Michael Howard, Secretary of State for Environment, United Kingdom, UNCED, Rio, 9th June 1992.

298. *Times* (London), 3rd June 1992, 1:5.

299. "Convention on Biological Diversity," British Information Services, News Release 48/92, 10th June 1992.

300. *Times* (London), 8th June 1992, 12:3.

301. *Ibid.*, 3rd June 1992, 12:1.

302. *Ibid.*, 15:2.

303. *Ibid.*, 1st June 1992, 15:2.

304. "Convention on Biological Diversity," British Information Services, News Release 48/92, 10th June 1992.

305. Reid testimony, 23rd November 1992, 47:62.

306. "Convention on Biological Diversity," British Information Services, News Release 48/92, 10th June 1992.

307. *Times* (London), 11th June 1992, 15:4.

308. *Ibid.*

309. *Ibid.*, 2nd June 1992, 10:3.

310. *Ibid.*

311. *Irish Times*, 11th June 1992, 11:1.

312. Usdin, "Biotech Industry Played Key Role," 8.

313. *New York Times*, 10th June 1992, A8:1.

314. *Ibid.*, 11th June 1992, A12:4.

315. *Ibid.*, 10th June 1992, A8:4.

316. *Ibid.*, 9th June 1992, 1:6.

317. *Ibid.*, 10th June 1992, A8:4.

318. *Japan Times*, 8th June 1992, 1:2–3.

319. *Earth Summit Bulletin*, 1st June 1992, 2:2.

320. *New York Times*, 10th June 1992, A8: 2.

321. *Ibid.*, 11th June 1992, A12:5.

322. *Christian Science Monitor*, 1st June 1992, 4:1.

323. *New York Times*, 10th June 1992, A8:3.

324. Statement by Shozaburo Nakamura, Minister in Charge of Global Environment, Japan, UNCED, Rio, 5th June 1992.

325. *Japan Times*, 12th June 1992, 3:1.

326. Statement by Shozaburo Nakamura, Minister in Charge of Global Environment, Japan, UNCED, Rio, 5th June 1992.

327. *Christian Science Monitor*, 26th May 1992, 3:1.

328. *Ibid.*, 3:3.

329. *World Monitor*, 30th June 1992, 33.

330. Arthur Campeau, Ambassador for Sustainable Development, Canada, telephone discussion with author, April 1993.

331. Wolfgang E. Burhenne, "Biodiversity—The Legal Aspects," *Environmental Policy and Law*, vol. 22, nos. 5/6 (December 1992): 325.

332. Statement by Rob Storey, Minister for the Environment, New Zealand, UNCED, Rio, 11th June 1992.

333. Statement by Per Stig Moller, Minister for the Environment, Denmark, UNCED, Rio, 9th June 1992.

334. Statement by Helmut Kohl, Chancellor, Federal Republic of Germany, UNCED, Rio, 12th June 1992.

335. Statement by Minister of Public Works and Transport, Spain, UNCED, Rio, 10th June 1992.

336. Statement by Ruth Feldgrill-Zankel, Minister of Environment, Austria, UNCED, Rio, 5th June 1992.

337. Statement by François Mitterrand, President, France, UNCED, Rio, 13th June 1992.

338. Statement by Segolene Royal, Minister for Environment, France, UNCED, Rio, 4th June 1992 (in French; translation by author).

The Kuala Lumpur Declaration, signed after the Second Ministerial Conference of Developing Countries on Environment and Development, 26th to 29th April 1992, contains the following provision on biodiversity: "We are of the view that providing a 'Global List of Biogeographic Areas of Global Importance' under the Convention on Biological Diversity is not necessary." Doc. No. SMCED/MC/DOC.2.

339. *Chronicle-Herald* (Halifax, N.S.), 5th June 1992, 1:3–4.

340. Statement by Michael Smith, Minister of the Environment, Ireland, UNCED, Rio, 10th June 1992.

341. *Japan Times*, 16th June 1992, 12:2.

342. Statement by François Mitterrand, President, France, UNCED, Rio, 13th June 1992.

343. Statement by Paavo Vayrynen, Minister for Foreign Affairs, Finland, UNCED, Rio, 10th June 1992.

344. Statement by Thorbjorn Berntsen, Minister of Environment, Norway, UNCED, Rio, 4th June 1992.

345. *Ibid.*

346. Statement by Andreas Gavrielides, Minister of Agriculture and Natural Resources, Cyprus, UNCED, Rio, 10th June 1992.

347. Statement by Olof Johansson, Minister of the Environment and Natural Resources, Sweden, UNCED, Rio, 8th June 1992.

348. Statement by Anibal Cavaco Silva, Prime Minister of Portugal, President of the European Community, UNCED, Rio, 12th June 1992.

349. "Convention on Biological Diversity Signed," *Environmental Policy and Law*, vol. 22, no. 4 (August 1992): 207.

350. Reid testimony, 23rd November 1992, 47:60.

351. Statement by Sir Wiwa Korowi, Governor-General, Papua New Guinea, UNCED, Rio, 13th June 1992.

352. Statement by Vincent Perera, Minister of Environment and Parliamentary Affairs, Sri Lanka, UNCED, Rio, June 1992.

353. *Ibid.*

354. Statement by Joeli Kalou, Minister of State for Environment, Fiji, UNCED, Rio, 11th June 1992.

355. Statement by Princess Sonam Chhoden Wangchuck, Representative of the King of Bhutan, UNCED, Rio, 11th June 1992.

356. *Ibid.*

357. Statement by Philip Leakey, Minister for Environment and Natural Resources, Kenya, UNCED, Rio, 12th June 1992.

358. Cited in Frye, "Uncle Sam at UNCED," 343.

359. *Christian Science Monitor*, 26th May 1992, 3:2.

360. Stone, "The Biodiversity Treaty."

361. Statement by Abdalla Ahmed Abdalla, Ambassador and Head of Delegation, Sudan, UNCED, Rio, 5th June 1992.

362. Statement by Resio S. Moses, Secretary of the Department of External Affairs, Federated States of Micronesia, UNCED, Rio, 10th June 1992.

363. Statement by Guy Willy Razanamasy, Prime Minister, Madagascar, UNCED, Rio, 12th June 1992.

364. R. Olembo, Deputy Assistant Executive Director, United Nations Environment Programme, to the Sub–Working Group on Biotechnology, Nairobi, 14th November 1990.

365. *Ibid.*

366. Centre for Science and Environment, "Statement on Global Environmental Democracy."

367. Reid testimony, 23rd November 1992, 47:61.

368. Statement by E-Hyock Kwon, Minister of Environment, Republic of Korea, UNCED, Rio, 11th June 1992.

369. *Ibid.*

370. *Ibid.*

371. Statement by Li Peng, Premier, People's Republic of China, UNCED, Rio, 12th June 1992.

372. Statement by Ahmad Mattar, Minister for the Environment, Singapore, UNCED, Rio, 11th June 1992.

373. Kuala Lumpur Declaration on Environment and Development, 26–29th April 1992, Kuala Lumpur, Malaysia, SMCED/MC/DOC.2.

374. Statement by William K. Reilly, Chief Administrator of the U.S. Environmental Protection Agency and Head of Delegation, United States of America, UNCED, Rio, 3rd June 1992.

375. Statement by Ali Hassan Mwinyi, President, United Republic of Tanzania, UNCED, Rio, 13th June 1992.

376. *Ibid.*

377. Cutter Information Corp., "Focus Report: UNCED's Crumbling Pillar: The Biodiversity Convention," *Global Environmental Change Report*, vol. 4, no. 11 (11th June 1992): 2.

378. *Times* (London), 3rd June 1992, 15:1.

379. Statement by Gro Harlem Brundtland, Prime Minister, Norway, UNCED, Rio, 3rd June 1992.

380. Statement by Ali Hassan Mwinyi, President, United Republic of Tanzania, UNCED, Rio, 13th June 1992.

381. *Ibid.*

382. Statement by Gro Harlem Brundtland, Prime Minister, Norway, UNCED, Rio, 3rd June 1992.

383. Tolba statement, Nairobi, 23rd September 1991, Speech 199/30.

384. Tolba statement, Geneva, 9th July 1990.

385. Tolba statement, Nairobi, 23rd September 1991, Speech 199/30.

386. Statement by Kiichi Miyazawa, Prime Minister, Japan, UNCED, Rio, 13th June 1992, and *Japan Times*, 2nd June 1992, 1.

387. *Japan Times*, 2nd June 1992, 1.

388. *New York Times*, 9th June 1992, 1:6.

389. *Japan Times*, 5th June 1992, 1.

390. *Christian Science Monitor*, 1st June 1992, 4:1.

391. *Ibid.*

392. *Ibid.*

393. Statement by George Bush, President, United States of America, UNCED, Rio, 12th June 1992.

394. *Times* (London), 2nd June 1992, 10:1.

395. Raeburn, "The Convention on Biological Diversity," 3.

396. Cutter Information Corp., "Forest Initiative Falters at UNCED," *Global Environmental Change Report*, vol. 4, no. 11 (11th June 1992): 3.

397. *Times* (London), 13th June 1992, 1:4.

398. *Ibid.*, 1:2–3.

399. Michael Howard, announcement, 9th June 1992, British Information Services, News Release 48/92, 10th June 1992.

400. Policy Statements by John Major, Prime Minister, United Kingdom, nos. 57/92 (12th June 1992) and 59/92 (15th June 1992), British Information Services.

401. Raeburn, "The Convention on Biological Diversity," 3.

402. *Ibid.*

403. "Funding for Darwin Initiative," Press Release 54/92, 19th November 1992, British Information Services.

404. *Times* (London), 2nd June 1992, 10:3.

405. Statement by Michael Howard, Secretary of State for Environment, United Kingdom, at the Natural History Museum, London, 24th June 1992.

406. Statement by Helmut Kohl, Chancellor, Federal Republic of Germany, UNCED, Rio, 12th June 1992.

407. Statement by François Mitterrand, President, France, UNCED, Rio, 13th June 1992.

408. Statement by Ruth Feldgrill-Zankel, Minister of Environment, Austria, UNCED, Rio, 5th June 1992.

409. Statement by Thorbjorn Berntsen, Minister of Environment, Norway, UNCED, Rio, 4th June 1992.

410. Statement by Olof Johansson, Minister of the Environment, Sweden, UNCED, Rio, 8th June 1992.

411. Statement by Michael Smith, Minister for the Environment, Ireland, UNCED, Rio, 10th June 1992.

412. *Ibid.*

413. *Christian Science Monitor*, 10th June 1992, 3:4.

414. *Globe and Mail*, 12th March 1993, A16:1. Note that the percentage for the United States includes strategic support to countries like Israel, Egypt, and Turkey.

415. *Ibid.*, A16:2.

416. *Times* (London), 12th June 1992, 15:1.

417. *Vancouver Sun*, 9th June 1992, A11:2.

418. *Chronicle-Herald*, 10th June 1992, A2:6.

419. Statement by Carl Bildt, Prime Minister, Sweden, UNCED, Rio, 13th June 1992.

420. *Vancouver Sun*, 9th June 1992, A11:1.

421. *Globe and Mail*, 9th June 1992, A8:4–5.

422. *Vancouver Sun*, 9th June 1992, A11:1.

423. Statement by Brian Mulroney, Prime Minister, Canada, UNCED, Rio, 12th June 1992.

424. *Globe and Mail*, 16th February 1993, 1:2.

425. *Ibid.*, 13th February 1993, A5:2.

426. *Ibid.*, 12th March 1993, A16:1.

427. *Ibid.*, 16th March 1993, A19:2.

428. *Ibid.*, 25th January 1993, A4:2.

429. *Vancouver Sun*, 9th June 1992, A11:3.

430. *Times* (London), 12th June 1992, 15:1, and *Globe and Mail*, 2nd June 1992, B26:1.

431. *Vancouver Sun*, 9th June 1992, A11:1.

432. *Times* (London), 12th June 1992, 17:1.

433. *Globe and Mail*, 1st March 1993, B1:1.

434. Statement by Dr. Mahathir bin Mohamad, Prime Minister, Malaysia, UNCED, Rio, 13th June 1992.

435. Raeburn, "The Convention on Biological Diversity," 6.

436. Campeau testimony, 23rd November 1992, 47:41.

437. *Ibid.*

438. Statement by Maurice Strong, Secretary-General, UNCED, Rio, 14th June 1992.

439. *Globe and Mail*, 17th November 1992, A6:5.

440. *Ibid.*, 10th December 1992, A8:1.

441. Raeburn, "The Convention on Biological Diversity," 4.

442. *Ibid.*, 4–5.

443. Reid testimony, 23rd November 1992, 47:60.

444. *Ibid.*

445. Campeau testimony, 23rd November 1992, 47:40.

446. Statement by Maurice Strong, Secretary-General, UNCED, Rio, 14th June 1992.

447. *Edmonton Journal*, 8th June 1992, A8:5.

Declaration of the United Nations Conference on the Human Environment

The United Nations Conference on the Human Environment,
Having met at Stockholm from 5 to 16 June 1972,
Having considered the need for a common outlook and for common principles to inspire and guide the peoples of the world in the preservation and enhancement of the human environment,

I

Proclaims that:

1. Man is both creature and moulder of his environment, which gives him physical sustenance and affords him the opportunity for intellectual, moral, social and spiritual growth. In the long and tortuous evolution of the human race on this planet a stage has been reached when, through the rapid acceleration of science and technology, man has acquired the power to transform his environment in countless ways and on an unprecedented scale. Both aspects of man's environment, the natural and the man-made, are essential to his well-being and to the enjoyment of basic human rights—even the right to life itself.

2. The protection and improvement of the human environment is a major issue which affects the well-being of peoples and economic development throughout the world; it is the urgent desire of the peoples of the whole world and the duty of all Governments.

3. Man has constantly to sum up experience and go on discovering, inventing, creating and advancing. In our time, man's capability to transform his surroundings, if used wisely, can bring to all peoples the benefits of development and the opportunity to enhance the quality of life. Wrongly or heedlessly applied, the same power can do incalculable harm to human beings and the human environment. We see around us growing evidence of man-made harm in many regions of the earth: dangerous levels of pollution in water, air, earth and living beings; major and undesirable disturbances to the ecological balance of the biosphere; destruction and depletion of irreplaceable resources; and gross deficiencies

harmful to the physical, mental and social health of man, in the man-made environment, particularly in the living and working environment.

4. In the developing countries most of the environmental problems are caused by under-development. Millions continue to live far below the minimum levels required for a decent human existence, deprived of adequate food and clothing, shelter and education, health and sanitation. Therefore, the developing countries must direct their efforts to development, bearing in mind their priorities and the need to safeguard and improve the environment. For the same purpose, the industrialized countries should make efforts to reduce the gap between themselves and the developing countries. In the industrialized countries, environmental problems are generally related to industrialization and technological development.

5. The natural growth of population continuously presents problems on the preservation of the environment, and adequate policies and measures should be adopted, as appropriate, to face these problems. Of all things in the world, people are the most precious. It is the people that propel social progress, create social wealth, develop science and technology and, through their hard work, continuously transform the human environment. Along with social progress and the advance of production, science and technology, the capability of man to improve the environment increases with each passing day.

6. A point has been reached in history when we must shape our actions throughout the world with a more prudent care for their environmental consequences. Through ignorance or indifference we can do massive and irreversible harm to the earthly environment on which our life and well-being depend. Conversely, through fuller knowledge and wiser action, we can achieve for ourselves and our posterity a better life in an environment more in keeping with human needs and hopes. There are broad vistas for the enhancement of environmental quality and the creation of a good life. What is needed is an enthusiastic but calm state of mind and intense but orderly work. For the purpose of attaining freedom in the world of nature, man must use knowledge to build, in collaboration with nature, a better environment. To defend and improve the human environment for present and future generations has become an imperative goal for mankind—a goal to be pursued together with, and in harmony with, the established and fundamental goals of peace and of world-wide economic and social development.

7. To achieve this environmental goal will demand the acceptance of responsibility by citizens and communities and by enterprises and institutions at every level, all sharing equitably in common efforts. Individuals in all walks of life as well as organizations in many fields, by their values and the sum of their actions, will shape the world environment of the future. Local and national governments will bear the greatest burden for large-scale environmental policy and action within their jurisdictions. International co-operation is also needed in order to

raise resources to support the developing countries in carrying out their responsibilities in this field. A growing class of environmental problems, because they are regional or global in extent or because they affect the common international realm, will require extensive co-operation among nations and action by international organizations in the common interest. The Conference calls upon Governments and peoples to exert common efforts for the preservation and improvement of the human environment, for the benefit of all the people and for their posterity.

II
Principles

States the common conviction that:

PRINCIPLE 1

Man has the fundamental right to freedom, equality and adequate conditions of life, in an environment of a quality that permits a life of dignity and well-being, and he bears a solemn responsibility to protect and improve the environment for present and future generations. In this respect, policies promoting or perpetuating *apartheid,* racial segregation, discrimination, colonial and other forms of oppression and foreign domination stand condemned and must be eliminated.

PRINCIPLE 2

The natural resources of the earth including the air, water, land, flora and fauna and epecially representative samples of natural ecosystems must be safeguarded for the benefit of present and future generations through careful planning or management, as appropriate.

PRINCIPLE 3

The capacity of the earth to produce vital renewable resources must be maintained and, wherever practicable, restored or improved.

PRINCIPLE 4

Man has a special responsibility to safeguard and wisely manage the heritage of wildlife and its habitat which are now gravely imperilled by a combination of adverse factors. Nature conservation including wildlife must therefore receive importance in planning for economic development.

PRINCIPLE 5

The non-renewable resources of the earth must be employed in such a way as to guard against the danger of their future exhaustion and to ensure that benefits from such employment are shared by all mankind.

PRINCIPLE 6

The discharge of toxic substances or of other substances and the release of heat, in such quantities or concentrations as to exceed the capacity of the environment to render them harmless, must be halted in order to ensure that serious or irreversible damage is not inflicted upon ecosystems. The just struggle of the peoples of all countries against pollution should be supported.

PRINCIPLE 7

States shall take all possible steps to prevent pollution of the seas by substances that are liable to create hazards to human health, to harm living resources and marine life, to damage amenities or to interfere with other legitimate uses of the sea.

PRINCIPLE 8

Economic and social development is essential for ensuring a favourable living and working environment for man and for creating conditions on earth that are necessary for the improvement of the quality of life.

PRINCIPLE 9

Environmental deficiencies generated by the conditions of underdevelopment and natural disasters pose grave problems and can best be remedied by accelerated development through the transfer of substantial quantities of financial and technological assistance as a supplement to the domestic effort of the developing countries and such timely assistance as may be required.

PRINCIPLE 10

For the developing countries, stability of prices and adequate earnings for primary commodities and raw materials are essential to environmental management since economic factors as well as ecological processes must be taken into account.

PRINCIPLE 11

The environmental policies of all States should enhance and not adversely affect the present or future development potential of developing countries, nor should they hamper the attainment of better living conditions for all, and appropriate steps should be taken by States and international organizations with a view to reaching agreement on meeting the possible national and international economic consequences resulting from the application of environmental measures.

PRINCIPLE 12

Resources should be made available to preserve and improve the environment, taking into account the circumstances and particular requirements of developing countries and any costs which may emanate from their incorporating environmental safeguards into their development planning and the need for making available to them, upon their request, additional international technical and financial assistance for this purpose.

PRINCIPLE 13

In order to achieve a more rational management of resources and thus to improve the environment, States should adopt an integrated and co-ordinated approach to their development planning so as to ensure that development is compatible with the need to protect and improve the human environment for the benefit of their population.

PRINCIPLE 14

Rational planning constitutes an essential tool for reconciling any conflict between the needs of development and the need to protect the environment.

PRINCIPLE 15

Planning must be applied to human settlements and urbanization with a view to avoiding adverse effects on the environment and obtaining maximum social, economic and environmental benefits for all. In this respect projects which are designed for colonialist and racist domination must be abandoned.

PRINCIPLE 16

Demographic policies, which are without prejudice to basic human rights and which are deemed appropriate by Governments concerned, should be applied in those regions where the rate of population growth or excessive population concentrations are likely to have adverse effects on the environment or development, or where low population density may prevent improvement of the human environment and impede development.

PRINCIPLE 17

Appropriate national institutions must be entrusted with the task of planning, managing or controlling the environmental resources of States with the view to enhancing environmental quality.

PRINCIPLE 18

Science and technology, as part of their contribution to economic and social development, must be applied to the identification, avoidance and control of environmental risks and the solution of environmental problems and for the common good of mankind.

PRINCIPLE 19

Education in environmental matters, for the younger generation as well as adults, giving due consideration to the underprivileged, is essential in order to broaden the basis for an enlightened opinion and responsible conduct by individuals, enterprises and communities in protecting and improving the environment in its full human dimension. It is also essential that mass media communications avoid contributing to the deterioration of the environment, but, on the contrary, disseminate information of an educational nature, on the need to protect and improve the environment in order to enable man to develop in every respect.

PRINCIPLE 20

Scientific research and development in the context of environmental problems, both national and multinational, must be promoted in all countries, especially the developing countries. In this connexion, the free flow of up-to-date scientific information and transfer of experience must be supported and assisted, to facilitate the solution of environmental problems; environmental technologies should be made available to developing countries on terms which would encourage their wide dissemination without constituting an economic burden on the developing countries.

PRINCIPLE 21

States have, in accordance with the Charter of the United Nations and the principles of international law, the sovereign right to exploit their own resources pursuant to their own environmental policies, and the responsibility to ensure that activities within their jurisdiction or control do not cause damage to the environment of other States or of areas beyond the limits of national jurisdiction.

PRINCIPLE 22

States shall co-operate to develop further the international law regarding liability and compensation for the victims of pollution and other environmental damage caused by activities within the jurisdiction or control of such States to areas beyond their jurisdiction.

PRINCIPLE 23

Without prejudice to such criteria as may be agreed upon by the international community, or to standards which will have to be determined nationally, it will be essential in all cases to consider the systems of values prevailing in each country, and the extent of the applicability of standards which are valid for the most advanced countries but which may be inappropriate and of unwarranted social cost for the developing countries.

PRINCIPLE 24

International matters concerning the protection and improvement of the environment should be handled in a co-operative spirit by all countries, big or small, on an equal footing. Co-operation through multilateral or bilateral arrangements or other appropriate means is essential to effectively control, prevent, reduce and eliminate adverse environmental effects resulting from activities conducted in all spheres, in such a way that due account is taken of the sovereignty and interests of all States.

PRINCIPLE 25

States shall ensure that international organizations play a co-ordinated, efficient and dynamic role for the protection and improvement of the environment.

PRINCIPLE 26

Man and his environment must be spared the effects of nuclear weapons and all other means of mass destruction. States must strive to reach prompt agreement, in the relevant international organs, on the elimination and complete destruction of such weapons.

The Rio Declaration on Environment and Development

Preamble

The United Nations Conference on Environment and Development,
Having met at Rio de Janeiro from 3 to 14 June 1992,
Reaffirming the Declaration of the United Nations Conference on the Human Environment, adopted at Stockholm on 16 June 1972, and seeking to build upon it,
With the goal of establishing a new and equitable global partnership through the creation of new levels of cooperation among States, key sectors of societies and people,
Working towards international agreements which respect the interests of all and protect the integrity of the global environment and developmental system,
Recognizing the integral and interdependent nature of the Earth, our home,
Proclaims that:

PRINCIPLE 1

Human beings are at the centre of concerns for sustainable development. They are entitled to a healthy and productive life in harmony with nature.

PRINCIPLE 2

States have, in accordance with the Charter of the United Nations and the principles of international law, the sovereign right to exploit their own resources pursuant to their own environmental and developmental policies, and the responsibility to ensure that activities within their jurisdiction or control do not cause damage to the environment of other States or of areas beyond the limits of national jurisdiction.

PRINCIPLE 3

The right to development must be fulfilled so as to equitably meet developmental and environmental needs of present and future generations.

PRINCIPLE 4

In order to achieve sustainable development, environmental protection shall constitute an integral part of the development process and cannot be considered in isolation from it.

PRINCIPLE 5

All States and all people shall cooperate in the essential task of eradicating poverty as an indispensable requirement for sustainable development, in order to decrease the disparities in standards of living and better meet the needs of the majority of the people of the world.

PRINCIPLE 6

The special situation and needs of developing countries, particularly the least developed and those most environmentally vulnerable, shall be given special priority. International actions in the field of environment and development should also address the interests and needs of all countries.

PRINCIPLE 7

States shall cooperate in a spirit of global partnership to conserve, protect and restore the health and integrity of the Earth's ecosystem. In view of the different contributions to global environmental degradation, States have common but differentiated responsibilities. The developed countries acknowledge the responsibility that they bear in the international pursuit of sustainable development in view of the pressures their societies place on the global environment and of the technologies and financial resources they command.

PRINCIPLE 8

To achieve sustainable development and a higher quality of life for all people, States should reduce and eliminate unsustainable patterns of production and consumption and promote appropriate demographic policies.

PRINCIPLE 9

States should cooperate to strengthen endogenous capacity-building for sustainable development by improving scientific understanding through exchanges of scientific and technological knowledge, and by enhancing the development, adaptation, diffusion and transfer of technologies, including new and innovative technologies.

PRINCIPLE 10

Environmental issues are best handled with the participation of all concerned citizens, at the relevant level. At the national level, each individual shall have appropriate access to information concerning the environment that is held by public authorities, including information on hazardous materials and activities in their communities, and the opportunity to participate in decision-making processes. States shall facilitate and encourage public awareness and participation by making information widely available. Effective access to judicial and administrative proceedings, including redress and remedy, shall be provided.

PRINCIPLE 11

States shall enact effective environmental legislation. Environmental standards, management objectives and priorities should reflect the environmental and developmental context to which they apply. Standards applied by some countries may be inappropriate and of unwarranted economic and social cost to other countries, in particular developing countries.

PRINCIPLE 12

States should cooperate to promote a supportive and open international economic system that would lead to economic growth and sustainable development in all countries, to better address the problems of environmental degradation. Trade policy measures for environmental purposes should not constitute a means of arbitrary or unjustifiable discrimination or a disguised restriction on international trade. Unilateral actions to deal with environmental challenges outside the jurisdiction of the importing country should be avoided. Environmental measures addressing transboundary or global environmental problems should, as far as possible, be based on an international consensus.

PRINCIPLE 13

States shall develop national law regarding liability and compensation for the victims of pollution and other environmental damage. States shall also cooperate in an expeditious and more determined manner to develop further international law regarding liability and compensation for adverse effects of environmental damage caused by activities within their jurisdiction or control to areas beyond their jurisdiction.

PRINCIPLE 14

States should effectively cooperate to discourage or prevent the relocation and transfer to other States of any activities and substances that cause severe environmental degradation or are found to be harmful to human health.

PRINCIPLE 15

In order to protect the environment, the precautionary approach shall be widely applied by States according to their capabilities. Where there are threats of serious or irreversible damage, lack of full scientific certainty shall not be used as a reason for postponing cost-effective measures to prevent environmental degradation.

PRINCIPLE 16

National authorities should endeavour to promote the internalization of environmental costs and the use of economic instruments, taking into account the approach that the polluter should, in principle, bear the cost of pollution, with due regard to the public interest and without distorting international trade and investment.

PRINCIPLE 17

Environmental impact assessment, as a national instrument, shall be undertaken for proposed activities that are likely to have a significant adverse impact on the environment and are subject to a decision of a competent national authority.

PRINCIPLE 18

States shall immediately notify other States of any natural disasters or other emergencies that are likely to produce sudden harmful effects on the environment of those States. Every effort shall be made by the international community to help States so afflicted.

PRINCIPLE 19

States shall provide prior and timely notification and relevant information to potentially affected States on activities that may have a significant adverse transboundary environmental effect and shall consult with those States at an early stage and in good faith.

PRINCIPLE 20

Women have a vital role in environmental management and development. Their full participation is therefore essential to achieve sustainable development.

PRINCIPLE 21

The creativity, ideals and courage of the youth of the world should be mobilized to forge a global partnership in order to achieve sustainable development and ensure a better future for all.

PRINCIPLE 22

Indigenous people and their communities, and other local communities, have a vital role in environmental management and development because of their knowledge and traditional practices. States should recognize and duly support their identity, culture and interests and enable their effective participation in the achievement of sustainable development.

PRINCIPLE 23

The environment and natural resources of people under oppression, domination and occupation shall be protected.

PRINCIPLE 24

Warfare is inherently destructive of sustainable development. States shall therefore respect international law providing protection for the environment in times of armed conflict and cooperate in its further development, as necessary.

PRINCIPLE 25

Peace, development and environmental protection are interdependent and indivisible.

PRINCIPLE 26

States shall resolve all their environmental disputes peacefully and by appropriate means in accordance with the Charter of the United Nations.

PRINCIPLE 27

States and people shall cooperate in good faith and in a spirit of partnership in the fulfilment of the principles embodied in this Declaration and in the further development of international law in the field of sustainable development.

Non–Legally Binding Authoritative Statement of Principles for a Global Consensus on the Management, Conservation and Sustainable Development of All Types of Forests

Preamble

(a) The subject of forests is related to the entire range of environmental and development issues and opportunities, including the right to socio-economic development on a sustainable basis.

(b) The guiding objective of these principles is to contribute to the management, conservation and sustainable development of forests and to provide for their multiple and complementary functions and uses.

(c) Forestry issues and opportunities should be examined in a holistic and balanced manner within the overall context of environment and development, taking into consideration the multiple functions and uses of forests, including traditional uses, and the likely economic and social stress when these uses are constrained or restricted, as well as the potential for development that sustainable forest management can offer.

(d) These principles reflect a first global consensus on forests. In committing themselves to the prompt implementation of these principles, countries also decide to keep them under assessment for their adequacy with regard to further international cooperation on forest issues.

(e) These principles should apply to all types of forests, both natural and planted, in all geographic regions and climatic zones, including austral, boreal, subtemperate, temperate, subtropical and tropical.

(f) All types of forests embody complex and unique ecological processes which are the basis for their present and potential capacity to provide resources to satisfy human needs as well as environmental values, and as such their sound management and conservation is of concern to the Governments of the countries to which they belong and are of value to local communities and to the environment as a whole.

(g) Forests are essential to economic development and the maintenance of all forms of life.

(h) Recognizing that the responsibility for forest management, conservation and sustainable development is in many States allocated among federal/national, state/provincial and local levels of government, each State, in accordance with its constitution and/or national legislation, should pursue these principles at the appropriate level of government.

Principles/Elements

1.   (a) "States have, in accordance with the Charter of the United Nations and the principles of international law, the sovereign right to exploit their own resources pursuant to their own environmental policies and have the responsibility to ensure that activities within their jurisdiction or control do not cause damage to the environment of other States or of areas beyond the limits of national jurisdiction."

(b) The agreed full incremental cost of achieving benefits associated with forest conservation and sustainable development requires increased international cooperation and should be equitably shared by the international community.

2.   (a) States have the sovereign and inalienable right to utilize, manage and develop their forests in accordance with their development needs and level of socio-economic development and on the basis of national policies consistent with sustainable development and legislation, including the conversion of such areas for other uses within the overall socio-economic development plan and based on rational land-use policies.

(b) Forest resources and forest lands should be sustainably managed to meet the social, economic, ecological, cultural and spiritual human needs of present and future generations. These needs are for forest products and services, such as wood and wood products, water, food, fodder, medicine, fuel, shelter, employment, recreation, habitats for wildlife, landscape diversity, carbon sinks and reservoirs, and for other forest products. Appropriate measures should be taken to protect forests against harmful effects of pollution, including air-borne pollution, fires, pests and diseases in order to maintain their full multiple value.

(c) The provision of timely, reliable and accurate information on forests and forest ecosystems is essential for public understanding and informed decision-making and should be ensured.

(d) Governments should promote and provide opportunities for the participation of interested parties, including local communities and indigenous people, industries, labour, non-governmental organizations and individuals, forest dwellers and women, in the development, implementation and planning of national forest policies.

3.   (a) National policies and strategies should provide a framework for in-

creased efforts, including the development and strengthening of institutions and programmes for the management, conservation and sustainable development of forests and forest lands.

(b) International institutional arrangements, building on those organizations and mechanisms already in existence, as appropriate, should facilitate international cooperation in the field of forests.

(c) All aspects of environmental protection and social and economic development as they relate to forests and forest lands should be integrated and comprehensive.

4. The vital role of all types of forests in maintaining the ecological processes and balance at the local, national, regional and global levels through, *inter alia*, their role in protecting fragile ecosystems, watersheds and freshwater resources and as rich storehouses of biodiversity and biological resources and sources of genetic material for biotechnology products, as well as photosynthesis, should be recognized.

5. (a) National forest policies should recognize and duly support the identity, culture and the rights of indigenous people, their communities and other communities and forest dwellers. Appropriate conditions should be promoted for these groups to enable them to have an economic stake in forest use, perform economic activities, and achieve and maintain cultural identity and social organization, as well as adequate levels of livelihood and well-being, through, *inter alia*, those land tenure arrangements which serve as incentives for the sustainable management of forests.

(b) The full participation of women in all aspects of the management, conservation and sustainable development of forests should be actively promoted.

6. (a) All types of forests play an important role in meeting energy requirements through the provision of a renewable source of bio-energy, particularly in developing countries, and the demands for fuelwood for household and industrial needs should be met through sustainable forest management, afforestation and reforestation. To this end, the potential contribution of plantations of both indigenous and introduced species for the provision of both fuel and industrial wood should be recognized.

(b) National policies and programmes should take into account the relationship, where it exists, between the conservation, management and sustainable development of forests and all aspects related to the production, consumption, recycling and/or final disposal of forest products.

(c) Decisions taken on the management, conservation and sustainable development of forest resources should benefit, to the extent practicable, from a comprehensive assessment of economic and non-economic values of forest goods and services and of the environmental costs and benefits. The development and improvement of methodologies for such evaluations should be promoted.

(d) The role of planted forests and permanent agricultural crops as sustainable and environmentally sound sources of renewable energy and industrial raw material should be recognized, enhanced and promoted. Their contribution to the maintenance of ecological processes, to offsetting pressure on primary/old-growth forest and to providing regional employment and development with the adequate involvement of local inhabitants should be recognized and enhanced.

(e) Natural forests also constitute a source of goods and services, and their conservation, sustainable management and use should be promoted.

7.   (a) Efforts should be made to promote a supportive international economic climate conducive to sustained and environmentally sound development of forests in all countries, which include, *inter alia*, the promotion of sustainable patterns of production and consumption, the eradication of poverty and the promotion of food security.

(b) Specific financial resources should be provided to developing countries with significant forest areas which establish programmes for the conservation of forests including protected natural forest areas. These resources should be directed notably to economic sectors which would stimulate economic and social substitution activities.

8.   (a) Efforts should be undertaken towards the greening of the world. All countries, notably developed countries, should take positive and transparent action towards reforestation, afforestation and forest conservation, as appropriate.

(b) Efforts to maintain and increase forest cover and forest productivity should be undertaken in ecologically, economically and socially sound ways through the rehabilitation, reforestation and re-establishment of trees and forests on unproductive, degraded and deforested lands, as well as through the management of existing forest resources.

(c) The implementation of national policies and programmes aimed at forest management, conservation and sustainable development, particularly in developing countries, should be supported by international financial and technical cooperation, including through the private sector, where appropriate.

(d) Sustainable forest management and use should be carried out in accordance with national development policies and priorities and on the basis of environmentally sound national guidelines. In the formulation of such guidelines, account should be taken, as appropriate and if applicable, of relevant internationally agreed methodologies and criteria.

(e) Forest management should be integrated with management of adjacent areas so as to maintain ecological balance and sustainable productivity.

(f) National policies and/or legislation aimed at management, conservation and sustainable development of forests should include the protection of ecologically viable representative or unique examples of forests, including primary/old-growth forests, cultural, spiritual, historical, religious and other unique and valued forests of national importance.

(g) Access to biological resources, including genetic material, shall be with due regard to the sovereign rights of the countries where the forests are located and to the sharing on mutually agreed terms of technology and profits from bio-technology products that are derived from these resources.

(h) National policies should ensure that environmental impact assessments should be carried out where actions are likely to have significant adverse impacts on important forest resources, and where such actions are subject to a decision of a competent national authority.

9.   (a) The efforts of developing countries to strengthen the management, conservation and sustainable development of their forest resources should be supported by the international community, taking into account the importance of redressing external indebtedness, particularly where aggravated by the net transfer of resources to developed countries, as well as the problem of achieving at least the replacement value of forests through improved market access for forest products, especially processed products. In this respect, special attention should also be given to the countries undergoing the process of transition to market economies.

(b) The problems that hinder efforts to attain the conservation and sustainable use of forest resources and that stem from the lack of alternative options available to local communities, in particular the urban poor and poor rural populations who are economically and socially dependent on forests and forest resources, should be addressed by Governments and the international community.

(c) National policy formulation with respect to all types of forests should take account of the pressures and demands imposed on forest ecosystems and resources from influencing factors outside the forest sector, and intersectoral means of dealing with these pressures and demands should be sought.

10.   New and additional financial resources should be provided to developing countries to enable them to sustainably manage, conserve and develop their forest resources, including through afforestation, reforestation and combating deforestation and forest and land degradation.

11.   In order to enable, in particular, developing countries to enhance their endogenous capacity and to better manage, conserve and develop their forest resources, the access to and transfer of environmentally sound technologies and corresponding know-how on favourable terms, including on concessional and preferential terms, as mutually agreed, in accordance with the relevant provisions of Agenda 21, should be promoted, facilitated and financed, as appropriate.

12.   (a) Scientific research, forest inventories and assessments carried out by national institutions which take into account, where relevant, biological, physical, social and economic variables, as well as technological development and its application in the field of sustainable forest management, conservation and development, should be strengthened through effective modalities, including international cooperation. In this context, attention should also be given to research and development of sustainably harvested non-wood products.

(b) National and, where appropriate, regional and international institutional capabilities in education, training, science, technology, economics, anthropology and social aspects of forests and forest management are essential to the conservation and sustainable development of forests and should be strengthened.

(c) International exchange of information on the results of forest and forest management research and development should be enhanced and broadened, as appropriate, making full use of education and training institutions, including those in the private sector.

(d) Appropriate indigenous capacity and local knowledge regarding the conservation and sustainable development of forests should, through institutional and financial support, and in collaboration with the people in local communities concerned, be recognized, respected, recorded, developed and, as appropriate, introduced in the implementation of programmes. Benefits arising from the utilization of indigenous knowledge should therefore be equitably shared with such people.

13. (a) Trade in forest products should be based on non-discriminatory and multilaterally agreed rules and procedures consistent with international trade law and practices. In this context, open and free international trade in forest products should be facilitated.

(b) Reduction or removal of tariff barriers and impediments to the provision of better market access and better prices for higher value-added forest products and their local processing should be encouraged to enable producer countries to better conserve and manage their renewable forest resources.

(c) Incorporation of environmental costs and benefits into market forces and mechanisms, in order to achieve forest conservation and sustainable development, should be encouraged both domestically and internationally.

(d) Forest conservation and sustainable development policies should be integrated with economic, trade and other relevant policies.

(e) Fiscal, trade, industrial, transportation and other policies and practices that may lead to forest degradation should be avoided. Adequate policies, aimed at management, conservation and sustainable development of forests, including where appropriate, incentives, should be encouraged.

14. Unilateral measures, incompatible with international obligations or agreements, to restrict and/or ban international trade in timber or other forest products should be removed or avoided, in order to attain long-term sustainable forest management.

15. Pollutants, particularly air-borne pollutants, including those responsible for acidic deposition, that are harmful to the health of forest ecosystems at the local, national, regional and global levels should be controlled.

Annex I to the Report of the Committee: United Nations Framework Convention On Climate Change

The Parties to this Convention,

Acknowledging that change in the Earth's climate and its adverse effects are a common concern of humankind,

Concerned that human activities have been substantially increasing the atmospheric concentrations of greenhouse gases, that these increases enhance the natural greenhouse effect, and that this will result on average in an additional warming of the Earth's surface and atmosphere and may adversely affect natural ecosystems and humankind,

Noting that the largest share of historical and current global emissions of greenhouse gases has originated in developed countries, that per capita emissions in developing countries are still relatively low and that the share of global emissions originating in developing countries will grow to meet their social and development needs,

Aware of the role and importance in terrestrial and marine ecosystems of sinks and reservoirs of greenhouse gases,

Noting that there are many uncertainties in predictions of climate change, particularly with regard to the timing, magnitude and regional patterns thereof,

Acknowledging that the global nature of climate change calls for the widest possible cooperation by all countries and their participation in an effective and appropriate international response, in accordance with their common but differential responsibilities and respective capabilities and their social and economic conditions,

Recalling the pertinent provisions of the Declaration of the United Nations Conference on the Human Environment, adopted at Stockholm on 16 June 1972,

Recalling also that States have, in accordance with the Charter of the United Nations and the principles of international law, the sovereign right to exploit their own resources pursuant to their own environmental and developmental policies, and the responsibility to ensure that activities within their jurisdiction or control do not cause damage to the environment of other States or of areas beyond the limits of national jurisdiction,

Reaffirming the principle of sovereignty of States in international cooperation to address climate change,

Recognizing that States should enact effective environmental legislation, that environmental standards, management objectives and priorities should reflect the environmental and developmental context to which they apply, and that standards applied by some countries may be inappropriate and of unwarranted economic and social cost to other countries, in particular developing countries,

Recalling the provisions of General Assembly resolution 44/228 of 22 December 1989 on the United Nations Conference on Environment and Development, and resolutions 43/53 of 6 December 1988, 44/207 of 22 December 1989, 45/212 of 21 December 1990 and 46/169 of 19 December 1991 on protection of global climate for present and future generations of mankind,

Recalling also the provisions of General Assembly resolution 44/206 of 22 December 1989 on the possible adverse effects of sea level rise on islands and coastal areas, particularly low-lying coastal areas and the pertinent provisions of General Assembly resolution 44/172 of 19 December 1989 on the implementation of the Plan of Action to Combat Desertification,

Recalling further the Vienna Convention for the Protection of the Ozone Layer, 1985, and the Montreal Protocol on Substances that Deplete the Ozone Layer, 1987, as adjusted and amended on 29 June 1990,

Noting the Ministerial Declaration of the Second World Climate Conference adopted on 7 November 1990,

Conscious of the valuable analytical work being conducted by many States on climate change and of the important contributions of the World Meteorological Organization, the United Nations Environment Programme and other organs, organizations and bodies of the United Nations system, as well as other international and intergovernmental bodies, to the exchange of results of scientific research and the coordination of research,

Recognizing that steps required to understand and address climate change will be environmentally, socially and economically most effective if they are based on relevant scientific, technical and economic considerations and continually re-evaluated in the light of new findings in these areas,

Recognizing that various actions to address climate change can be justified economically in their own right and can also help in solving other environmental problems,

Recognizing also the need for developed countries to take immediate action in a flexible manner on the basis of clear priorities, as a first step towards comprehensive response strategies at the global, national and, where agreed, regional levels that take into account all greenhouse gases, with due consideration of their relative contributions to the enhancement of the greenhouse effect,

Recognizing further that low-lying and other small island countries, countries with low-lying coastal, arid and semi-arid areas or areas liable to floods, drought and desertification, and developing countries with fragile mountainous ecosystems are particularly vulnerable to the adverse effects of climate change,

Recognizing the special difficulties of those countries, especially developing countries, whose economies are particularly dependent on fossil fuel production, use and exportation, as a consequence of action taken on limiting greenhouse gas emissions,

Affirming that responses to climate change should be coordinated with social and economic development in an integrated manner with a view to avoiding adverse impacts on the latter, taking into full account the legitimate priority needs of developing countries for the achievement of sustained economic growth and the eradication of poverty,

Recognizing that all countries, especially developing countries, need access to resources required to achieve sustainable social and economic development and that, in order for developing countries to progress towards that goal, their energy consumption will need to grow taking into account the possibilities for achieving greater energy efficiency and for controlling greenhouse gas emissions in general, including through the application of new technologies on terms which make such an application economically and socially beneficial,

Determined to protect the climate system for present and future generations,

Have agreed as follows:

Article 1
*Definitions**

For the purposes of this Convention:

1. "Adverse effects of climate change" means changes in the physical environment or biota resulting from climate change which have significant deleterious effects on the composition, resilience or productivity of natural and managed ecosystems or on the operation of socio-economic systems or on human health and welfare.

2. "Climate change" means a change of climate which is attributed directly or indirectly to human activity that alters the composition of the global atmosphere and which is in addition to natural climate variability observed over comparable time periods.

3. "Climate system" means the totality of the atmosphere, hydrosphere, biosphere and geosphere and their interactions.

4. "Emissions" means the release of greenhouse gases and/or their precursors into the atmosphere over a specified area and period of time.

*Titles of articles are included solely to assist the reader.

5.   "Greenhouse gases" means those gaseous constituents of the atmosphere, both natural and anthropogenic, that absorb and re-emit infrared radiation.

6.   "Regional economic integration organization" means an organization constituted by sovereign States of a given region which has competence in respect of matters governed by this Convention or its protocols and has been duly authorized, in accordance with its internal procedures, to sign, ratify, accept, approve or accede to the instruments concerned.

7.   "Reservoir" means a component or components of the climate system where a greenhouse gas or a precursor of a greenhouse gas is stored.

8.   "Sink" means any process, activity or mechanism which removes a greenhouse gas, an aerosol or a precursor of a greenhouse gas from the atmosphere.

9.   "Source" means any process or activity which releases a greenhouse gas, an aerosol or a precursor of a greenhouse gas into the atmosphere.

Article 2
Objective

The ultimate objective of this Convention and any related legal instruments that the Conference of the Parties may adopt is to achieve, in accordance with the relevant provisions of the Convention, stabilization of greenhouse gas concentrations in the atmosphere at a level that would prevent dangerous anthropogenic interference with the climate system. Such a level should be achieved within a time frame sufficient to allow ecosystems to adapt naturally to climate change, to ensure that food production is not threatened and to enable economic development to proceed in a sustainable manner.

Article 3
Principles

In their actions to achieve the objective of the Convention and to implement its provisions, the Parties shall be guided, *inter alia,* by the following:

1.   The Parties should protect the climate system for the benefit of present and future generations of humankind, on the basis of equity and in accordance with their common but differentiated responsibilities and respective capabilities. Accordingly, the developed country Parties should take the lead in combating climate change and the adverse effects thereof.

2.   The specific needs and special circumstances of developing country Parties, especially those that are particularly vulnerable to the adverse effects of climate change, and of those Parties, especially in developing country Parties, that would

have to bear a disproportionate or abnormal burden under the Convention, should be given full consideration.

3.   The Parties should take precautionary measures to anticipate, prevent or minimize the causes of climate change and mitigate its adverse effects. Where there are threats of serious or irreversible damage, lack of full scientific certainty should not be used as a reason for postponing such measures, taking into account that policies and measures to deal with climate change should be cost-effective so as to ensure global benefits at the lowest possible cost. To achieve this, such policies and measures should take into account different socio-economic contexts, be comprehensive, cover all relevant sources, sinks and reservoirs of greenhouse gases and adaptation, and comprise all economic sectors. Efforts to address climate change may be carried out cooperatively by interested Parties.

4.   The Parties have a right to, and should, promote sustainable development. Policies and measures to protect the climate system against human-induced change should be appropriate for the specific conditions of each Party and should be integrated with national development programmes, taking into account that economic development is essential for adopting measures to address climate change.

5.   The Parties should cooperate to promote a supportive and open international economic system that would lead to sustainable economic growth and development in all Parties, particularly developing country Parties, thus enabling them better to address the problems of climate change. Measures taken to combat climate change, including unilateral ones, should not constitute a means of arbitrary or unjustifiable discrimination or a disguised restriction on international trade.

Article 4
Commitments

1.   All Parties, taking into account their common but differentiated responsibilities and their specific national and regional development priorities, objectives and circumstances, shall:

    (a)   Develop, periodically update, publish and make available to the Conference of the Parties, in accordance with Article 12, national inventories of anthropogenic emissions by sources and removals by sinks of all greenhouse gases not controlled by the Montreal Protocol, using comparable methodologies to be agreed upon by the Conference of the Parties;

    (b)   Formulate, implement, publish and regularly update national and, where appropriate, regional programmes containing measures to mitigate climate change by addressing anthropogenic emissions by sources and removals by sinks of all greenhouse gases not controlled by the Montreal Protocol, and measures to facilitate adequate adaptation to climate change;

(c) Promote and cooperate in the development, application and diffusion, including transfer, of technologies, practices and processes that control, reduce or prevent anthropogenic emissions of greenhouse gases not controlled by the Montreal Protocol in all relevant sectors, including the energy, transport, industry, agriculture, forestry and waste management sectors;

(d) Promote sustainable management, and promote and cooperate in the conservation and enhancement, as appropriate, of sinks and reservoirs of all greenhouse gases not controlled by the Montreal Protocol, including biomass, forests and oceans as well as other terrestrial, coastal and marine ecosystems;

(e) Cooperate in preparing for adaptation to the impacts of climate change; develop and elaborate appropriate and integrated plans for coastal zone management, water resources and agriculture, and for the protection and rehabilitation of areas, particularly in Africa, affected by drought and desertification, as well as floods;

(f) Take climate change considerations into account, to the extent feasible, in their relevant social, economic and environmental policies and actions, and employ appropriate methods, for example impact assessments, formulated and determined nationally, with a view to minimizing adverse effects on the economy, on public health and on the quality of the environment, of projects or measures undertaken by them to mitigate or adapt to climate change;

(g) Promote and cooperate in scientific, technological, technical, socio-economic and other research, systematic observation and development of data archives related to the climate system and intended to further the understanding and to reduce or eliminate the remaining uncertainties regarding the causes, effects, magnitude and timing of climate change and the economic and social consequences of various response strategies;

(h) Promote and cooperate in the full, open and prompt exchange of relevant scientific, technological, technical, socio-economic and legal information related to the climate system and climate change, and to the economic and social consequences of various response strategies;

(i) Promote and cooperate in education, training and public awareness related to climate change and encourage the widest participation in this process, including that of non-governmental organizations; and

(j)   Communicate to the Conference of the Parties information related to implementation, in accordance with Article 12.

2.   The developed country Parties and other Parties included in annex I commit themselves specifically as provided for in the following:

(a)   Each of these Parties shall adopt national[1] policies and take corresponding measures on the mitigation of climate change, by limiting its anthropogenic emissions of greenhouse gases and protecting and enhancing its greenhouse gas sinks and reservoirs. These policies and measures will demonstrate that developed countries are taking the lead in modifying longer-term trends in anthropogenic emissions consistent with the objective of the Convention, recognizing that the return by the end of the present decade to earlier levels of anthropogenic emissions of carbon dioxide and other greenhouse gases not controlled by the Montreal Protocol would contribute to such modification, and taking into account the differences in these Parties' starting points and approaches, economic structures and resource bases, the need to maintain strong and sustainable economic growth, available technologies and other individual circumstances, as well as the need for equitable and appropriate contributions by each of these Parties to the global effort regarding that objective. These Parties may implement such policies and measures jointly with other Parties and may assist other Parties in contributing to the achievement of the objective of the Convention and, in particular, that of this subparagraph;

(b)   In order to promote progress to this end, each of these Parties shall communicate, within six months of the entry into force of the Convention for it and periodically thereafter, and in accordance with Article 12, detailed information on its policies and measures referred to in subparagraph (a) above, as well as on its resulting projected anthropogenic emissions by sources and removals by sinks of greenhouse gases not controlled by the Montreal Protocol for the period referred to in subparagraph (a), with the aim of returning individually or jointly to their 1990 levels these anthropogenic emissions of carbon dioxide and other greenhouse gases not controlled by the Montreal Protocol. This information will be reviewed by the Conference of the Parties, at its first session and periodically thereafter, in accordance with Article 7;

(c)   Calculations of emissions by sources and removals by sinks of greenhouse gases for the purposes of subparagraph (b) above should take into account the best available scientific knowledge, including of the

1. This includes policies and measures adopted by regional economic integration organizations.

effective capacity of sinks and the respective contributions of such gases to climate change. The Conference of the Parties shall consider and agree on methodologies for these calculations at its first session and review them regularly thereafter;

(d) The Conference of the Parties shall, at its first session, review the adequacy of subparagraphs (a) and (b) above. Such review shall be carried out in the light of the best available scientific information and assessment on climate change and its impacts, as well as relevant technical, social and economic information. Based on this review, the Conference of the Parties shall take appropriate action, which may include the adoption of amendments to the commitments in subparagraphs (a) and (b) above. The Conference of the Parties, at its first session, shall also take decisions regarding criteria for joint implementation as indicated in subparagraph (a) above. A second review of subparagraphs (a) and (b) shall take place not later than 31 December 1998, and thereafter at regular intervals determined by the Conference of the Parties, until the objective of the Convention is met;

(e) Each of these Parties shall:

(i) coordinate as appropriate with other such Parties, relevant economic and administrative instruments developed to achieve the objective of the Convention; and

(ii) identify and periodically review its own policies and practices which encourage activities that lead to greater levels of anthropogenic emissions of greenhouse gases not controlled by the Montreal Protocol than would otherwise occur;

(f) The Conference of the Parties shall review, not later than 31 December 1998, available information with a view to taking decisions regarding such amendments to the lists in annexes I and II as may be appropriate, with the approval of the Party concerned;

(g) Any Party not included in annex I may, in its instrument of ratification, acceptance, approval or accession, or at any time thereafter, notify the Depositary that it intends to be bound by subparagraphs (a) and (b) above. The Depositary shall inform the other signatories and Parties of any such notification.

3. The developed country Parties and other developed Parties included in annex II shall provide new and additional financial resources to meet the agreed full costs incurred by developing country Parties in complying with their obligations under Article 12, paragraph 1. They shall also provide such financial resources, including for the transfer of technology, needed by the developing country Parties to meet the agreed full incremental costs of implementing measures that are covered by paragraph 1 of this Article and that are agreed between

a developing country Party and the international entity or entities referred to in Article 11, in accordance with that Article. The implementation of these commitments shall take into account the need for adequacy and predictability in the flow of funds and the importance of appropriate burden sharing among the developed country Parties.

4. The developed country Parties and other developed Parties included in annex II shall also assist the developing country Parties that are particularly vulnerable to the adverse effects of climate change in meeting costs of adaptation to those adverse effects.

5. The developed country Parties and other developed Parties included in annex II shall take all practicable steps to promote, facilitate and finance, as appropriate, the transfer of, or access to, environmentally sound technologies and know-how to other Parties, particularly developing country Parties, to enable them to implement the provisions of the Convention. In this process, the developed country Parties shall support the development and enhancement of endogenous capacities and technologies of developing country Parties. Other Parties and organizations in a position to do so may also assist in facilitating the transfer of such technologies.

6. In the implementation of their commitments under paragraph 2 above, a certain degree of flexibility shall be allowed by the Conference of the Parties to the Parties included in annex I undergoing the process of transition to a market economy, in order to enhance the ability of these Parties to address climate change, including with regard to the historical level of anthropogenic emissions of greenhouse gases not controlled by the Montreal Protocol chosen as a reference.

7. The extent to which developing country Parties will effectively implement their commitments under the Convention will depend on the effective implementation by developed country Parties of their commitments under the Convention related to financial resources and transfer of technology and will take fully into account that economic and social development and poverty eradication are the first and overriding priorities of the developing country Parties.

8. In the implementation of the commitments in this Article, the Parties shall give full consideration to what actions are necessary under the Convention, including actions related to funding, insurance and the transfer of technology, to meet the specific needs and concerns of developing country Parties arising from the adverse effects of climate change and/or the impact of the implementation of response measures, especially on:

(a) Small island countries;

(b) Countries with low-lying coastal areas;

(c) Countries with arid and semi-arid areas, forested areas and areas liable to forest decay;

(d)   Countries with areas prone to natural disasters;

(e)   Countries with areas liable to drought and desertification;

(f)   Countries with areas of high urban atmospheric pollution;

(g)   Countries with areas with fragile ecosystems, including mountainous ecosystems;

(h)   Countries whose economies are highly dependent on income generated from the production, processing and export, and/or on consumption of fossil fuels and associated energy-intensive products; and

(i)   Land-locked and transit countries.

Further, the Conference of the Parties may take actions, as appropriate, with respect to this paragraph.

9.   The Parties shall take full account of the specific needs and special situations of the least developed countries in their actions with regard to funding and transfer of technology.

10.   The Parties shall, in accordance with Article 10, take into consideration in the implementation of the commitments of the Convention the situation of Parties, particularly developing country Parties, with economies that are vulnerable to the adverse effects of the implementation of measures to respond to climate change. This applies notably to Parties with economies that are highly dependent on income generated from the production, processing and export, and/or consumption of fossil fuels and associated energy-intensive products and/or the use of fossil fuels for which such Parties have serious difficulties in switching to alternatives.

Article 5
Research and Systematic Observation

In carrying out their commitments under Article 4, paragraph 1(g), the Parties shall:

(a)   Support and further develop, as appropriate, international and intergovernmental programmes and networks or organizations aimed at defining, conducting, assessing and financing research, data collection and systematic observation, taking into account the need to minimize duplication of effort;

(b)   Support international and intergovernmental efforts to strengthen systematic observation and national scientific and technical research capacities and capabilities, particularly in developing countries, and to promote access to, and the exchange of, data and analyses thereof obtained from areas beyond national jurisdiction; and

(c)   Take into account the particular concerns and needs of developing countries and cooperate in improving their endogenous capacities and capabilities to participate in the efforts referred to in subparagraphs (a) and (b) above.

Article 6
Education, Training and Public Awareness

In carrying out their commitments under Article 4, paragraph 1(i), the Parties shall:

(a) Promote and facilitate at the national and, as approriate, subregional and regional levels, and in accordance with national laws and regulations, and within their respective capacities:

(i) the development and implementation of educational and public awareness programmes on climate change and its effects;

(ii) public access to information on climate change and its effects;

(iii) public participation in addressing climate change and its effects and developing adequate responses; and

(iv) training of scientific, technical and managerial personnel.

(b) Cooperate in and promote, at the international level, and, where appropriate, using existing bodies:

(i) the development and exchange of educational and public awareness material on climate change and its effects; and

(ii) the development and implementation of education and training programmes, including the strengthening of national institutions and the exchange or secondment of personnel to train experts in this field, in particular for developing countries.

Article 7
Conference of the Parties

1. A Conference of the Parties is hereby established.

2. The Conference of the Parties, as the supreme body of this Convention, shall keep under regular review the implementation of the Convention and any related legal instruments that the Conference of the Parties may adopt, and shall make, within its mandate, the decisions necessary to promote the effective implementation of the Convention. To this end, it shall:

(a) Periodically examine the obligations of the Parties and the institutional arrangements under the Convention, in the light of the objective of the Convention, the experience gained in its implementation and the evolution of scientific and technological knowledge;

(b) Promote and facilitate the exchange of information on measures adopted by the Parties to address climate change and its effects, taking into account the differing circumstances, responsibilities and capabilities of the Parties and their respective commitments under the Convention;

(c)   Facilitate, at the request of two or more Parties, the coordination of measures adopted by them to address climate change and its effects, taking into account the differing circumstances, responsibilities and capabilities of the Parties and their respective commitments under the Convention;

(d)   Promote and guide, in accordance with the objective and provisions of the Convention, the development and periodic refinement of comparable methodologies, to be agreed on by the Conference of the Parties, *inter alia*, for preparing inventories of greenhouse gas emissions by sources and removals by sinks, and for evaluating the effectiveness of measures to limit the emissions and enhance the removals of these gases;

(e)   Assess, on the basis of all information made available to it in accordance with the provisions of the Convention, the implementation of the Convention by the Parties, the overall effects of the measures taken pursuant to the Convention, in particular environmental, economic and social effects as well as their cumulative impacts and the extent to which progress towards the objective of the Convention is being achieved;

(f)   Consider and adopt regular reports on the implementation of the Convention and ensure their publication;

(g)   Make recommendations on any matters necessary for the implementation of the Convention;

(h)   Seek to mobilize financial resources in accordance with Article 4, paragraphs 3, 4 and 5, and Article 11;

(i)   Establish such subsidiary bodies as are deemed necessary for the implementation of the Convention;

(j)   Review reports submitted by its subsidiary bodies and provide guidance to them;

(k)   Agree upon and adopt, by consensus, rules of procedure and financial rules for itself and for any subsidiary bodies;

(l)   Seek and utilize, where appropriate, the services and cooperation of, and information provided by, competent international organizations and intergovernmental and non-governmental bodies; and

(m)   Exercise such other functions as are required for the achievement of the objective of the Convention as well as all other functions assigned to it under the Convention.

3.   The Conference of the Parties shall, at its first session, adopt its own rules of procedure as well as those of the subsidiary bodies established by the Convention, which shall include decision-making procedures for matters not already covered by decision-making procedures stipulated in the Convention. Such procedures may include specified majorities required for the adoption of particular decisions.

4.   The first session of the Conference of the Parties shall be convened by the interim secretariat referred to in Article 21 and shall take place not later than one year after the date of entry into force of the Convention. Thereafter, ordinary sessions of the Conference of the Parties shall be held every year unless otherwise decided by the Conference of the Parties.

5.   Extraordinary sessions of the Conference of the Parties shall be held at such other times as may be deemed necessary by the Conference, or at the written request of any Party, provided that, within six months of the request being communicated to the Parties by the secretariat, it is supported by at least one-third of the Parties.

6.   The United Nations, its specialized agencies and the International Atomic Energy Agency, as well as any State member thereof or observers thereto not Party to the Convention, may be represented at sessions of the Conference of the Parties as observers. Any body or agency, whether national or international, governmental or non-governmental, which is qualified in matters covered by the Convention, and which has informed the secretariat of its wish to be represented at a session of the Conference of the Parties as an observer, may be so admitted unless at least one-third of the Parties present object. The admission and participation of observers shall be subject to the rules of procedure adopted by the Conference of the Parties.

Article 8
Secretariat

1.   A secretariat is hereby established.

2.   .The functions of the secretariat shall be:
  (a)   To make arrangements for sessions of the Conference of the Parties and its subsidiary bodies established under the Convention and to provide them with services as required;
  (b)   To compile and transmit reports submitted to it;
  (c)   To facilitate assistance to the Parties, particularly developing country Parties, on request, in the compilation and communication of information required in accordance with the provisions of the Convention;
  (d)   To prepare reports on its activities and present them to the Conference of the Parties;

(e)   To ensure the necessary coordination with the secretariats of other relevant international bodies;

(f)   To enter, under the overall guidance of the Conference of the Parties, into such administrative and contractual arrangements as may be required for the effective discharge of its functions; and

(g)   To perform the other secretariat functions specified in the Convention and in any of its protocols and such other functions as may be determined by the Conference of the Parties.

3.   The Conference of the Parties, at its first session, shall designate a permanent secretariat and make arrangements for its functioning.

Article 9
Subsidiary Body for Scientific and Technological Advice

1.   A subsidiary body for scientific and technological advice is hereby established to provide the Conference of the Parties and, as appropriate, its other subsidiary bodies with timely information and advice on scientific and technological matters relating to the Convention. This body shall be open to participation by all Parties and shall be multidisciplinary. It shall comprise government representatives competent in the relevant field of expertise. It shall report regularly to the Conference of the Parties on all aspects of its work.

2.   Under the guidance of the Conference of the Parties, and drawing upon existing competent international bodies, this body shall:

(a)   Provide assessments of the state of scientific knowledge relating to climate change and its effects;

(b)   Prepare scientific assessments on the effects of measures taken in the implementation of the Convention;

(c)   Identify innovative, efficient and state-of-the-art technologies and know-how and advise on the ways and means of promoting development and/or transferring such technologies;

(d)   Provide advice on scientific programmes, international cooperation in research and development related to climate change, as well as on ways and means of supporting endogenous capacity-building in developing countries; and

(e)   Respond to scientific, technological and methodological questions that the Conference of the Parties and its subsidiary bodies may put to the body.

3.   The functions and terms of reference of this body may be further elaborated by the Conference of the Parties.

Article 10
Subsidiary Body for Implementation

1. A subsidiary body for implementation is hereby established to assist the Conference of the Parties in the assessment and review of the effective implementation of the Convention. This body shall be open to participation by all Parties and comprise government representatives who are experts on matters related to climate change. It shall report regularly to the Conference of the Parties on all aspects of its work.

2. Under the guidance of the Conference of the Parties, this body shall:

    (a) Consider the information communicated in accordance with Article 12, paragraph 1, to assess the overall aggregated effect of the steps taken by the Parties in the light of the latest scientific assessments concerning climate change;

    (b) Consider the information communicated in accordance with Article 12, paragraph 2, in order to assist the Conference of the Parties in carrying out the reviews required by Article 4, paragraph 2(d); and

    (c) Assist the Conference of the Parties, as appropriate, in the preparation and implementation of its decisions.

Article 11
Financial Mechanism

1. A mechanism for the provision of financial resources on a grant or concessional basis, including for the transfer of technology, is hereby defined. It shall function under the guidance of and be accountable to the Conference of the Parties, which shall decide on its policies, programme priorities and eligibility criteria related to this Convention. Its operation shall be entrusted to one or more existing international entities.

2. The financial mechanism shall have an equitable and balanced representation of all Parties within a transparent system of governance.

3. The Conference of the Parties and the entity or entities entrusted with the operation of the financial mechanism shall agree upon arrangements to give effect to the above paragraphs, which shall include the following:

    (a) Modalities to ensure that the funded projects to address climate change are in conformity with the policies, programme priorities and eligibility criteria established by the Conference of the Parties;

    (b) Modalities by which a particular funding decision may be reconsidered in light of these policies, programme priorities and eligibility criteria;

    (c) Provision by the entity or entities of regular reports to the Conference of the Parties on its funding operations, which is consistent with the requirement for accountability set out in paragraph 1 above; and

(d)   Determination in a predictable and identifiable manner of the amount of funding necessary and available for the implementation of this Convention and the conditions under which that amount shall be periodically reviewed.

4.   The Conference of the Parties shall make arrangements to implement the above mentioned provisions at its first session, reviewing and taking into account the interim arrangements referred to in Article 21, paragraph 3, and shall decide whether these interim arrangements shall be maintained. Within four years thereafter, the Conference of the Parties shall review the financial mechanism and take appropriate measures.

5.   The developed country Parties may also provide and developing country Parties avail themselves of financial resources related to the implementation of the Convention through bilateral, regional and other multilateral channels.

Article 12
Communication of Information Related to Implementation

1.   In accordance with Article 4, paragraph 1, each Party shall communicate to the Conference of the Parties, through the secretariat, the following elements of information:

(a)   A national inventory of anthropogenic emissions by sources and removals by sinks of all greenhouse gases not controlled by the Montreal Protocol, to the extent its capacities permit, using comparable methodologies to be promoted and agreed upon by the Conference of the Parties;

(b)   A general description of steps taken or envisaged by the Party to implement the Convention; and

(c)   Any other information that the Party considers relevant to the achievement of the objective of the Convention and suitable for inclusion in its communciation, including, if feasible, material relevant for calculations of global emission trends.

2.   Each developed country Party and each other Party included in annex I shall incorporate in its communication the following elements of information:

(a)   A detailed description of the policies and measures that it has adopted to implement its commitment under Article 4, paragraphs 2(a) and 2(b); and

(b)   A specific estimate of the effects that the policies and measures referred to in subparagraph (a) immediately above will have on anthropogenic emissions by its sources and removals by its sinks of greenhouse gases during the period referred to in Article 4, paragraph 2(a).

3.   In addition, each developed country Party and each other developed Party included in annex II shall incorporate details of measures taken in accordance with Article 4, paragraphs 3, 4 and 5.

4.   Developing country Parties may, on a voluntary basis, propose projects for financing, including specific technologies, materials, equipment, techniques or practices that would be needed to implement such projects, along with, if possible, an estimate of all incremental costs, of the reductions of emissions and increments of removals of greenhouse gases, as well as an estimate of the consequent benefits.

5.   Each developed country Party and each other Party included in annex I shall make its initial communication within six months of the entry into force of the Convention for that Party. Each Party not so listed shall make its initial communication within three years of the entry into force of the Convention for that Party, or of the availability of financial resources in accordance with Article 4, paragraph 3. Parties that are least developed countries may make their initial communication at their discretion. The frequency of subsequent communications by all Parties shall be determined by the Conference of the Parties, taking into account the differentiated timetable set by this paragraph.

6.   Information communicated by Parties under this Article shall be transmitted by the secretariat as soon as possible to the Conference of the Parties and to any subsidiary bodies concerned. If necessary, the procedures for the communication of information may be further considered by the Conference of the Parties.

7.   From its first session, the Conference of the Parties shall arrange for the provision to developing country Parties of technical and financial support, on request, in compiling and communicating information under this Article, as well as in identifying the technical and financial needs associated with proposed projects and response measures under Article 4. Such support may be provided by other Parties, by competent international organizations and by the secretariat, as appropriate.

8.   Any group of Parties may, subject to guidelines adopted by the Conference of the Parties, and to prior notification to the Conference of the Parties, make a joint communication in fulfilment of their obligations under this Article, provided that such a communication includes information on the fulfilment by each of these Parties of its individual obligations under the Convention.

9.   Information received by the secretariat that is designated by a Party as confidential, in accordance with criteria to be established by the Conference of the Parties, shall be aggregated by the secretariat to protect its confidentiality before being made available to any of the bodies involved in the communication and review of information.

10. Subject to paragraph 9 above, and without prejudice to the ability of any Party to make public its communication at any time, the secretariat shall make communications by Parties under this Article publicly available at the time they are submitted to the Conference of the Parties.

Article 13
Resolution of Questions Regarding Implementation

The Conference of the Parties shall, at its first session, consider the establishment of a multilateral consultative process, available to Parties on their request, for the resolution of questions regarding the implementation of the Convention.

Article 14
Settlement of Disputes

1.   In the event of a dispute between any two or more Parties concerning the interpretation or application of the Convention, the Parties concerned shall seek a settlement of the dispute through negotiation or any other peaceful means of their own choice.

2.   When ratifying, accepting, approving or acceding to the Convention, or at any time thereafter, a Party which is not a regional economic integration organization may declare in a written instrument submitted to the Depositary that, in respect of any dispute concerning the interpretation or application of the Convention, it recognizes as compulsory *ipso facto* and without special agreement, in relation to any party accepting the same obligation:

   (a)   Submission of the dispute to the International Court of Justice, and/or

   (b)   Arbitration in accordance with procedures to be adopted by the Conference of the Parties as soon as practicable, in an annex on arbitration.

A Party which is a regional economic integration organization may make a declaration with like effect in relation to arbitration in accordance with the procedures referred to in subparagraph (b) above.

3.   A declaration made under paragraph 2 above shall remain in force until it expires in accordance with its terms or until three months after written notice of its revocation has been deposited with the Depositary.

4.   A new declaration, a notice of revocation or the expiry of a declaration shall not in any way affect proceedings pending before the International Court of Justice or the arbitral tribunal, unless the parties to the dispute otherwise agree.

5.   Subject to the operation of paragraph 2 above, if after twelve months following notification by one Party to another that a dispute exists between them, the Parties concerned have not been able to settle their dispute through the means

mentioned in paragraph 1 above, the dispute shall be submitted, at the request of any of the parties to the dispute, to conciliation.

6.   A conciliation commission shall be created upon the request of one of the parties to the dispute. The commission shall be composed of an equal number of members appointed by each party concerned and a chairman chosen jointly by the members appointed by each party. The commission shall render a recommendatory award, which the parties shall consider in good faith.

7.   Additional procedures relating to conciliation shall be adopted by the Conference of the Parties, as soon as practicable, in an annex on conciliation.

8.   The provisions of this Article shall apply to any related legal instrument which the Conference of the Parties may adopt, unless the instrument provides otherwise.

Article 15
Amendments to the Convention

1.   Any party may propose amendments to the Convention.

2.   Amendments to the Convention shall be adopted at an ordinary session of the Conference of the Parties. The text of any proposed amendment to the Convention shall be communicated to the Parties by the secretariat at least six months before the meeting at which it is proposed for adoption. The secretariat shall also communicate proposed amendments to the signatories to the Convention and, for information, to the Depositary.

3.   The Parties shall make every effort to reach agreement on any proposed amendment to the Convention by consensus. If all efforts at consensus have been exhausted, and no agreement reached, the amendment shall as a last resort be adopted by a three-fourths majority vote of the Parties present and voting at the meeting. The adopted amendment shall be communicated by the secretariat to the Depositary, who shall circulate it to all Parties for their acceptance.

4.   Instruments of acceptance in respect of an amendment shall be deposited with the Depositary. An amendment adopted in accordance with paragraph 3 above shall enter into force for those Parties having accepted it on the ninetieth day after the date of receipt by the Depositary of an instrument of acceptance by at least three-fourths of the Parties to the Convention.

5.   The amendment shall enter into force for any other Party on the ninetieth day after the date on which that Party deposits with the Depositary its instrument of acceptance of the said amendment.

6.   For the purposes of this Article, "Parties present and voting" means Parties present and casting an affirmative or negative vote.

Article 16
Adoption and Amendment of Annexes to the Convention

1.   Annexes to the Convention shall form an integral part thereof and, unless otherwise expressly provided, a reference to the Convention constitutes at the same time a reference to any annexes thereto. Without prejudice to the provisions of Article 14, paragraphs 2(b) and 7, such annexes shall be restricted to lists, forms and any other material of a descriptive nature that is of a scientific, technical, procedural or administrative character.

2.   Annexes to the Convention shall be proposed and adopted in accordance with the procedure set forth in Article 15, paragraphs 2, 3, and 4.

3.   An annex that has been adopted in accordance with paragraph 2 above shall enter into force for all Parties to the Convention six months after the date of the communication by the Depositary to such Parties of the adoption of the annex, except for those Parties that have notified the Depositary, in writing, within that period of their non-acceptance of the annex. The annex shall enter into force for Parties which withdraw their notification of non-acceptance on the ninetieth day after the date on which withdrawal of such notification has been received by the Depositary.

4.   The proposal, adoption and entry into force of amendments to annexes to the Convention shall be subject to the same procedure as that for the proposal, adoption and entry into force of annexes to the Convention in accordance with paragraphs 2 and 3 above.

5.   If the adoption of an annex or an amendment to an annex involves an amendment to the Convention, that annex or amendment to an annex shall not enter into force until such time as the amendment to the Convention enters into force.

Article 17
Protocols

1.   The Conference of the Parties may, at any ordinary session, adopt protocols to the Convention.

2.   The text of any proposed protocol shall be communicated to the Parties by the secretariat at least six months before such a session.

3.   The requirements for the entry into force of any protocol shall be established by that instrument.

4.   Only Parties to the Convention may be Parties to a protocol.

5.   Decisions under any protocol shall be taken only by the Parties to the protocol concerned.

Article 18
Right to Vote

1.   Each Party to the Convention shall have one vote, except as provided for in paragraph 2 below.

2.   Regional economic integration organizations, in matters within their competence, shall exercise their right to vote with a number of votes equal to the number of their member States that are Parties to the Convention. Such an organization shall not exercise its right to vote if any of its member States exercises its right, and vice versa.

Article 19
Depositary

The Secretary-General of the United Nations shall be the Depositary of the Convention and of protocols adopted in accordance with Article 17.

Article 20
Signature

This Convention shall be open for signature by States Members of the United Nations or of any of its specialized agencies or that are Parties to the Statute of the International Court of Justice and by regional economic integration organizations at Rio de Janeiro, during the United Nations Conference on Environment and Development, and thereafter at United Nations Headquarters in New York from 20 June 1992 to 19 June 1993.

Article 21
Interim Arrangements

1.   The secretariat functions referred to in Article 8 will be carried out on an interim basis by the secretariat established by the General Assembly of the United Nations in its resolution 45/212 of 21 December 1990, until the completion of the first session of the Conference of the Parties.

2.   The head of the interim secretariat referred to in paragraph 1 above will cooperate closely with the Intergovernmental Panel on Climate Change to ensure that the Panel can respond to the need for objective scientific and technical advice. Other relevant scientific bodies could also be consulted.

3.   The Global Environment Facility of the United Nations Development Programme, the United Nations Environmental Programme and the International Bank for Reconstruction and Development shall be the international entity entrusted with the operation of the financial mechanism referred to in Article 11

on an interim basis. In this connection, the Global Environment Facility should be appropriately restructured and its membership made universal to enable it to fulfil the requirements of Article 11.

Article 22
Ratification, Acceptance, Approval or Accession

1.   The Convention shall be subject to ratification, acceptance, approval or accession by States and by regional economic integration organizations. It shall be open for accession from the day after the date on which the Convention is closed for signature. Instruments of ratification, acceptance, approval or accession shall be deposited with the Depositary.

2.   Any regional economic integration organization which becomes a Party to the Convention without any of its member States being a Party shall be bound by all the obligations under the Convention. In the case of such organizations, one or more of whose member States is a Party to the Convention, the organization and its member States shall decide on their respective responsibilities for the performance of their obligations under the Convention. In such cases, the organization and the member States shall not be entitled to exercise rights under the Convention concurrently.

3.   In their instruments of ratification, acceptance, approval or accession, regional economic integration organizations shall declare the extent of their competence with respect to the matters governed by the Convention. These organizations shall also inform the Depositary, who shall in turn inform the Parties, of any substantial modification in the extent of their competence.

Article 23
Entry into Force

1.   The Convention shall enter into force on the ninetieth day after the date of deposit of the fiftieth instrument of ratification, acceptance, approval or accession.

2.   For each State or regional economic integration organization that ratifies, accepts or approves the Convention or accedes thereto after the deposit of the fiftieth instrument of ratification, acceptance, approval or accession, the Convention shall enter into force on the ninetieth day after the date of deposit by such State or regional economic integration organization of its instrument of ratification, acceptance, approval or accession.

3.   For the purposes of paragraphs 1 and 2 above, any instrument deposited by a regional economic integration organization shall not be counted as additional to those deposited by States members of the organization.

Article 24
Reservations

No reservations may be made to the Convention.

Article 25
Withdrawal

1.　At any time after three years from the date on which the Convention has entered into force for a Party, that Party may withdraw from the Convention by giving written notification to the Depositary.

2.　Any such withdrawal shall take effect upon expiry of one year from the date of receipt by the Depositary of the notification of withdrawal, or on such later date as may be specified in the notification of withdrawal.

3.　Any Party that withdraws from the Convention shall be considered as also having withdrawn from any protocol to which it is a Party.

Article 26
Authentic Texts

The original of this Convention, of which the Arabic, Chinese, English, French, Russian and Spanish texts are equally authentic, shall be deposited with the Secretary-General of the United Nations.

IN WITNESS WHEREOF the undersigned, being duly authorized to that effect, have signed this Convention.

DONE at New York this ninth day of May one thousand nine hundred and ninety-two.

Annex I

Australia
Austria
Belarus[a]
Belgium
Bulgaria[a]
Canada
Czechoslovakia[a]
Denmark
European Community

[a]Countries that are undergoing the process of transition to a market economy.

Estonia[a]
Finland
France
Germany
Greece
Hungary[a]
Iceland
Ireland
Italy
Japan
Latvia[a]
Lithuania[a]
Luxembourg
Netherlands
New Zealand
Norway
Poland[a]
Portugal
Romania[a]
Russian Federation[a]
Spain
Sweden
Switzerland
Turkey
Ukraine[a]
United Kingdom of Great Britain and Northern Ireland
United States of America

Annex II

Australia
Austria
Belgium
Canada
Denmark
European Community
Finland
France
Germany
Greece
Iceland
Ireland

Italy
Japan
Luxembourg
Netherlands
New Zealand
Norway
Portugal
Spain
Sweden
Switzerland
Turkey
United Kingdom of Great Britain and Northern Ireland
United States of America

Annex II to the Report of the Committee: Resolution Adopted by the Intergovernmental Negotiating Committee for a Framework Convention on Climate Change

INC/1992/1.   INTERIM ARRANGEMENTS

The Intergovernmental Negotiating Committee for a Framework Convention on Climate Change,

Having agreed upon and adopted the text of the United Nations Framework Convention on Climate Change,

Considering that preparations are required for an early and effective operation of the Convention once it has entered into force,

Further considering that, in the interim arrangements, involvement in the negotiations of all participants in the Committee is essential,

Recalling General Assembly resolutions 45/212 of 21 December 1990 and 46/169 of 19 December 1991,

1.   *Calls upon* all States and regional economic integration organizations entitled to do so to sign the Convention during the United Nations Conference on Environment and Development in Rio de Janeiro or at the earliest subsequent opportunity and thereafter to ratify, accept, approve or accede to the Convention;

2.   *Requests* the Secretary-General to make the necessary arrangements for convening a session of the Committee, in accordance with paragraph 4 of General Assembly resolution 46/169, to prepare for the first session of the Conference of the Parties as specified in the Convention;

3.   *Requests further* the Secretary-General to make recommendations to the General Assembly at its forty-seventh session regarding arrangements for further sessions of the Committee until the entry into force of the Convention;

4.   *Invites* the Secretary-General to include in his report to the General Assembly, as required in paragraphs 4 and 9 of resolution 46/169, proposals that would enable the secretariat established under resolution 45/212 to continue its activities until the designation of the secretariat of the Convention by the Conference of the Parties;

5.   *Appeals* to Governments and organizations to make voluntary contributions to the extrabudgetary funds established under General Assembly resolution 45/212 in order to contribute to the costs of the interim arrangements, and to ensure full and effective participation of developing countries, in particular the least developed countries and small island developing countries, as well as developing countries stricken by drought and desertification, in all the sessions of the Committee;

6.   *Invites* States and regional economic integration organizations entitled to sign the Convention to communicate as soon as feasible to the head of the secretariat information regarding measures consistent with the provisions of the Convention pending its entry into force.

9 May 1992

United Nations Convention on Biological Diversity, 5 June 1992

Preamble

The Contracting Parties,

Conscious of the intrinsic value of biological diversity and of the ecological, genetic, social, economic, scientific, educational, cultural, recreational and aesthetic values of biological diversity and its components,

Conscious also of the importance of biological diversity for evolution and for maintaining life sustaining systems of the biosphere,

Affirming that the conservation of biological diversity is a common concern of humankind,

Reaffirming that States have sovereign rights over their own biological resources,

Reaffirming also that States are responsible for conserving their biological diversity and for using their biological resources in a sustainable manner,

Concerned that biological diversity is being significantly reduced by certain human activities,

Aware of the general lack of information and knowledge regarding biological diversity and of the urgent need to develop scientific, technical and institutional capacities to provide the basic understanding upon which to plan and implement appropriate measures,

Noting that it is vital to anticipate, prevent and attack the causes of significant reduction or loss of biological diversity at source,

Noting also that where there is a threat of significant reduction or loss of biological diversity, lack of full scientific certainty should not be used as a reason for postponing measures to avoid or minimize such a threat,

Noting further that the fundamental requirement for the conservation of biological diversity is the *in-situ* conservation of ecosystems and natural habitats and the maintenance and recovery of viable populations of species in their natural surroundings,

Noting further that *ex-situ* measures, preferably in the country of origin, also have an important role to play,

Recognizing the close and traditional dependence of many indigenous and local communities embodying traditional lifestyles on biological resources, and the desirability of sharing equitably benefits arising from the use of traditional knowledge, innovations and practices relevant to the conservation of biological diversity and the sustainable use of its components,

Recognizing also the vital role that women play in the conservation and sustainable use of biological diversity and affirming the need for the full participation of women at all levels of policy-making and implementation for biological diversity conservation,

Stressing the importance of, and the need to promote, international, regional and global cooperation among States and intergovernmental organizations and the non-governmental sector for the conservation of biological diversity and the sustainable use of its components,

Acknowledging that the provision of new and additional financial resources and appropriate access to relevant technologies can be expected to make a substantial difference in the world's ability to address the loss of biological diversity,

Acknowledging further that special provision is required to meet the needs of developing countries, including the provision of new and additional financial resources and appropriate access to relevant technologies,

Noting in this regard the special conditions of the least developed countries and small island States,

Acknowledging that substantial investments are required to conserve biological diversity and that there is the expectation of a broad range of environmental, economic and social benefits from those investments,

Recognizing that economic and social development and poverty eradication are the first and overriding priorities of developing countries,

Aware that conservation and sustainable use of biological diversity is of critical importance for meeting the food, health and other needs of the growing world population, for which purpose access to and sharing of both genetic resources and technologies are essential,

Noting that, ultimately, the conservation and sustainable use of biological diversity will strengthen friendly relations among States and contribute to peace for humankind,

Desiring to enhance and complement existing international arrangements for the conservation of biological diversity and sustainable use of its components, and

Determined to conserve and sustainably use biological diversity for the benefit of present and future generations,

Have agreed as follows:

Article 1
Objectives

The objectives of this Convention, to be pursued in accordance with its relevant provisions, are the conservation of biological diversity, the sustainable use of its components and the fair and equitable sharing of the benefits arising out of the utilization of genetic resources, including by appropriate access to genetic resources and by appropriate transfer of relevant technologies, taking into account all rights over those resources and to technologies, and by appropriate funding.

Article 2
Use of Terms

For the purposes of this Convention:

"Biological diversity" means the variability among living organisms from all sources including, *inter alia*, terrestrial, marine and other aquatic ecosystems and the ecological complexes of which they are part; this includes diversity within species, between species and of ecosystems.

"Biological resources" includes genetic resources, organisms or parts thereof, populations, or any other biotic component of ecosystems with actual or potential use or value for humanity.

"Biotechnology" means any technological application that uses biological systems, living organisms, or derivatives thereof, to make or modify products or processes for specific use.

"Country of origin of genetic resources" means the country which possesses those genetic resources in *in situ* conditions.

"Country providing genetic resources" means the country supplying genetic resources collected from *in-situ* sources, including populations of both wild and domesticated species, or taken from *ex-situ* sources, which may or may not have originated in that country.

"Domesticated or cultivated species" means species in which the evolutionary process has been influenced by humans to meet their needs.

"Ecosystem" means a dynamic complex of plant, animal and micro-organism communities and their non-living environment interacting as a functional unit.

"Ex-situ conservation" means the conservation of components of biological diversity outside their natural habitats.

"Genetic material" means any material of plant, animal, microbial or other origin containing functional units of heredity.

"Genetic resources" means genetic material of actual or potential value.

"Habitat" means the place or type of site where an organism or population naturally occurs.

"In-situ conditions" means conditions where genetic resources exist within eco-systems and natural habitats, and, in the case of domesticated or cultivated spe-cies, in the surroundings where they have developed their distinctive properties.

"In-situ conservation" means the conservation of ecosystems and natural habitats and the maintenance and recovery of viable populations of species in their natu-ral surroundings and, in the case of domesticated or cultivated species, in the surroundings where they have developed their distinctive properties.

"Protected area" means a geographically defined area which is designated or regulated and managed to achieve specific conservation objectives.

"Regional economic integration organization" means an organization constituted by sovereign States of a given region, to which its member States have transferred competence in respect of matters governed by this Convention and which has been duly authorized, in accordance with its internal procedures, to sign, ratify, accept, approve or accede to it.

"Sustainable use" means the use of components of biological diversity in a way and at a rate that does not lead to the long-term decline of biological diversity, thereby maintaining its potential to meet the needs and aspirations of present and future generations.

"Technology" includes biotechnology.

Article 3
Principle

States have, in accordance with the Charter of the United Nations and the princi-ples of international law, the sovereign right to exploit their own resources pursu-ant to their own environmental policies, and the responsibility to ensure that activities within their jurisdiction or control do not cause damage to the environ-ment of other States or of areas beyond the limits of national jurisdiction.

Article 4
Jurisdictional Scope

Subject to the rights of other States, and except as otherwise expressly provided in this Convention, the provisions of this Convention apply, in relation to each Contracting Party:

(a) In the case of components of biological diversity, in areas within the limits of its national jurisdiction; and

(b) In the case of processes and activities, regardless of where their effects occur, carried out under its jurisdiction or control, within the area of its national jurisdiction or beyond the limits of national jurisdiction.

Article 5
Cooperation

Each Contracting Party shall, as far as possible and as appropriate, cooperate with other Contracting Parties, directly or, where appropriate, through competent international organizations, in respect of areas beyond national jurisdiction and on other matters of mutual interest, for the conservation and sustainable use of biological diversity.

Article 6
General Measures for Conservation and Sustainable Use

Each Contracting Party shall, in accordance with its particular conditions and capabilities:

(a) Develop national strategies, plans or programmes for the conservation and sustainable use of biological diversity or adapt for this purpose existing strategies, plans or programmes which shall reflect, *inter alia*, the measures set out in this Convention relevant to the Contracting Party concerned; and

(b) Integrate, as far as possible and as appropriate, the conservation and sustainable use of biological diversity into relevant sectoral or cross-sectoral plans, programmes and policies.

Article 7
Identification and Monitoring

Each Contracting Party shall, as far as possible and as appropriate, in particular for the purposes of Articles 8 to 10:

(a) Identify components of biological diversity important for its conservation and sustainable use having regard to the indicative list of categories set down in Annex I;

(b) Monitor, through sampling and other techniques, the components of biological diversity identified pursuant to subparagraph (a) above, paying particular attention to those requiring urgent conservation measures and those which offer the greatest potential for sustainable use;

(c) Identify processes and categories of activities which have or are likely to have significant adverse impacts on the conservation and sustainable use of biological diversity, and monitor their effects through sampling and other techniques; and

(d) Maintain and organize, by any mechanism, data derived from identification and monitoring activities pursuant to subparagraphs (a), (b) and (c) above.

Article 8
In-situ *Conservation*

Each Contracting Party shall, as far as possible and as appropriate:

(a) Establish a system of protected areas or areas where special measures need to be taken to conserve biological diversity;

(b) Develop, where necessary, guidelines for the selection, establishment and management of protected areas or areas where special measures need to be taken to conserve biological diversity;

(c) Regulate or manage biological resources important for the conservation of biological diversity whether within or outside protected areas, with a view to ensuring their conservation and sustainable use;

(d) Promote the protection of ecosystems, natural habitats and the maintenance of viable populations of species in natural surroundings;

(e) Promote environmentally sound and sustainable development in areas adjacent to protected areas with a view to furthering protection of these areas;

(f) Rehabilitate and restore degraded ecosystems and promote the recovery of threatened species, *inter alia*, through the development and implementation of plans or other management strategies;

(g) Establish or maintain means to regulate, manage or control the risks associated with the use and release of living modified organisms resulting from biotechnology which are likely to have adverse environmental impacts that could affect the conservation and sustainable use of biological diversity, taking also into account the risks to human health;

(h) Prevent the introduction of, control or eradicate those alien species which threaten ecosystems, habitats or species;

(i) Endeavour to provide the conditions needed for compatibility between present uses and the conservation of biological diversity and the sustainable use of its components;

(j) Subject to its national legislation, respect, preserve and maintain knowledge, innovations and practices of indigenous and local communities embodying traditional lifestyles relevant for the conservation and sustainable use of biological diversity and promote their wider application with the approval and involvement of the holders of such knowledge, innovations and practices and encourage the equitable sharing of the benefits arising from the utilization of such knowledge, innovations and practices;

(k) Develop or maintain necessary legislation and/or other regulatory provisions for the protection of threatened species and populations;

(l) Where a significant adverse effect on biological diversity has been determined pursuant to Article 7, regulate or manage the relevant processes and categories of activities; and

(m) Cooperate in providing financial and other support for *in-situ* conservation outlined in subparagraphs (a) to (l) above, particularly to developing countries.

Article 9
Ex-situ *Conservation*

Each Contracting Party shall, as far as possible and as appropriate, and predominantly for the purpose of complementing *in-situ* measures:

(a) Adopt measures for the *ex-situ* conservation of components of biological diversity, preferably in the country of origin of such components;

(b) Establish and maintain facilities for *ex-situ* conservation of and research on plants, animals and micro-organisms, preferably in the country of origin of genetic resources;

(c) Adopt measures for the recovery and rehabilitation of threatened species and for their reintroduction into their natural habitats under appropriate conditions;

(d) Regulate and manage collection of biological resources from natural habitats for *ex-situ* conservation purposes so as not to threaten ecosystems and *in-situ* populations of species, except where special temporary *ex-situ* measures are required under subparagraph (c) above; and

(e) Cooperate in providing financial and other support for *ex-situ* conservation outlined in subparagraphs (a) to (d) above and in the establishment and maintenance of *ex-situ* conservation facilities in developing countries.

Article 10
Sustainable Use of Components of Biological Diversity

Each Contracting Party shall, as far as possible and as appropriate:

(a) Integrate consideration of the conservation and sustainable use of biological resources into national decision-making;

(b) Adopt measures relating to the use of biological resources to avoid or minimize adverse impacts on biological diversity;

(c) Protect and encourage customary use of biological resources in accordance with traditional cultural practices that are compatible with conservation or sustainable use requirements;

(d) Support local populations to develop and implement remedial action in degraded areas where biological diversity has been reduced; and

(e) Encourage cooperation between its governmental authorities and its private sector in developing methods for sustainable use of biological resources.

Article 11
Incentive Measures

Each Contracting Party shall, as far as possible and as appropriate, adopt economically and socially sound measures that act as incentives for the conservation and sustainable use of components of biological diversity.

Article 12
Research and Training

The Contracting Parties, taking into account the special needs of developing countries, shall:

(a) Establish and maintain programmes for scientific and technical education and training in measures for the identification, conservation and sustainable use of biological diversity and its components and provide support for such education and training for the specific needs of developing countries;

(b) Promote and encourage research which contributes to the conservation and sustainable use of biological diversity, particularly in developing countries, *inter alia*, in accordance with decisions of the Conference of the Parties taken in consequence of recommendations of the Subsidiary Body on Scientific, Technical and Technological Advice; and

(c) In keeping with the provisions of Articles 16, 18 and 20, promote and cooperate in the use of scientific advances in biological diversity research in developing methods for conservation and sustainable use of biological resources.

Article 13
Public Education and Awareness

The Contracting Parties shall:

(a) Promote and encourage understanding of the importance of, and the measures required for, the conservation of biological diversity, as well as its propagation through media, and the inclusion of these topics in educational programmes; and

(b) Cooperate, as appropriate, with other States and international organizations in developing educational and public awareness programmes, with respect to conservation and sustainable use of biological diversity.

Article 14
Impact Assessment and Minimizing Adverse Impacts

1.   Each Contracting Party, as far as possible and as appropriate, shall:

(a) Introduce appropriate procedures requiring environmental impact assessment of its proposed projects that are likely to have significant adverse effects on biological diversity with a view to avoiding or minimizing such effects and, where appropriate, allow for public participation in such procedures;

(b) Introduce appropriate arrangements to ensure that the environmental consequences of its programmes and policies that are likely to have significant adverse impacts on biological diversity are duly taken into account;

(c) Promote, on the basis of reciprocity, notification, exchange of information and consultation on activities under their jurisdiction or control which are likely to significantly affect adversely the biological diversity of other States or areas beyond the limits of national jurisdiction, by encouraging the conclusion of bilateral, regional or multilateral arrangements, as appropriate;

(d) In the case of imminent or grave danger or damage, originating under its jurisdiction or control, to biological diversity within the area under jurisdiction of other States or in areas beyond the limits of national jurisdiction, notify immediately the potentially affected States of such danger or damage, as well as initiate action to prevent or minimize such danger or damage; and

(e) Promote national arrangements for emergency responses to activities or events, whether caused naturally or otherwise, which present a grave and imminent danger to biological diversity and encourage international cooperation to supplement such national efforts and, where appropriate and agreed by the States or regional economic integration organizations concerned, to establish joint contingency plans.

2.   The Conference of the Parties shall examine, on the basis of studies to be carried out, the issue of liability and redress, including restoration and compensation, for damage to biological diversity, except where such liability is a purely internal matter.

Article 15
Access to Genetic Resources

1.   Recognizing the sovereign rights of States over their natural resources, the authority to determine access to genetic resources rests with the national governments and is subject to national legislation.

2.   Each Contracting Party shall endeavour to create conditions to facilitate access to genetic resources for environmentally sound uses by other Contracting

Parties and not to impose restrictions that run counter to the objectives of this Convention.

3.   For the purpose of this Convention, the genetic resources being provided by a Contracting Party, as referred to in this Article and Articles 16 and 19, are only those that are provided by Contracting Parties that are countries of origin of such resources or by the Parties that have acquired the genetic resources in accordance with this Convention.

4.   Access, where granted, shall be on mutually agreed terms and subject to the provisions of this Article.

5.   Access to genetic resources shall be subject to prior informed consent of the Contracting Party providing such resources, unless otherwise determined by that Party.

6.   Each Contracting Party shall endeavour to develop and carry out scientific research based on genetic resources provided by other Contracting Parties with the full participation of, and where possible in, such Contracting Parties.

7.   Each Contracting Party shall take legislative, administrative or policy measures, as appropriate, and in accordance with Articles 16 and 19 and, where necessary, through the financial mechanism established by Articles 20 and 21 with the aim of sharing in a fair and equitable way the results of research and development and the benefits arising from the commercial and other utilization of genetic resources with the Contracting Party providing such resources. Such sharing shall be upon mutually agreed terms.

Article 16
Access to and Transfer of Technology

1.   Each Contracting Party, recognizing that technology includes biotechnology, and that both access to and transfer of technology among Contracting Parties are essential elements for the attainment of the objectives of this Convention, undertakes subject to the provisions of this Article to provide and/or facilitate access for and transfer to other Contracting Parties of technologies that are relevant to the conservation and sustainable use of biological diversity or make use of genetic resources and do not cause significant damage to the environment.

2.   Access to and transfer of technology referred to in paragraph 1 above to developing countries shall be provided and/or facilitated under fair and most favourable terms, including on concessional and preferential terms where mutually agreed, and, where necessary, in accordance with the financial mechanism established by Articles 20 and 21. In the case of technology subject to patents and other intellectual property rights, such access and transfer shall be provided on terms which recognize and are consistent with the adequate and effective

protection of intellectual property rights. The application of this paragraph shall be consistent with paragraphs 3, 4 and 5 below.

3.   Each Contracting Party shall take legislative, administrative or policy measures, as appropriate, with the aim that Contracting Parties, in particular those that are developing countries, which provide genetic resources are provided access to and transfer of technology which makes use of those resources, on mutually agreed terms, including technology protected by patents and other intellectual property rights, where necessary, through the provisions of Articles 20 and 21 and in accordance with international law and consistent with paragraphs 4 and 5 below.

4.   Each Contracting Party shall take legislative, administrative or policy measures, as appropriate, with the aim that the private sector facilitates access to, joint development and transfer of technology referred to in paragraph 1 above for the benefit of both governmental institutions and the private sector of developing countries and in this regard shall abide by the obligations included in paragraphs 1, 2 and 3 above.

5.   The Contracting Parties, recognizing that patents and other intellectual property rights may have an influence on the implementation of this Convention, shall cooperate in this regard subject to national legislation and international law in order to ensure that such rights are supportive of and do not run counter to its objectives.

Article 17
Exchange of Information

1.   The Contracting Parties shall facilitate the exchange of information, from all publicly available sources, relevant to the conservation and sustainable use of biological diversity, taking into account the special needs of developing countries.

2.   Such exchange of information shall include exchange of results of technical, scientific and socio-economic research, as well as information on training and surveying programmes, specialized knowledge, indigenous and traditional knowledge as such and in combination with the technologies referred to in Article 16, paragraph 1. It shall also, where feasible, include repatriation of information.

Article 18
Technical and Scientific Cooperation

1.   The Contracting Parties shall promote international technical and scientific cooperation in the field of conservation and sustainable use of biological diversity, where necessary, through the appropriate international and national institutions.

2.   Each Contracting Party shall promote technical and scientific cooperation with other Contracting Parties, in particular developing countries, in implementing this Convention, *inter alia*, through the development and implementation of national policies. In promoting such cooperation, special attention should be given to the development and strengthening of national capabilities, by means of human resources development and institution building.

3.   The Conference of the Parties, at its first meeting, shall determine how to establish a clearing-house mechanism to promote and facilitate technical and scientific cooperation.

4.   The Contracting Parties shall, in accordance with national legislation and policies, encourage and develop methods of cooperation for the development and use of technologies, including indigenous and traditional technologies, in pursuance of the objectives of this Convention. For this purpose, the Contracting Parties shall also promote cooperation in the training of personnel and exchange of experts.

5.   The Contracting Parties shall, subject to mutual agreement, promote the establishment of joint research programmes and joint ventures for the development of technologies relevant to the objectives of this Convention.

Article 19
Handling of Biotechnology and Distribution of Its Benefits

1.   Each Contracting Party shall take legislative, administrative or policy measures, as appropriate, to provide for the effective participation in biotechnological research activities by those Contracting Parties, especially developing countries, which provide the genetic resources for such research, and where feasible in such Contracting Parties.

2.   Each Contracting Party shall take all practicable measures to promote and advance priority access on a fair and equitable basis by Contracting Parties, especially developing countries, to the results and benefits arising from biotechnologies based upon genetic resources provided by those Contracting Parties. Such access shall be on mutually agreed terms.

3.   The Parties shall consider the need for and modalities of a protocol setting out appropriate procedures, including, in particular, advance informed agreement, in the field of the safe transfer, handling and use of any living modified organism resulting from biotechnology that may have adverse effect on the conservation and sustainable use of biological diversity.

4.   Each Contracting Party shall, directly or by requiring any natural or legal person under its jurisdiction providing the organisms referred to in paragraph 3 above, provide any available information about the use and safety regulations

required by that Contracting Party in handling such organisms, as well as any available information on the potential adverse impact of the specific organisms concerned to the Contracting Party into which those organisms are to be introduced.

Article 20
Financial Resources

1.   Each Contracting Party undertakes to provide, in accordance with its capabilities, financial support and incentives in respect of those national activities which are intended to achieve the objectives of this Convention, in accordance with its national plans, priorities and programmes.

2.   The developed country Parties shall provide new and additional financial resources to enable developing country Parties to meet the agreed full incremental costs to them of implementing measures which fulfil the obligations of this Convention and to benefit from its provisions and which costs are agreed between a developing country Party and the institutional structure referred to in Article 21, in accordance with policy, strategy, programme priorities and eligibility criteria and an indicative list of incremental costs established by the Conference of the Parties. Other Parties, including countries undergoing the process of transition to a market economy, may voluntarily assume the obligations of the developed country Parties. For the purpose of this Article, the Conference of the Parties, shall at its first meeting establish a list of developed country Parties and other Parties which voluntarily assume the obligations of the developed country Parties. The Conference of the Parties shall periodically review and if necessary amend the list. Contributions from other countries and sources on a voluntary basis would also be encouraged. The implementation of these commitments shall take into account the need for adequacy, predictability and timely flow of funds and the importance of burden-sharing among the contributing Parties included in the list.

3.   The developed country Parties may also provide, and developing country Parties avail themselves of, financial resources related to the implementation of this Convention through bilateral, regional and other multilateral channels.

4.   The extent to which developing country Parties will effectively implement their commitments under this Convention will depend on the effective implementation by developed country Parties of their commitments under this Convention related to financial resources and transfer of technology and will take fully into account the fact that economic and social development and eradication of poverty are the first and overriding priorities of the developing country Parties.

5.   The Parties shall take full account of the specific needs and special situation of least developed countries in their actions with regard to funding and transfer of technology.

6.   The Contracting Parties shall also take into consideration the special conditions resulting from the dependence on, distribution and location of, biological diversity within developing country Parties, in particular small island States.

7.   Consideration shall also be given to the special situation of developing countries, including those that are most environmentally vulnerable, such as those with arid and semi-arid zones, coastal and mountainous areas.

Article 21
Financial Mechanism

1.   There shall be a mechanism for the provision of financial resources to developing country Parties for purposes of this Convention on a grant or concessional basis the essential elements of which are described in this Article. The mechanism shall function under the authority and guidance of, and be accountable to, the Conference of the Parties for purposes of this Convention. The operations of the mechanism shall be carried out by such institutional structure as may be decided upon by the Conference of the Parties at its first meeting. For purposes of this Convention, the Conference of the Parties shall determine the policy, strategy, programme priorities and eligibility criteria relating to the access to and utilization of such resources. The contributions shall be such as to take into account the need for predictability, adequacy and timely flow of funds referred to in Article 20 in accordance with the amount of resources needed to be decided periodically by the Conference of the Parties and the importance of burden-sharing among the contributing Parties included in the list referred to in Article 20, paragraph 2. Voluntary contributions may also be made by the developed country Parties and by other countries and sources. The mechanism shall operate within a democratic and transparent system of governance.

2.   Pursuant to the objectives of this Convention, the Conference of the Parties shall at its first meeting determine the policy, strategy and programme priorities, as well as detailed criteria and guidelines for elegibility for access to and utilization of the financial resources including monitoring and evaluation on a regular basis of such utilization. The Conference of the Parties shall decide on the arrangements to give effect to paragraph 1 above after consultation with the institutional structure entrusted with the operation of the financial mechanism.

3.   The Conference of the Parties shall review the effectiveness of the mechanism established under this Article, including the criteria and guidelines referred to in paragraph 2 above, not less than two years after the entry into force of this Convention and thereafter on a regular basis. Based on such review, it shall take appropriate action to improve the effectiveness of the mechanism if necessary.

4.   The Contracting Parties shall consider strengthening existing financial in-

stitutions to provide financial resources for the conservation and sustainable use of biological diversity.

Article 22
Relationship with Other International Conventions

1.   The provisions of this Convention shall not affect the rights and obligations of any Contracting Party deriving from any existing international agreement, except where the exercise of those rights and obligations would cause a serious damage or threat to biological diversity.

2.   Contracting Parties shall implement this Convention with respect to the marine environment consistently with the rights and obligations of States under the law of the sea.

Article 23
Conference of the Parties

1.   A Conference of the Parties is hereby established. The first meeting of the Conference of the Parties shall be convened by the Executive Director of the United Nations Environment Programme not later than one year after the entry into force of this Convention. Thereafter, ordinary meetings of the Conference of the Parties shall be held at regular intervals to be determined by the Conference at its first meeting.

2.   Extraordinary meetings of the Conference of the Parties shall be held at such other times as may be deemed necessary by the Conference, or at the written request of any Party, provided that, within six months of the request being communicated to them by the Secretariat, it is supported by at least one third of the Parties.

3.   The Conference of the Parties shall by consensus agree upon and adopt rules of procedure for itself and for any subsidiary body it may establish, as well as financial rules governing the funding of the Secretariat. At each ordinary meeting, it shall adopt a budget for the financial period until the next ordinary meeting.

4.   The Conference of the Parties shall keep under review the implementation of this Convention, and, for this purpose, shall:

(a) Establish the form and the intervals for transmitting the information to be submitted in accordance with Article 26 and consider such information as well as reports submitted by any subsidiary body;

(b) Review scientific, technical and technological advice on biological diversity provided in accordance with Article 25;

(c) Consider and adopt, as required, protocols in accordance with Article 28;

(d) Consider and adopt, as required, in accordance with Articles 29 and 30, amendments to this Convention and its annexes;

(e) Consider amendments to any protocol, as well as to any annexes thereto, and, if so decided, recommend their adoption to the parties to the protocol concerned;

(f) Consider and adopt, as required, in accordance with Article 30, additional annexes to this Convention;

(g) Establish such subsidiary bodies, particularly to provide scientific and technical advice, as are deemed necessary for the implementation of this Convention;

(h) Contact, through the Secretariat, the executive bodies of conventions dealing with matters covered by this Convention with a view to establishing appropriate forms of cooperation with them; and

(i) Consider and undertake any additional action that may be required for the achievement of the purposes of this Convention in the light of experience gained in its operation.

5.   The United Nations, its specialized agencies and the International Atomic Energy Agency, as well as any State not Party to this Convention, may be represented as observers at meetings of the Conference of the Parties. Any other body or agency, whether governmental or non-governmental, qualified in fields relating to conservation and sustainable use of biological diversity, which has informed the Secretariat of its wish to be represented as an observer at a meeting of the Conference of the Parties, may be admitted unless at least one third of the Parties present object. The admission and participation of observers shall be subject to the rules of procedure adopted by the Conference of the Parties.

Article 24
Secretariat

1.   A secretariat is hereby established. Its function shall be:

(a) To arrange for and service meetings of the Conference of the Parties provided for in Article 23;

(b) To perform the functions assigned to it by any protocol;

(c) To prepare reports on the execution of its functions under this Convention and present them to the Conference of the Parties;

(d) To coordinate with other relevant international bodies and, in particular to enter into such administrative and contractual arrangements as may be required for the effective discharge of its functions; and

(e) To perform such other functions as may be determined by the Conference of the Parties.

2.   At its first ordinary meeting, the Conference of the Parties shall designate the secretariat from amongst those existing competent international organizations which have signified their willingness to carry out the secretariat functions under this Convention.

Article 25
Subsidiary Body on Scientific, Technical and Technological Advice

1.   A subsidiary body for the provision of scientific, technical and technological advice is hereby established to provide the Conference of the Parties and, as appropriate, its other subsidiary bodies with timely advice relating to the implementation of this Convention. This body shall be open to participation by all Parties and shall be multidisciplinary. It shall comprise government representatives competent in the relevant field of expertise. It shall report regularly to the Conference of the Parties on all aspects of its work.

2.   Under the authority of and in accordance with guidelines laid down by the Conference of the Parties, and upon its request, this body shall:

(a) Provide scientific and technical assessments of the status of biological diversity;

(b) Prepare scientific and technical assessments of the effects of types of measures taken in accordance with the provisions of this Convention;

(c) Identify innovative, efficient and state-of-the-art technologies and know-how relating to the conservation and sustainable use of biological diversity and advise on the ways and means of promoting development and/or transferring such technologies;

(d) Provide advice on scientific programmes and international cooperation in research and development related to conservation and sustainable use of biological diversity; and

(e) Respond to scientific, technical, technological and methodological questions that the Conference of the Parties and its subsidiary bodies may put to the body.

3.   The functions, terms of reference, organization and operation of this body may be further elaborated by the Conference of the Parties.

Article 26
Reports

Each Contracting Party shall, at intervals to be determined by the Conference of the Parties, present to the Conference of the Parties, reports on measures which it has taken for the implementation of the provisions of this Convention and their effectiveness in meeting the objectives of this Convention.

Article 27
Settlement of Disputes

1.   In the event of a dispute between Contracting Parties concerning the interpretation or application of this Convention, the parties concerned shall seek solution by negotiation.

2.   If the parties concerned cannot reach agreement by negotiation, they may jointly seek the good offices of, or request mediation by, a third party.

3.   When ratifying, accepting, approving or acceding to this Convention, or at any time thereafter, a State or regional economic integration organization may declare in writing to the Depositary that for a dispute not resolved in accordance with paragraph 1 or paragraph 2 above, it accepts one or both of the following means of dispute settlement as compulsory:

(a)   Arbitration in accordance with the procedure laid down in Part 1 of Annex II;

(b)   Submission of the dispute to the International Court of Justice.

4.   If the parties to the dispute have not, in accordance with paragraph 3 above, accepted the same or any procedure, the dispute shall be submitted to conciliation in accordance with Part 2 of Annex II unless the parties otherwise agree.

5.   The provisions of this Article shall apply with respect to any protocol except as otherwise provided in the protocol concerned.

Article 28
Adoption of Protocols

1.   The Contracting Parties shall cooperate in the formulation and adoption of protocols to this Convention.

2.   Protocols shall be adopted at a meeting of the Conference of the Parties.

3.   The text of any proposed protocol shall be communicated to the Contracting Parties by the Secretariat at least six months before such a meeting.

Article 29
Amendment of the Convention or Protocols

1.   Amendments to this Convention may be proposed by any Contracting Party. Amendments to any protocol may be proposed by any Party to that protocol.

2.   Amendments to this Convention shall be adopted at a meeting of the Conference of the Parties. Amendments to any protocol shall be adopted at a meeting

of the Parties to the Protocol in question. The text of any proposed amendment to this Convention or to any protocol, except as may otherwise be provided in such protocol, shall be communicated to the Parties to the instrument in question by the secretariat at least six months before the meeting at which it is proposed for adoption. The secretariat shall also communicate proposed amendments to the signatories to this Convention for information.

3. The Parties shall make every effort to reach agreement on any proposed amendment to this Convention or to any protocol by consensus. If all efforts at consensus have been exhausted, and no agreement reached, the amendment shall as a last resort be adopted by a two-third majority vote of the Parties to the instrument in question present and voting at the meeting, and shall be submitted by the Depositary to all Parties for ratification, acceptance or approval.

4. Ratification, acceptance or approval of amendments shall be notified to the Depositary in writing. Amendments adopted in accordance with paragraph 3 above shall enter into force among Parties having accepted them on the ninetieth day after the deposit of instruments of ratification, acceptance or approval by at least two thirds of the Contracting Parties to this Convention or of the Parties to the protocol concerned, except as may otherwise be provided in such protocol. Thereafter the amendments shall enter into force for any other Party on the ninetieth day after that Party deposits its instrument of ratification, acceptance or approval of the amendments.

5. For the purposes of this Article, "Parties present and voting" means Parties present and casting an affirmative or negative vote.

Article 30
Adoption and Amendment of Annexes

1. The annexes to this Convention or to any protocol shall form an integral part of the Convention or of such protocol, as the case may be, and, unless expressly provided otherwise, a reference to this Convention or its protocols constitutes at the same time a reference to any annexes thereto. Such annexes shall be restricted to procedural, scientific, technical and administrative matters.

2. Except as may be otherwise provided in any protocol with respect to its annexes, the following procedure shall apply to the proposal, adoption and entry into force of additional annexes to this Convention or of annexes to any protocol:

(a) Annexes to this Convention or to any protocol shall be proposed and adopted according to the procedure laid down in Article 29;

(b) Any Party that is unable to approve an additional annex to this Convention or an annex to any protocol to which it is Party shall so notify the Depositary, in writing, within one year from the date of the communication of the adoption by the Depositary. The Depositary shall without delay notify all Parties of any such notification received. A Party may at any time withdraw a previous declaration of objection and the annexes shall thereupon enter into force for that Party subject to subparagraph (c) below;

(c) On the expiry of one year from the date of the communication of the adoption by the Depositary, the annex shall enter into force for all Parties to this Convention or to any protocol concerned which have not submitted a notification in accordance with the provisions of subparagraph (b) above.

3.   The proposal, adoption and entry into force of amendments to annexes to this Convention or to any protocol shall be subject to the same procedure as for the proposal, adoption and entry into force of annexes to the Convention or annexes to any protocol.

4.   If an additional annex or an amendment to an annex is related to an amendment to this Convention or to any protocol, the additional annex or amendment shall not enter into force until such time as the amendment to the Convention or to the protocol concerned enters into force.

Article 31
Right to Vote

1.   Except as provided for in paragraph 2 below, each Contracting Party to this Convention or to any protocol shall have one vote.

2.   Regional economic integration organizations, in matters within their competence, shall exercise their right to vote with a number of votes equal to the number of their member States which are Contracting Parties to this Convention or the relevant protocol. Such organizations shall not exercise their right to vote if their member States exercise theirs, and vice versa.

Article 32
Relationship Between This Convention and Its Protocols

1.   A State or a regional economic integration organization may not become a Party to a protocol unless it is, or becomes at the same time, a Contracting Party to this Convention.

2.   Decisions under any protocol shall be taken only by the Parties to the protocol concerned. Any Contracting Party that has not ratified, accepted or approved a protocol may participate as an observer in any meeting of the parties to that protocol.

Article 33
Signature

This Convention shall be open for signature at Rio de Janeiro by all States and any regional economic integration organization from 5 June 1992 until 14 June

1992, and at the United Nations Headquarters in New York from 15 June 1992 to 4 June 1993.

Article 34
Ratification, Acceptance or Approval

1.   This Convention and any protocol shall be subject to ratification, acceptance or approval by States and by regional economic integration organizations. Instruments of ratification, acceptance or approval shall be deposited with the Depositary.

2.   Any organization referred to in paragraph 1 above which becomes a Contracting Party to this Convention or any protocol without any of its member States being a Contracting Party shall be bound by all the obligations under the Convention or the protocol, as the case may be. In the case of such organizations, one or more of whose member States is a Contracting Party to this Convention or relevant protocol, the organization and its member States shall decide on their respective responsibilities for the performance of their obligations under the Convention or protocol, as the case may be. In such cases, the organization and the member States shall not be entitled to exercise rights under the Convention or relevant protocol concurrently.

3.   In their instruments of ratification, acceptance or approval, the organizations referred to in paragraph 1 above shall declare the extent of their competence with respect to the matters governed by the Convention or the relevant protocol. These organizations shall also inform the Depositary of any relevant modification in the extent of their competence.

Article 35
Accession

1.   This Convention and any protocol shall be open for accession by States and by regional economic integration organizations from the date on which the Convention or the protocol concerned is closed for signature. The instruments of accession shall be deposited with the Depositary.

2.   In their instruments of accession, the organizations referred to in paragraph 1 above shall declare the extent of their competence with respect to the matters governed by the Convention or the relevant protocol. These organizations shall also inform the Depositary of any relevant modification in the extent of their competence.

3.   The provisions of Article 34, paragraph 2, shall apply to regional economic integration organizations which accede to this Convention or any protocol.

Article 36
Entry Into Force

1.   This Convention shall enter into force on the ninetieth day after the date of deposit of the thirtieth instrument of ratification, acceptance, approval or accession.

2.   Any protocol shall enter into force on the ninetieth day after the date of deposit of the number of instruments of ratification, acceptance, approval or accession, specified in that protocol, has been deposited.

3.   For each Contracting Party which ratifies, accepts or approves this Convention or accedes thereto after the deposit of the thirtieth instrument of ratification, acceptance, approval or accession, it shall enter into force on the ninetieth day after the date of deposit by such Contracting Party of its instrument of ratification, acceptance, approval or accession.

4.   Any protocol, except as otherwise provided in such protocol, shall enter into force for a Contracting Party that ratifies, accepts or approves that protocol or accedes thereto after its entry into force pursuant to paragraph 2 above, on the ninetieth day after the date on which that Contracting Party deposits its instrument of ratification, acceptance, approval or accession, or on the date on which this Convention enters into force for that Contracting Party, whichever shall be the later.

5.   For the purpose of paragraphs 1 and 2 above, any instrument deposited by a regional economic integration organization shall not be counted as additional to those deposited by member States of such organization.

Article 37
Reservations

No reservations may be made to this Convention.

Article 38
Withdrawals

1.   At any time after two years from the date on which this Convention has entered into force for a Contracting Party, that Contracting Party may withdraw from the Convention by giving written notification to the Depositary.

2.   Any such withdrawal shall take place upon expiry of one year after the date of its receipt by the Depositary, or on such later date as may be specified in the notification of the withdrawal.

3.   Any Contracting Party which withdraws from this Convention shall be considered as also having withdrawn from any protocol to which it is party.

Article 39
Financial Interim Arrangements

Provided that it has been fully restructured in accordance with the requirements of Article 21, the Global Environment Facility of the United Nations Development Programme, the United Nations Environment Programme and the International Bank for Reconstruction and Development shall be the institutional structure referred to in Article 21 on an interim basis, for the period between the entry into force of this Convention and the first meeting of the Conference of the Parties or until the Conference of the Parties decides which institutional structure will be designated in accordance with Article 21.

Article 40
Secretariat Interim Arrangements

The secretariat to be provided by the Executive Director of the United Nations Environment Programme shall be the secretariat referred to in Article 24, paragraph 2, on an interim basis for the period between the entry into force of this Convention and the first meeting of the Conference of the Parties.

Article 41
Depositary

The Secretary-General of the United Nations shall assume the functions of Depositary of this Convention and any protocols.

Article 42
Authentic Texts

The original of this Convention, of which the Arabic, Chinese, English, French, Russian and Spanish texts are equally authentic, shall be deposited with the Secretary-General of the United Nations.

IN WITNESS WHEREOF the undersigned, being duly authorized to that effect, have signed this Convention.

Done at Rio de Janeiro on this fifth day of June, one thousand nine hundred and ninety-two.

Annex I
Identification and Monitoring

1. Ecosystems and habitats: containing high diversity, large numbers of en-

demic or threatened species, or wilderness; required by migratory species; of social, economic, cultural or scientific importance; or, which are representative, unique or associated with key evolutionary or other biological processes;

2.   Species and communities which are: threatened; wild relatives of domesticated or cultivated species; of medicinal, agricultural or other economic value; or social, scientific or cultural importance; or importance for research into the conservation and sustainable use of biological diversity, such as indicator species; and

3.   Described genomes and genes of social, scientific or economic importance.

Annex II

PART 1   ARBITRATION
ARTICLE 1

The claimant party shall notify the secretariat that the parties are referring a dispute to arbitration pursuant to Article 27. The notification shall state the subject-matter of arbitration and include, in particular, the articles of the Convention or the protocol, the interpretation or application of which are at issue. If the parties do not agree on the subject-matter of the dispute before the President of the tribunal is designated, the arbitral tribunal shall determine the subject matter. The secretariat shall forward the information thus received to all Contracting Parties to this Convention or to the protocol concerned.

ARTICLE 2

1.   In disputes between two parties, the arbitral tribunal shall consist of three members. Each of the parties to the dispute shall appoint an arbitrator and the two arbitrators so appointed shall designate by common agreement the third arbitrator who shall be the President of the tribunal. The latter shall not be a national of one of the parties to the dispute, nor have his or her usual place of residence in the territory of one of these parties, nor be employed by any of them, nor have dealt with the case in any other capacity.

2.   In disputes between more than two parties, parties in the same interest shall appoint one arbitrator jointly by agreement.

3.   Any vacancy shall be filled in the manner prescribed for the initial appointment.

ARTICLE 3

1.   If the President of the arbitral tribunal has not been designated within two months of the appointment of the second arbitrator, the Secretary-General of the United Nations shall, at the request of a party, designate the President within a further two-month period.

2.   If one of the parties to the dispute does not appoint an arbitrator within two months of receipt of the request, the other party may inform the Secretary-General who shall make the designation within a further two-month period.

ARTICLE 4

The arbitral tribunal shall render its decisions in accordance with the provisions of this Convention, any protocols concerned, and international law.

ARTICLE 5

Unless the parties to the dispute otherwise agree, the arbitral tribunal shall determine its own rules of procedure.

ARTICLE 6

The arbitral tribunal may, at the request of one of the parties, recommend essential interim measures of protection.

ARTICLE 7

The parties to the dispute shall facilitate the work of the arbitral tribunal and, in particular, using all means at their disposal, shall:

(a) Provide it with all relevant documents, information and facilities; and

(b) Enable it, when necessary, to call witnesses or experts and receive their evidence.

ARTICLE 8

The parties and the arbitrators are under an obligation to protect the confidentiality of any information they receive in confidence during the proceedings of the arbitral tribunal.

ARTICLE 9

Unless the arbitral tribunal determines otherwise because of the particular circumstances of the case, the costs of the tribunal shall be borne by the parties to the dispute in equal shares. The tribunal shall keep a record of all its costs, and shall furnish a final statement thereof to the parties.

ARTICLE 10

Any Contracting Party that has an interest of a legal nature in the subject-matter of the dispute which may be affected by the decision in the case, may intervene in the proceedings with the consent of the tribunal.

ARTICLE 11

The tribunal may hear and determine counterclaims arising directly out of the subject-matter of the dispute.

ARTICLE 12

Decisions both on procedure and substance of the arbitral tribunal shall be taken by a majority vote of its members.

ARTICLE 13

If one of the parties to the dispute does not appear before the arbitral tribunal or fails to defend its case, the other party may request the tribunal to continue the proceedings and to make its award. Absence of a party or a failure of a party to defend its case shall not constitute a bar to the proceedings. Before rendering its final decision, the arbitral tribunal must satisfy itself that the claim is well founded in fact and law.

ARTICLE 14

The tribunal shall render its final decision within five months of the date on which it is fully constituted unless it finds it necessary to extend the time-limit for a period which should not exceed five more months.

ARTICLE 15

The final decision of the arbitral tribunal shall be confined to the subject-matter of the dispute and shall state the reasons on which it is based. It shall contain the names of the members who have participated and the date of the final decision. Any member of the tribunal may attach a separate or dissenting opinion to the final decision.

ARTICLE 16

The award shall be binding on the parties to the dispute. It shall be without appeal unless the parties to the dispute have agreed in advance to an appellate procedure.

ARTICLE 17

Any controversy which may arise between the parties to the dispute as regards the interpretation or manner of implementation of the final decision may be submitted by either party for decision to the arbitral tribunal which rendered it.

PART 2   CONCILIATION
ARTICLE 1

A conciliation commission shall be created upon the request of one of the parties to the dispute. The commission shall, unless the parties otherwise agree, be composed of five members, two appointed by each Party concerned and a President chosen jointly by those members.

ARTICLE 2

In disputes between more than two parties, parties in the same interest shall appoint their members of the commission jointly by agreement. Where two or more parties have separate interests or there is a disagreement as to whether they are of the same interest, they shall appoint their members separately.

ARTICLE 3

If any appointments by the parties are not made within two months of the date of the request to create a conciliation commission, the Secretary-General of the United Nations shall, if asked to do so by the party that made the request, make those appointments within a further two-month period.

ARTICLE 4

If a President of the conciliation commission has not been chosen within two months of the last of the members of the commission being appointed, the Secretary-General of the United Nations shall, if asked to do so by a party, designate a President within a further two-month period.

ARTICLE 5

The conciliation commission shall take its decisions by majority vote of its members. It shall, unless the parties to the dispute otherwise agree, determine its own procedure. It shall render a proposal for resolution of the dispute, which the parties shall consider in good faith.

ARTICLE 6

A disagreement as to whether the conciliation commission has competence shall be decided by the commission.

Index

Abzug, Bella, 78
Adams, Patricia, 112, 203
Adede, A. O., 30
Africa, 4, 36, 104, 197, 232
Agenda 21, 14–15, 31, 131, 140–41, 198;
  costs of, 15; and Middle East peace
  process, 37; and oceanic pollution, 73–
  74. *See also* Rio Declaration
Alders, Hans, 81
Alesana, Tofilau Eti, 195
American Convention on Human Rights,
  94
American Forests for the Future Initiative,
  284
Antarctica Treaty, 71–72
Asia, 104, 227, 232
Atmospheric Environment Service, 163
Austria, 18, 82, 175–76, 285
Awoonor, Kofi N., 190

Babbitt, Bruce, 273–74
Badawi, Abdullah Haji Ahmad, 184
Bangladesh, 107, 110, 195
Benedick, Richard E., 207
Berntsen, Thorbjorn, 73, 78, 206, 275, 285
Bhutan, 277
Bildt, Carl, 285
Biodiversity, 10–11; action plans, 231; bi-
  oremediation, 223; and biotechnology,
  245–46; and corporations, 232–33,
  255–59, 272–73; and deforestation,
  221, 225; and habitat, 225–26; impor-
  tance of, 222–24; as national resource,
  231, 255, 260–61; wildlife, 74–76
Biodiversity Convention, 13–14, 18, 75–
  76, 129, 131, 134–39; and biological

resources, 221, 236, 257, 260–61; and
  Conference of the Parties, 268, 270–71,
  282, 293–94; and conservation, 236,
  265, 279; controversial clauses, 259–
  61, 279–80; costs of protection, 282–
  84, 287–88; definitions, 220–22; and
  environmental cleanup, 263–64; financial provisions of, 137–38, 237–38,
  248, 264–71, 281; as morally binding,
  233–34, 280; and nationality of species,
  136–37; and Northern financial fears,
  267–68; Northern pledges, 283–85; objectives of, 234–35, 265; obligations of,
  289–94; and poverty, 137; and prior
  ownership, 260–61; and sanctions,
  273–74, 294; Southern advantage in,
  236–37, 255–57, 278, 279; Southern
  apprehensions, 281; and sovereignty,
  135; and technology transfer, 236–37,
  257–60; weak language of, 265
Biodiversity Research Initiative, 243
Biotechnology, 14, 134–35; regulation of,
  245–46; and trade agreements, 247; in
  United States, 254–63. *See also* Biodiv-
  ersity Convention; Intellectual property
  rights; Technology
Blix, Hans, 197–98
Blunkett, David, 269
Bohlen, E. U. Curtis, 245
Boom, Brian, 254
Borrego, Carlos, 82, 204, 207
Boutros-Ghali, Boutros, 11, 206
Bradshaw, Stephen, 100, 112, 114, 132
Brady Plan, 119
Brazil, 56, 105–6, 116, 221, 250
Brinkhorst, Laurens Jan, 272